Digital Technologies for Sustainable Manufacturing

Arturo Molina
Instituto Tecnológico de Tláhuac III, Tecnológico Nacional de México, Mexico

Juana Isabel Méndez
Tecnologico de Monterrey, School of Architecture, Art and Design, México
and
Tecnologico de Monterrey, School of Engineering and Sciences, México

Dora Iliana Medina
Department of Industrial Engineering and Operations
Instituto Tecnológico Autónomo de México (ITAM), Mexico

Fernando Gómez-Zaldívar
Tecnologico de Monterrey, School of Social Sciences and Government, México

Alex Elías-Zúñiga
Tecnologico de Monterrey
Institute of Advanced Materials for Sustainable Manufacturing, México

Pedro Ponce
Tecnologico de Monterrey
Institute of Advanced Materials for Sustainable Manufacturing, México

Rita Q. Fuentes-Aguilar
Tecnologico de Monterrey
Institute of Advanced Materials for Sustainable Manufacturing, México

Alejandro Montesinos-Castellanos
Tecnologico de Monterrey
Institute of Advanced Materials for Sustainable Manufacturing, México

CRC Press
Taylor & Francis Group
Boca Raton London New York

CRC Press is an imprint of the
Taylor & Francis Group, an **informa** business
A SCIENCE PUBLISHERS BOOK

Cover credit: **Cover illustrations reproduced by kind courtesy of the authors.**

First edition published 2026
by CRC Press
2385 NW Executive Center Drive, Suite 320, Boca Raton FL 33431

and by CRC Press
4 Park Square, Milton Park, Abingdon, Oxon, OX14 4RN

CRC Press is an imprint of Taylor & Francis Group, LLC

Library of Congress Cataloging-in-Publication Data (applied for)

ISBN: 978-1-032-72888-9 (hbk)
ISBN: 978-1-032-81837-5 (pbk)
ISBN: 978-1-003-50161-9 (ebk)

DOI: 10.1201/9781003501619

Typeset in Euclid
by Prime Publishing Services

To all my friends and colleagues at Tec de Monterrey,
to my sons Julio and Charlie, and to my wife Silvia.
—A. Molina

To the readers of this book, with gratitude
for their time and curiosity.
—D.I. Medina

To my parents, Gandhi and Isabel, my sister, and my spouse,
Vic, who guided me at every step of my life.
—J.I. Méndez

To my beloved wife, Norma, and my wonderful sons,
Pedro and Jamie.
—P. Ponce

To my daughter SE, my husband, and both
moms—thanks for having my back (and sometimes my dishes).
Without your help, I wouldn't have had the
time to chase my passion for research.
—RQ. Fuentes-Aguilar

Foreword

In an era marked by a critical need for sustainability, "*Digital Technologies for Sustainable Manufacturing*" stands as a seminal text, guiding us through the transformative powers of digitalization in manufacturing. Rooted deeply in the Life Cycle Engineering (LCE) framework, this book addresses sustainable practices through our meticulous integration of the 5S principles—Social, Sustainable, Sensing, Smart, and Safe—and the strategic leverage of Competitive Intelligence (CI).

As global manufacturing faces unprecedented challenges, the urgency for a sustainable overhaul is undeniable. We respond by offering a comprehensive blueprint combining economic viability with ecological responsibility. Through the lens of LCE, we methodically unfold the complexities and potentials of embedding digital technologies across Material Development, Product Development, Processes Development, Manufacturing System Development, and Supply Chain Development entities. Each stage—from Idea to Basic Development, Advanced Development, and Launching through to end-of-life—ensures a sustainable LCE for manufactured products.

Our exploration into the 5S principles as an integral component of the LCE framework highlights a revolutionary approach to manufacturing. These principles advocate for operations that are not only environmentally aligned but are also socially responsible, intelligently integrated, and inherently safe. The inclusion of CI throughout the book further enriches this narrative, presenting a foresighted method of harnessing market and environmental data to fortify our manufacturing processes against future challenges and competitive pressures.

"*Digital Technologies for Sustainable Manufacturing*" is a call to action for industry leaders, policymakers, and academics to adopt the forward-thinking strategies detailed in this text. It is a roadmap designed to inspire and equip stakeholders with the knowledge and tools necessary to transition effectively into more sustainable manufacturing operations.

This book is not just an academic treatise but a practical guide that connects theory with real-world applications, illustrated through detailed case studies and empirical data. As we navigate towards a more sustainable future, the insights offered here are both timely and timeless, poised to make a significant impact on how we think about and implement manufacturing processes in the digital age.

Preface

In an era where environmental sustainability is no longer a choice but an imperative, this book, Digital Technologies for Sustainable Manufacturing, provides a comprehensive framework for transforming industrial practices. Rooted in Life Cycle Engineering (LCE) approach, it aims to merge economic goals with ecological responsibility by embracing digital innovations that optimize manufacturing processes from conception to end-of-life management. The text outlines how emerging technologies—such as Artificial Intelligence, Internet of Things, digital twins, among others—facilitate this transformation, ensuring that every step, from material selection to supply chain, contributes to a sustainable future.

The book is organized based on a LCE framework, illustrating how entities like materials, products, processes, manufacturing systems, and supply chains can be developed, optimized, and maintained with sustainability as a core priority. Each chapter delves into one entity, showing how 5S principles—Social, Sustainable, Sensing, Smart, and Safe—are integral to this life cycle approach. The book showcases real-world applications that bridge theoretical concepts with practical solutions by highlighting specific technologies and methods, such as high-throughput screening in materials science or augmented reality in process design.

Detailed case studies provide readers with examples that bring the theoretical frameworks to life. For instance, developing modular CNC machines demonstrates how digital technologies can elevate sustainability in manufacturing systems. Likewise, discussions on the future of sustainable manufacturing emphasize a need for an adaptable approach that anticipates technological advances while minimizing environmental impacts.

Authored by experts in the field, this book serves as both a roadmap and an inspiration for engineers, scientists, and industry leaders who seek to align manufacturing with sustainability goals. By combining cutting-edge digital tools with a holistic life cycle perspective, Digital Technologies for Sustainable Manufacturing presents a vision for an industry that not only meets the demands of the present but also paves the way for a greener, more resilient future.

Mexico
November 2024

Arturo Molina
Juana Isabel Méndez
Dora Iliana Medina
Fernando Gómez-Zaldívar
Alex Elías-Zúñiga
Pedro Ponce
Rita Q. Fuentes-Aguilar
Alejandro Montesinos-Castellanos

How to Read This Book

Digital Technologies for Sustainable Manufacturing introduces an integrative framework to reshape manufacturing aligned with sustainability goals. Through the lens of Life Cycle Engineering (LCE), this book explores how digital technologies—such as Artificial Intelligence, the Internet of Things, and digital twins—enhance sustainable manufacturing processes, from design and production to end-of-life management. The book emphasizes the 5S principles—Social, Sustainable, Sensing, Smart, and Safe—as essential components for creating resilient, efficient, and sustainable manufacturing systems.

Each chapter in this book addresses a different stage or element of the manufacturing life cycle, illustrated with real-world case studies and applications. Below is a guide to the key topics covered in each chapter to help readers navigate the concepts most relevant to their interests.

Chapter 1 is the ideal starting point for understanding the fundamentals of Life Cycle Engineering and the role of digital technologies in sustainable manufacturing. This chapter presents the LCE framework and explores the transformative role of digital technologies. Through a case study of a modular CNC micromachine, readers will see how the 5S principles apply across a product's life cycle, from conception to end-of-life.

Chapter 2 provides an in-depth look at the 5S principles that underpin the LCE framework. This chapter breaks down each of the 5S principles—Social, Sustainable, Sensing, Smart, and Safe—demonstrating how they contribute to system efficiency, ethical responsibility, and adaptability. These principles are essential for those aiming to integrate sustainability into every phase of the life cycle of systems.

Chapter 3 focuses on materials discovery, creation, and development for those interested in sustainable materials science. This chapter covers how digital technologies, like computational materials science and machine learning, are revolutionizing sustainable material development. A case study on Honeywell Aerospace illustrates how additive manufacturing of turbine blades supports sustainable practices, reducing both energy consumption and emissions.

Chapter 4 is the best choice to learn about sustainable product design and prototyping. This chapter discusses product development from ideation through prototyping, covering tools such as CAD, CAM, and digital twins.

The case study of the Mio-Tracker bioengineering device highlights the application of eco-design principles and energy-efficient manufacturing methods in product development.

Chapter 5 provides a comprehensive overview of sustainable process development and optimization. This chapter explores how process development can minimize resource usage and waste using technologies like Computer-Aided Process Planning (CAPP) and predictive analytics. A case study on the manufacturing of micromachine components demonstrates the importance of LCE in process design.

Chapter 6 focuses on system design, modularity, and predictive maintenance to understand sustainable manufacturing systems and their integration. Manufacturers can monitor systems in real-time through tools such as digital twins and the Industrial Internet of Things (IIoT), allowing for operational efficiency and minimizing environmental impact.

For those interested in sustainable supply chain management, Chapter 7 covers the development of supply chains from sourcing to end-of-life recycling. This chapter highlights the role of blockchain and cloud computing in enhancing transparency, traceability, and sustainability throughout global supply chains.

Finally, Chapter 8 is crucial in examining to exploring how competitive intelligence strengthens sustainable manufacturing. This chapter discusses how businesses can leverage the LCE and 5S frameworks to gain strategic insights, anticipate sustainability trends, and enhance their competitive edge.

Each chapter stands alone, the readers can focus on specific areas of interest or read the book sequentially for a complete understanding of digital technologies used to achieve sustainable manufacturing. By merging practical applications with cutting-edge digital solutions, Digital Technologies for Sustainable Manufacturing presents a vision for a more sustainable, resilient, and innovative future in manufacturing.

Contents

1

Life Cycle Engineering as a Framework for Digital Technologies for Sustainable Manufacturing

This chapter explores the transformative impact of digital technologies in sustainable manufacturing, focusing on the Life Cycle Engineering (LCE) framework as a catalyst for sustainable and economically viable industrial practices. It discusses integrating advanced technologies such as Artificial Intelligence (AI), the Internet of Things (IoT), and digital twins, which optimize manufacturing processes from material selection to end-of-life management, aligning with the Sustainable Development Goals for sustainable industrialization. Special emphasis is placed on the 5S principles—Social, Sustainable, Sensing, Smart, and Safe—integral to enhancing product life cycle management and fostering innovation in manufacturing practices. The chapter includes a detailed case study on the Micromachine, a modular CNC machine tool, illustrating the application of LCE and the 5S principles across its development life cycle. This machine's evolution exemplifies how digital technologies can enhance manufacturing systems' adaptability, efficiency, and sustainability. Looking ahead, the chapter forecasts a future where digital transformation in manufacturing not only drives economic benefits but also significantly reduces environmental impacts, advocating for a manufacturing paradigm that is as sustainable as it is transformational. This future-oriented perspective underscores the imperative for a holistic approach to manufacturing that integrates sustainability at every stage of the system life cycle.

1.1 Introduction

In the face of ever-growing environmental concerns and the urgent need for sustainable practices, the manufacturing industry stands at a crucial juncture. Traditional manufacturing methods, often resource-intensive and environmentally burdensome, are gradually giving way to innovative, sustainable practices that promise to reduce environmental impact and enhance efficiency and economic viability. This shift is not merely a response

DOI: 10.1201/9781003501619-1

to regulatory pressures but a proactive move towards a more responsible and forward-thinking approach to industrial production.

The burgeoning role of digital technologies in this transformation cannot be overstated. As we venture further into the digital age, integrating advanced technologies in manufacturing processes—ranging from Artificial Intelligence (AI) and Internet of Things (IoT) to smart automation and beyond—presents unprecedented opportunities to redefine the very essence of manufacturing. This chapter aims to explore these dynamic changes, offering insights into sustainable manufacturing practices that harmonize economic goals with environmental stewardship.

Hence, sustainable manufacturing emerges as a critical paradigm in an era marked by significant ecological challenges and a global push towards sustainability. It transcends traditional production techniques, focusing on minimizing environmental degradation while enhancing resource efficiency. The imperative for such practices is clear: to mitigate the adverse impacts of industrial activities on the planet while meeting the increasing demands of a growing global population.

Sustainable manufacturing also addresses the ethical concerns of resource usage. It promotes the conservation of non-renewable resources, the reduction of waste, and the minimization of carbon footprints. By adopting a life cycle approach to product development—considering the environmental impact from raw material extraction to end-of-life disposal—manufacturers can contribute significantly to the global sustainability agenda.

In that sense, digital technologies are at the forefront of revolutionizing manufacturing processes. These technologies enable manufacturers to leverage big data analytics, cloud computing, and sophisticated sensing technologies to optimize production lines, reduce waste, and improve product quality. For instance, AI and machine learning algorithms can predict failures before they occur, reducing downtime and maintenance costs.

Moreover, digital twins—virtual replicas of physical devices—allow manufacturers to simulate production processes and identify inefficiencies without the costs associated with physical testing. IoT devices facilitate real-time monitoring of manufacturing operations, enhancing operational transparency and enabling rapid responses to operational hiccups. These technologies improve efficiency and drive innovations that lead to smarter, more sustainable manufacturing practices.

1.2 Life Cycle Engineering Description

Life Cycle Engineering (LCE) is an integrative approach that focuses on optimizing the entire life cycle of materials, products, processes, systems, and supply chains. It considers all phases, from the extraction of raw materials to end-of-life disposal or recycling, aiming for sustainability, cost-effectiveness, and high performance. By considering the whole life cycle, LCE helps make informed decisions that lead to sustainable and economically viable outcomes. FIGURE 1.1 describes our framework for using LCE to develop materials, products, manufacturing processes, manufacturing systems and supply chains.

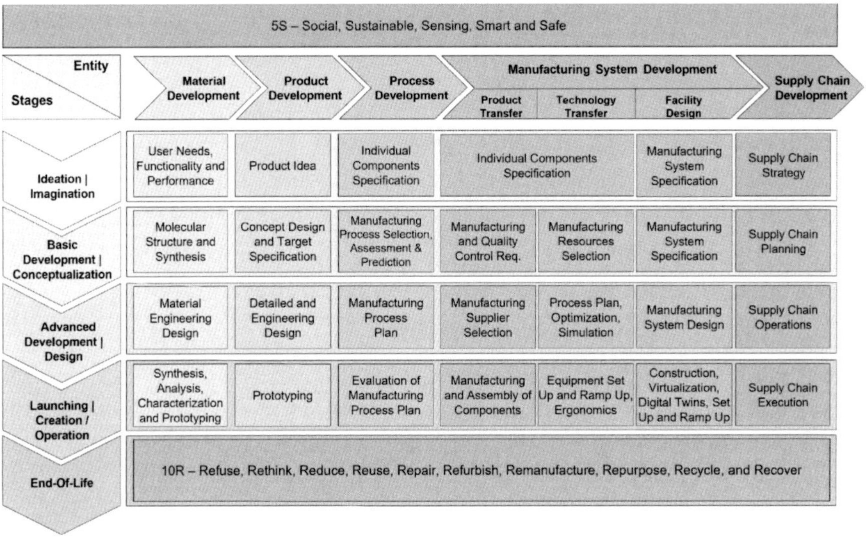

FIGURE 1.1
Framework for life cycle engineering.

1.2.1 Benefits and Limitations of Applying Life Cycle Engineering

Incorporating an LCE approach emphasizes minimizing environmental and societal impacts throughout a system's lifespan, from the initial extraction of raw materials to the end-of-life phase, which may include recycling and upcycling strategies. The United Nations Environment Programme has been a proponent of such integrated approaches to sustainability, recognizing their importance in achieving Sustainable Development Goals (SDGs) [26].

Integrating an LCE perspective encourages the creation of processing methods that reduce waste and conserve energy, which can significantly decrease the environmental burden. Furthermore, by intertwining ecodesign principles with process optimization and digital technology, LCE fosters innovation in various sectors, including aerospace and healthcare, enhancing the sustainability of materials and technologies [138]. This approach can also lead to market leadership and competitive advantage, particularly as consumers increasingly value sustainability, and can ensure compliance with environmental management standards like those in the ISO 14000 series [123].

However, the adoption of LCE practices is not devoid of challenges. Significant investment is often required to implement these principles, and the complexity of life cycle assessments can present barriers, particularly for small and medium-sized enterprises (SMEs) [70]. Detailed data collection and analysis require specific expertise and can be resource-intensive [57].

Besides, relying on sophisticated digital technologies for implementing LCE also comes with difficulties. The need for a skilled workforce to navigate complex digital systems, the possibility of rapid technological obsolescence, and the threats posed by cyber vulnerabilities are significant considerations [112]. Moreover, the complexities of applying LCE principles across global supply chains can introduce additional challenges in maintaining transparency and consistency due to diverse regulatory environments and the intricacies involved in international operations [92].

1.2.2 How Life Cycle Engineering Contributes to Sustainable Manufacturing

The United Nations Environment Programme (UNEP) strongly recommends that sustainable product development requires a comprehensive view of the production process, considering every stage from raw material extraction to final disposal or recycling [140].

LCE ensures that the environmental impacts are minimized during the raw material extraction phase. For example, promoting less wasteful processing methods aligns with the eco-design principles, which, decreases the environmental burden by early intervention in the design stage of product development [87]. Using strategies like recycling and upcycling at the product's end-of-life stage also supports a more circular economy [40].

Digital technologies further enhance LCE by streamlining the development of sustainable materials and manufacturing processes. The use of AI and machine learning allows for the rapid development of new materials and innovation in material science [112]. These advancements are crucial in reducing time-to-market and enhancing efficiency, which are vital for the competitiveness of various industries.

LCE principles extend into manufacturing, where optimization processes focus on reducing waste and energy use. The International Energy Agency (IEA) informs that the optimization of energy use in industrial processes can result in significant energy savings and reduce environmental impacts [67]. For example, tools like Computer-Aided Manufacturing (CAM) and simulations provided by Finite Element Analysis (FEA) minimize the need for physical prototypes, which conserves resources and accelerates development [92].

In manufacturing systems, digital technologies like Digital Twins and the Industrial Internet of Things (IIoT) facilitate real-time monitoring and predictive maintenance, significantly enhancing system longevity and performance [112].

Hence, applying LCE throughout the material development to supply chain process promotes the use of predictive modeling and real-time feedback loops [92].

Furthermore, the 5S principles—Social, Sustainable, Sensing, Smart, and Safe—are crucial in enhancing the life cycle of systems in terms of sustainability, efficiency, and cost-effectiveness. Following a breakdown of how these principles can be seamlessly integrated into the LCE framework for developing systems:

- **Social:** This principle emphasizes the importance of positive social impacts and ethical practices throughout the system's life cycle. It involves engaging with communities, promoting fair labor practices, and ensuring that the benefits of technological advancements are equitably shared among all stakeholders.
- **Sustainable:** Sustainability within the LCE framework focuses on minimizing environmental impacts at every stage of the system's life—from design and manufacturing to usage and disposal. This involves using renewable resources, reducing emissions, and designing recyclable or biodegradable products.
- **Sensing:** The integration of sensors and data collection technologies is crucial for monitoring and improving system performance and sustainability. Sensing technologies enable real-time data gathering that informs decision-making and enhances systems' responsiveness to environmental and operational changes.
- **Smart:** Smart technologies such as AI and machine learning make systems more adaptive and efficient. These technologies enable systems to analyze data, optimize operations, and predict maintenance needs, thus reducing downtime and improving performance.
- **Safe:** Safety is paramount in the LCE approach, where systems are designed to be safe for users and the environment throughout their life cycle. This involves implementing robust safety protocols, ensuring data security, and mitigating system operations and disposal risks.

By incorporating these 5S principles, the LCE framework ensures that systems are optimized for performance and adhere to high ethical, environmental, and safety standards. This integration supports the creation of systems that are sustainable, efficient, and capable of adapting to future challenges while minimizing their environmental footprint.

Chapter 2 examines in greater detail how the 5S principles can be implemented throughout a system's life cycle, from conception to end-of-life management, to enhance sustainability, efficiency, and cost-effectiveness. By doing so, it facilitates the development of systems that comply with current technological and operational standards and adapt to future changes, ensuring long-term sustainability and resilience.

1.3 Life Cycle Diagram Flow

This section describes the whole process of Life cycle Engineering. Figure 1.1 outlines the LCE into different entities and stages with a multifaceted and interrelated activities required to develop and manufacture a product. It integrates various components of LCE, from initial idea generation to the delivery of the final product, emphasizing the interconnectedness of each stage and entity in the overarching process [93].

TABLE 1.1 outlines the stages and entities involved in the LCE framework. It is structured into several columns, each representing a distinct entity in the process, including Material Development, Product Development, Process Development, Manufacturing System Development, Product Transfer, Technology Transfer, Facility Design, and Supply Chain Development.

Each entity column is further divided into rows corresponding to specific stages of that entity's role in the overall process. For instance, the Material Development Entity includes stages such as *Ideation—Imagination* that considers the user needs, functionality and performance, following the *Basic Development—Conceptualization* for the molecular structure and synthesis to achieve the desired functions. The *Advanced Development* follows this—Design stage, explaining the need for material engineering design to ensure that the materials match the production. Finally, the fourth stage is the Launching—Creation/Operation stage, in which the synthesis, analysis, characterization, and prototyping are needed for testing real-world conditions before full-scale production. This indicates a progression from the conceptual understanding of user requirements to the practical application in material synthesis and design, culminating in the prototyping stage.

Similarly, the Product Development Entity progresses from *Product Idea* to *Conceptual Design and Target Specification*, then to *Detailed and Engineering Design*, and finally to *Prototyping*. This sequence indicates a flow from initial idea generation through detailed design to the creation of prototypes.

The Process Development Entity and other entities follow a similar structure, each with stages pertinent to their focus areas. For example, the Supply Chain Development Entity includes *Supply Chain Strategy*, *Supply Chain Planning*, *Supply Chain Operations*, and *Supply Chain Execution*, which describe the strategic planning and operational execution of the supply chain necessary to support the product life cycle.

TABLE 1.1
LCE Framework: Stage and Entity Description

	ENTITY						
				Manufacturing System			
STAGES	**Material**	**Product**	**Process**	**Product Transfer**	**Technology Transfer**	**Facility Design**	**Supply Chain**
Ideation—Imagination	User Needs, Functionality and Performance. User applications, environments, and performance needs analysis. Material choice hinges on functionality, meeting product purposes like structural strength or conductivity. Performance criteria cover durability, environmental resistance, longevity, and standards adherence.	Product Idea. Systematic search for selection and development of promising product ideas. This process ensures that the chosen ideas are both technically feasible and economically viable. The scope of the project and the project plan are defined.	Individual Component Specification. Product design is received and decomposed to identify all individual components. Customer requirements are identified mainly in three aspects: geometry, material and production rates. Finally, the scope of the project and project plan are defined.	Individual Component Specification. Product information is collected. Bill of Materials (BOM) is carried out to identify materials, standard components, quality requirements and delivery times.	Individual component specification. The information about components or families of components is captured in three aspects: geometry (drawings), materials (specifications), and production rates (batch size).	Product and Manufacturing Process specification. The technical information of the product and manufacturing processes must be used to begin the design of a new facility.	Supply Chain Strategy. Plan the efficient flow of materials and information from sourcing to consumer, ensuring timely delivery and risk mitigation, with a focus on technology integration and adaptability to demand and market shifts.
Basic Development—Conceptualization	Molecular Structure and Synthesis. It is the design and engineering of the molecular makeup of materials to achieve desired properties and functions. The process is focused on creating materials with specific characteristics to match the product's requirements. Scalability, cost-effectiveness, and environmental impact must be considered.	Conceptual Design and Target Specification. This stage involves collecting information about constraints and customer requirements for the solution.	Manufacturing Process Selection. Components are classified as standard parts or manufactured parts. A component could be transferred or could be manufactured by the company.	Manufacturing and quality control requirements. Manufacturing and Quality control specifications and requirements are defined in terms of volume, cost, and time. Product attributes (material, dimension, tolerances, surface finish). Control variables are defined for acceptance or rejection.	Manufacturing Resources Specification and selection. Available technology (machines, Material Handling Equipment, Tooling, Jig/Fixture) is evaluated to select the best equipment to manufacture this component.	Manufacturing System Specification. Available technology (machines, material handling equipment, tooling, jig/fixture) is evaluated to select the best product manufacturing equipment.	Supply Chain Planning. Integrates forecasting, inventory, production, and distribution to meet customer demand. Uses data to predict sales and manages stock levels against costs and service goals. Ensures raw materials and components are secured from suppliers for uninterrupted production.

	ENTITY						
				Manufacturing System			
STAGES	**Material**	**Product**	**Process**	**Product Transfer**	**Technology Transfer**	**Facility Design**	**Supply Chain**
Advanced Development—Design	Material Engineering Design. The form, structure, and dimensions of the materials are established. Creation of precise drawings and specifications for manufacturing, ensuring that the materials match the production.	Detailed and Engineering Design. This stage develops the individual parts' arrangement, form, dimensions, and surface properties. Drawings and manufacturing documents are produced.	Manufacturing Process Plan. The production plan is developed for each component. The manufacturing process and plan is designed specifying time and costs. It is necessary to define the manufacturing process plan to achieve production volume, costs, times, quality and flexibilities.	Manufacturing Supplier Selection. The partners are evaluated and selected based on capacities and capabilities from different companies to integrate their competencies to develop the project.	Process Plan. The process chart is elaborated to define raw materials, tools, fixtures, gauges, and other devices necessary to manufacture the component. Also, documents for control quality in the process and for standard operation are elaborated.	Manufacturing System Design. The Manufacturing Strategies must be defined, and the station, cells, shop-floor, and Factory must be configured. The capacity plan must be evaluated using simulation tools.	Supply Chain Operations. Addresses daily supply chain management, focusing on procurement, production, distribution, and logistics to streamline processes, manage inventory, ensure product delivery, and adapt to market changes.
Launching—Creation/ Operation	Synthesis, analysis, characterization and prototyping. Synthesizing new materials or improving existing ones to achieve desired properties. Analyzing these materials to understand their composition and structure for potential applications. Characterizing the materials to evaluate their physical, chemical, and mechanical properties, ensuring they meet design specifications. Prototyping to create sample or trial versions of the materials for testing in real-world conditions before full-scale production.	Prototyping. This stage verifies any remaining problems in the product. Tests are carried out to check the functionality and potential design modifications.	Evaluation of Manufacturing Process Plan. The process must be evaluated using simulation tools in order to verified that the product can be produced and volumes could be achieved.	Manufacturing and Assembly of Components. Components are manufactured, assembled, and delivered to the facility, and finally, quality controls are checked and documented	Equipment set up and ramp of Production. Equipment is checked, tested and set-up. Ramp up and production is started-up.	Facility construction, set-up, and ramp-up of Production. The manufacturing system is configured, constructed, and installed. Ramp-up production began, and the facility's performance was evaluated and adjusted.	Supply Chain Execution. The established supply chain is executed, including order fulfillment, logistics management, and the monitoring of supplier performance to ensure timely delivery and quality. It relies on real-time tracking and collaboration with all supply chain partners to ensure efficient operations and customer satisfaction, leveraging technology for automation and smooth information exchange.

1.3.1 Entity Description: Material, Product, Process, Manufacturing Systems, and Supply Chain

Here is a breakdown of how LCE can be applied to various domains:

1.3.1.1 Material Development

At the core of this approach is assessing and reducing environmental and social impacts during raw material extraction, developing processing methods that minimize waste and energy consumption, and promoting end-of-life strategies like recycling and upcycling. TABLE 1.2 depicts each stage of the LCE framework corresponding to this entity.

Digital technologies have catalyzed advancements in materials engineering, with techniques like computational materials science, high-throughput screening, and AI and machine learning leading to the rapid prediction and development of new materials. 3D printing facilitates the creation of complex material geometries, while digital twins and advanced characterization tools aid in understanding material behavior in real-world conditions. Additionally, using data repositories, IoT, robotics, and automation enhances collaborative efforts and expedites material research and development. Emerging technologies like blockchain and quantum computing further secure data and promise solutions to complex problems. This confluence of LCE principles and digital technology accelerates the creation and application of advanced materials and drives innovation across multiple industries, from aerospace to healthcare, thereby reducing time-to-market and fostering sector-wide advancements.

- **Raw Material Extraction**: Assess the environmental and social impact of extracting raw materials.
- **Material Processing**: Develop sustainable processes that reduce waste and energy consumption.
- **Material End-of-Life**: Encourage recycling, upcycling, or biodegradability.

Digital technologies have greatly accelerated the development of advanced materials. These materials, whether new alloys, polymers, ceramics, or composites, often possess unique properties that can be optimized for specific applications (see TABLE 1.3)

By integrating these digital technologies, material scientists and engineers can accelerate the discovery, understanding, and application of advanced materials. This rapid development reduces the time-to-market for new materials and fosters innovation in various sectors, including aerospace, electronics, energy, and healthcare.

Chapter 3 focuses on Materials: Discovery, Creation, and Development, emphasizing sustainability in materials science by integrating Life Cycle Engineering (LCE) and advanced digital technologies. The chapter explores

TABLE 1.2
Material Development Entity Description by Stage

Stage	Material Development Entity
Ideation—Imagination	***User Needs, Functionality and Performance.*** User applications, environments, and performance needs analysis. Material choice hinges on functionality, meeting product purposes like structural strength or conductivity. Performance criteria cover durability, environmental resistance, longevity, and standards adherence.
Basic Development—Conceptualization	***Molecular Structure and Synthesis.*** It is the design and engineering of the molecular makeup of materials to achieve desired properties and functions. The process is focused on creating materials with specific characteristics to match the product's requirements. Scalability, cost-effectiveness, and environmental impact must be considered.
Advanced Development—Design	***Material Engineering Design.*** The materials' form, structure, and dimensions are established. Creation of precise drawings and specifications for manufacturing, ensuring that the materials match with the production.
Launching—Creation/Operation	***Synthesis, Analysis, Characterization and Prototyping.*** Synthesizing new materials or improving existing ones to achieve desired properties. Analyzing these materials to understand their composition and structure for potential applications. Characterizing the materials to evaluate their physical, chemical, and mechanical properties, ensuring they meet design specifications. Prototyping to create sample or trial versions of the materials for testing in real-world conditions before full-scale production.

the complete life cycle of material development, covering stages from ideation to end-of-life management, focusing on minimizing waste and conserving resources. Key technologies discussed include machine learning, materials informatics, and high-throughput screening, which accelerate material discovery and optimize properties for specific applications.

A case study on Honeywell Aerospace illustrates the application of LCE principles in manufacturing turbine blades using additive manufacturing. This example highlights sustainable practices that reduce energy use and carbon emissions. The chapter also presents future trends in sustainable

TABLE 1.3
Examples of Digital Technologies Employed during the Material Development Entity

Digital Technology	Application
Computational Materials Science	Uses computational methods and simulations to predict the properties of new materials without the need for experimental testing. Density Functional Theory (DFT) and Molecular Dynamics (MD) simulations are commonly used techniques.
High-throughput Screening	Combines automated synthesis methods with rapid characterization techniques. Enables the evaluation of many material compositions in parallel.
Artificial Intelligence and Machine Learning	Analyzes vast material properties and performance datasets. Helps identify patterns, predict material behavior, and suggest potential material compositions.
3D Printing/Additive Manufacturing	Enables rapid prototyping of materials with complex geometries. Allows for the layer-by-layer construction of materials with varied compositions, leading to graded or functionally graded materials.
Digital Twins	Digital representations of materials and their behavior under different conditions. Facilitates the understanding and prediction of material performance in real-world applications.
Advanced Characterization Tools	Techniques such as advanced electron microscopy, spectroscopy, and tomography generate digital data that can be analyzed to understand microstructures and properties.
Data Repositories and Cloud Computing	Storing and analyzing material datasets in centralized or cloud databases. Facilitates collaboration among researchers and provides access to vast amounts of material data.
Virtual Reality (VR) and Augmented Reality (AR)—Extended Reality	Provides an immersive experience for visualizing material structures and properties at different scales, from atomic to macroscopic.
Internet of Things (IoT)	Sensors embedded in manufacturing or testing setups provide real-time data on material synthesis and performance. Enables a feedback loop for optimizing material processing parameters.

Digital Technology	Application
Robotics and Automation	Automated synthesis and testing labs (often called "self-driving labs") can operate around the clock. Increases the pace of material experimentation and discovery.
Blockchain Technology	Ensures data authenticity, traceability, and security in collaborative material research projects.
Quantum Computing	Although still in its nascent stages, quantum computing holds the potential for solving complex material science problems intractable for classical computers.

manufacturing, including digital twins, circular economy principles, and smart manufacturing, positioning industries to thrive sustainably in an evolving market landscape.

1.3.1.2 Product Development

TABLE 1.4 depicts each stage of the LCE framework corresponding to this entity.

Digital technologies have redefined the product development landscape, guided by the LCE stages and 5S principles (social, sustainable, sensing, smart and safe). Manufacturers can mitigate environmental impacts and elevate product value by deploying these technologies within a structured LCE framework.

- **Design Phase:** Incorporate eco-design principles to minimize environmental impacts. Consider durability, recyclability, and minimal use of hazardous substances.
- **Manufacturing:** Optimize processes to reduce waste, energy, and resource consumption.
- **Usage:** Make products energy-efficient and educate consumers about sustainable usage.
- **End-of-Life:** Develop and promote recycling and disposal programs.

Digital technologies have significantly enhanced the speed and efficiency of product development. These tools and techniques allow for rapid prototyping, enhanced collaboration, and better design accuracy. TABLE 1.5 describes some of the key digital technologies used for rapid product development.

By integrating these digital technologies, companies can accelerate product development, reduce costs, increase accuracy, and ensure that the final product meets or exceeds market demands.

TABLE 1.4
Product Development Entity Description by Stage

Stage	Product Development Entity
Ideation—Imagination	***Product Idea.*** Systematic search for selection and development of promising product ideas using patents, benchmarking and competitive intelligent analysis. This process ensures that the chosen ideas are both technically feasible and economically viable. The scope of the project and the project plan are defined.
Basic Development—Conceptualization	***Conceptual Design and Target Specification.*** This stage involves collecting information about customer requirements (restriction, usage, pains, etc.) to be embodied in the solution.
Advanced Development—Design	***Detailed and Engineering Design.*** This stage develops the structure, form, dimensions, functions, and mechanical properties of all individual parts of a Product. Drawings and manufacturing documents are produced.
Launching—Creation/ Operation	***Prototyping.*** This stage verifies any remaining problems in the product. Tests are carried out to check the functionality and potential design modifications using virtual and physical prototypes.

TABLE 1.5
Examples of Digital Technologies Employed during the Product Development Entity

Digital Technology	Application
Computer-Aided Design (CAD)	Allows designers to create detailed 2D or 3D models of products. Offers simulation capabilities to assess a product's functionality before it is physically produced.
Computer-Aided Manufacturing (CAM)	Directly uses CAD outputs to generate manufacturing instruction (e.g. CNC code) for manufacturing machinery. Enables the evaluation of alternative manufacturing processes to transform different material compositions.
Computer-Aided Process Planning (CAPP)	Bridges the gap between CAD and CAM by translating design models into actionable manufacturing process plans. Automates the preparation of machine setup instructions and tooling requirements.

Chapter 4 details the Product Development entity and its stages from Ideation to Launching. Besides, a Mio-Tracker is given as a study case. This

Digital Technology	Application
3D Printing	Enables rapid prototyping of product designs. Facilitates quick iterations based on physical testing and feedback.
Digital Twins	Virtual representations of a physical product or its components. Allows for real-time simulation and emulation of products to predict performance under various conditions.
Simulation and Analysis Software	Tools like Finite Element Analysis (FEA) and Computational Fluid Dynamics (CFD) help in predicting how a product will react to real-world forces and conditions. Reduces the need for extensive physical testing.
Product Life cycle Management (PLM) Software	Manages the entire product life cycle, from ideation to disposal by integrating data, information and knowledge models. Enhances collaboration and data management across different stages of product development.
Augmented Reality (AR) and Virtual Reality (VR)	AR can overlay digital information on the real world, aiding in design visualization and modifications. VR allows stakeholders to immerse themselves in a digital environment, helping in product testing and user experience evaluations.
Cloud Computing	Offers scalable data storage, collaboration, and software deployment resources. Allows team members across the globe to access and collaborate on product designs seamlessly.
Internet of Things (IoT)	Embeds product sensors to gather real-time data. Data collected can inform product´s performance feedback, improvements and updates.
Artificial Intelligence (AI) and Machine Learning (ML)	Helps in analyzing vast amounts of data for insights. Can predict consumer behavior, optimize designs, and improve manufacturing processes.
Collaborative Robots (Cobots)	Robots designed to work alongside humans. Can assist in tasks like decision making, engineering evaluations, and rapid prototyping.
Version Control Systems	Tools like Git help track and manage product design changes. Enhances collaboration and ensures everyone is working on the most recent design version.
Blockchain	Increases transparency and traceability of a product's performance during its complete life cycle. Ensures authenticity and security in product development processes.

bioengineering product serves as a successful approach for this entity's development.

1.3.1.3 Processes Development Entity

Applying LCE principles aims to enhance resource efficiency, minimize waste, and transition towards sustainable energy sources. Process development focuses on creating environmentally friendly manufacturing methods, while process optimization involves analytics and continuous improvement to refine efficiency and sustainability. Moreover, as manufacturing processes reach the end of their life cycle, LCE principles guide sustainability assessments during process closures and facilitate the transition to newer, more sustainable technologies. TABLE 1.6 depicts each stage of the LCE framework corresponding to this entity.

TABLE 1.6
Processes Development Entity Description by Stage

Stage	Product Development Entity
Ideation—Imagination	***Individual Component Specification.*** Product design is received, and it is decomposed to identify all individual components. Customer requirements are identified mainly in three aspects: dimensions, material properties, tolerances, surface finish, and volume. The scope of the project and project plan are defined.
Basic Development—Conceptualization	***Manufacturing Process Selection.*** Components are classified as standard components (from catalogues) or non-standard components (to be manufactured by specification). A component could be outsourced or can be manufactured in-house.
Advanced Development— Design	***Manufacturing Process Plan.*** The production plan is developed for each individual component. The manufacturing process and plan is designed and generated according to manufacturing specifications. The manufacturing process plan must be defined to achieve production volume, costs, times, quality, and flexibility (Key Performance Indicators—KPIs).
Launching—Creation/ Operation	***Evaluation of Manufacturing Process Plan.*** The process must be evaluated using simulation tools to verify that the product and its components can be produced and KPIs can be achieved.

Digital technologies are vital in this context, enabling rapid and efficient development of manufacturing processes. Computer-Aided Manufacturing (CAM) and 3D Printing/Additive Manufacturing streamline production, offering precision and reducing time from concept to market. Simulation software, such as Finite Element Analysis (FEA) and Computational Fluid Dynamics (CFD), allows for virtual modeling to perfect processes before physical execution. Digital Twins mirror the manufacturing process digitally for real-time optimization, while the Industrial Internet of Things (IIoT) gathers essential data for process enhancement.

Machine Learning and Artificial Intelligence analyze process data to refine manufacturing operations and predict potential restrictions or risks. Augmented and Virtual Reality assist in manufacturing process validation including assembly, and provide immersive training, while robotics and automation execute repetitive tasks, enhancing speed and precision. Cloud computing supports remote process control and improves collaborative decision-making. Advanced Process Control (APC) maintains product quality, and smart sensors, alongside vision systems, ensuring continuous monitoring and quality control.

The Digital Thread integrates data across production stages for consistency, and blockchain technology bolsters life cycle and supply chain transparency. By employing these digital tools, manufacturers can forge processes that are agile, adaptable, and aligned with sustainability principles, effectively balancing rapid development with environmental stewardship.

- **Process Development:** Emphasize resource efficiency, waste minimization, and sustainable energy sources.
- **Process Optimization:** Use analytics and continuous improvement methodologies to ensure efficient and sustainable processes.
- **End-of-Process Life:** Assess the sustainability of process shutdowns or transitions to newer technologies.

Rapid manufacturing process development is essential for companies aiming to bring products to market quickly while maintaining quality and reducing costs. Digital technologies play a pivotal role in streamlining and enhancing the manufacturing process. Some of the key digital technologies used for rapid manufacturing process development are displayed in TABLE 1.7.

Leveraging these digital technologies allows manufacturers to develop more agile, flexible, and efficient production processes. The insights and optimizations made possible by these tools lead to faster process development, reduced lead times, and higher-quality outputs.

Chapter 5 details the Processes Development Entity. This chapter details each entity stage and presents a case study that applies Life Cycle Engineering (LCE) principles to manufacturing a rigid coupling for a Micromachine. This component, essential for rotational functions, is designed with

TABLE 1.7
Examples of Digital Technologies Employed during the Processes Development Entity

Digital Technology	Application
Computer-Aided X (CAX)	Use of software to support ideation, design, engineering and manufacturing activities to be more precise, efficient and effective using engineering data and rules.
Allows for more precise and efficient production processes.	
3D Printing/Additive Manufacturing	Builds prototypes (components, molds, tooling) for rapid production of physical parts and tooling, to validate physical properties and performance. Before scaling up the production of such components and tools. Can significantly shorten the production cycle and allows for on-demand manufacturing.
Simulation and Analysis Software	Digital tools like Finite Element Analysis (FEA) and Computational Fluid Dynamics (CFD) can model and predict the performance of manufacturing processes. These tools help in optimizing parameters before actual production begins.
Digital Twins	Creates a digital replica of the physical manufacturing process. Can be used to monitor, analyze, and optimize the real-world manufacturing process in real-time.
Industrial Internet of Things (IIoT)	Embeds sensors in machinery to collect data during the manufacturing process to provide feedback to engineers to optimize the production process and improve KPIs. Data-driven insights can lead to process optimization, predictive maintenance, and reduced downtime.
Machine Learning and Artificial Intelligence	Analyzes large datasets from the manufacturing process to optimize operations. Predictive analytics can foresee issues and optimize parameters for improved efficiency.
Augmented Reality (AR) and Virtual Reality (VR)	AR can assist workers by overlaying digital information on real-world scenarios, aiding in production, assembly, maintenance, or training. VR can be used for immersive training and process visualization.

Digital Technology	Application
Robotics and Automation	Robots can perform repetitive tasks with high precision and speed. Cobots (collaborative robots) can work alongside humans, aiding in tasks and improving efficiency.
Cloud Computing	Enables remote monitoring and control of manufacturing processes. Facilitates data storage and sharing, improving collaboration and decision-making.
Advanced Process Control (APC)	Uses statistical methods and control strategies to optimize the manufacturing process. Helps maintain product quality and reduce variations.
Smart Sensors and Vision Systems	Monitors and collects data on various aspects of the manufacturing process. Vision systems can inspect products for defects, ensuring quality control.
Digital Thread	Integrates data flows across the manufacturing life cycle. Connects various stages of production, ensuring continuity and consistency.
Blockchain Technology	Enhances traceability and transparency in the supply chain. Ensures data integrity and security in the manufacturing process.

standardization to reduce production costs and simplify assembly. The case study demonstrates how each LCE phase—design, manufacturing, operation, and end-of-life—aligns with the 5S principles, emphasizing Social, Sustainable, Sensing, Smart, and Safe aspects.

1.3.1.4 Manufacturing System Development Entity

The design and operation of manufacturing systems prioritize efficiency, modularity, and adaptability, with sustainable practices integral to daily operations, including proactive energy and waste management. Maintenance approaches are geared towards predictive and preventive strategies to enhance system longevity and performance, while sustainable practices are also crucial during system decommissioning. TABLE 1.8 depicts each stage of the LCE framework corresponding to this entity.

TABLE 1.8
Manufacturing System Development Entity Description by Stage

Stage	Product Transfer	Technology Transfer	Facility Design
Ideation—Imagination	Individual Component Specification. Product information is collected. Bill of Materials (BOM) is carried out to identify materials, standard components, quality requirements and delivery times.	Individual component specification. The information about components or families of components is captured in three aspects: geometry (drawings), materials (specifications), and production volumes (batch size).	Product and Manufacturing Process specification. Technical information on the product and manufacturing processes must be used to begin the design of a new facility.
Basic Development—Conceptualization	Manufacturing and quality control requirements are defined to be shared to suppliers. Manufacturing and Quality control specifications and requirements are defined in terms of volume, cost, and time. Product attributes (material, dimension, tolerances, surface finish). Control variables are defined for acceptance or rejection	Manufacturing Resources Specification and selection. Available technology (machines, Material Handling Equipment, Tooling, Jig/Fixture) is evaluated to select the best equipment to manufacture this component.	Manufacturing System Specification. Available technology (machines, Material Handling Equipment, Tooling, Jig/Fixture) is evaluated to select the best equipment to manufacture the product based on KPIs (costs, volume, production times)

Stage	Product Transfer	Technology Transfer	Facility Design
Advanced Development—Design	Manufacturing Supplier Selection. The partners are evaluated and selected based on capacities and capabilities from different manufacturing companies to integrate their expertise to develop the manufacturing project.	Process Plan. The process chart is elaborated to define raw materials, tools, fixtures, gauges and other devices necessary to manufacture the component. Also, documents for quality control in the process and for standard operation are elaborated.	Manufacturing System Design. The Manufacturing Strategies must be defined, and the station, cells, shop-floor, lines and factory layout must be configured. The capacity plan must be evaluated using simulation tools or digital twins.
Launching—Creation/ Operation	Manufacturing and Assembly of components. Components are manufactured, assembled, and delivered to the facility, and finally, quality controls are checked and documented.	All the information related to product/components and manufacturing technology is transfer to the supplier. Technical advice is given to suppliers for equipment setup, and production ramp-up. Equipment is checked, tested, and set up. Ramp-up, and production is started.	Facility construction, set-up and ramp-up of production. The manufacturing system is configured, constructed, and installed. Ramp-up production began, and the facility's performance is evaluated, tuned and readjusted.

Digital technology innovations are key to the rapid development of these manufacturing systems. Computer-aided design (CAD) and Engineering (CAE) facilitate the creation of optimized system layouts and virtual testing of manufacturing processes, streamlining the path from design to production. Simulation software enables scenario testing to fine-tune system designs. Digital Twins offer real-time system replicas for ongoing analysis and optimization, while the Industrial Internet of Things (IIoT) allows for efficient system monitoring and predictive maintenance through embedded sensors.

Artificial Intelligence (AI) and Machine Learning (ML) provide data-driven insights for system design and operation optimization, including production scheduling and quality control. Augmented Reality (AR) and Virtual Reality (VR) support operators with digital overlays for machine maintenance and immersive system simulations for training purposes. Cloud and Edge Computing offer scalable resources for data analysis, with Edge Computing enabling rapid, localized decision-making.

Blockchain technology ensures secure, transparent data exchanges within the supply chain. Advanced robotics and collaborative robots (Cobots) automate and enhance tasks such as material handling and assembly, working safely alongside humans. The Digital Thread maintains consistent data flow throughout the product life cycle, and Model-Based Systems Engineering (MBSE) provides a comprehensive view of system optimization.

One key manufacturing technology is a Flexible Manufacturing System (FMS) that quickly adjusts to production changes, supported by Product Life cycle Management (PLM), Enterprise Resource Planning (ERP), Manufacturing Execution System (MES), Shop-Floor control and Factory Automation (PLC, sensors). Advanced Analytics and Big Data provide performance insights for continuous improvement. Cyber-Physical Systems (CPS) integrate computing with physical operations, facilitating adaptability and real-time monitoring. Modular and scalable designs enabled by digital tools allow systems to evolve with changing business needs.

- **System Design:** Develop systems that emphasize efficiency, modularity, and adaptability.
- **Operation:** Implement sustainable practices in daily operations, including energy management, waste reduction, emission control and water usage.
- **Maintenance:** Adopt predictive and preventive maintenance strategies to extend system lifespan and efficiency.
- **Decommissioning:** Ensure that decommissioning processes are sustainable and have minimal environmental impact.

Rapid manufacturing system design is essential for industries aiming to respond swiftly to market changes, reduce costs, and boost overall productivity. Digital technologies enhance the design, analysis, and implementation of manufacturing systems. Some of the primary digital technologies used for rapid manufacturing system design are described in TABLE 1.9.

TABLE 1.9
Examples of Digital Technologies Employed during the Manufacturing System Development Entity

Digital Technology	Application
Computer-Aided Design (CAD) for Facility Development	Enables designers to create and optimize manufacturing facilities, workflows, and factory layouts. Simulates manufacturing processes to identify bottlenecks or inefficiencies. Allows for designing, simulating, and testing manufacturing systems in a virtual environment before physical implementation. CAM integrates with CAD to streamline the transition from design to production.
Simulation Software	Tools like Discrete Event Simulation (DES) model the flow of materials, work-in-progress, and finished goods through a system. Allows for "what-if" analyses to test different scenarios and optimize system design.
Digital Twins	Provides a real-time digital representation of a manufacturing system. Helps monitor, analyze, and optimize the manufacturing system's performance in real-time.
Industrial Internet of Things (IIoT)	Embeds sensors in machinery and equipment to collect data during the manufacturing process. Enables real-time monitoring and feedback, optimizing system performance and maintenance. Facilitates predictive maintenance, real-time monitoring, and efficient resource utilization.
Artificial Intelligence (AI) and Machine Learning (ML)	Uses data-driven insights to optimize manufacturing system design and operation. Can predict system disruptions, optimize workflows, and improve resource allocation. AI algorithms can optimize production schedules, enhance quality control, and predict system failures. ML can analyze vast datasets to improve system efficiency, throughput, and flexibility.
Augmented Reality (AR) and Virtual Reality (VR)	AR provides digital overlays to assist operators in machine setup, maintenance, and training. VR allows for immersive simulations of manufacturing system designs, aiding in training and system evaluation.

Digital Technology	**Application**
	AR assists in system layout planning, maintenance, and training by overlaying digital information on the real environment. VR provides immersive simulations for training, system design evaluation, and optimization.
Cloud Computing and Edge Computing	Provides scalable computing resources to analyze vast amounts of data from manufacturing systems. Enables real-time processing and decision-making at the source of data generation. Offers scalable computational resources for data storage, analysis, and collaboration. Edge computing processes data at the source, reducing latency and enabling faster decision-making.
Blockchain Technology	Enhances transparency, traceability, and security in manufacturing systems. Can be used for tracking components throughout the supply chain and verifying the authenticity of parts. Facilitates secure data exchange and ensures authenticity in multi-stakeholder environments.
Advanced Robotics and Cobots (Collaborative Robots)	Robots optimize tasks such as material handling, assembly, and inspections. Cobots work alongside humans, bringing automation to tasks while ensuring safety and collaboration. Automated guided vehicles (AGVs) and robotic arms improve material handling and assembly tasks. Enhances system flexibility, scalability, and adaptability.
Digital Thread	Integrates data across the entire product life cycle, from design to manufacturing to after-sales service. Enables seamless data flow, reducing errors and accelerating system design.
Model-Based Systems Engineering (MBSE)	Uses a systems approach to design, considering both the components and their interactions. Provides a holistic view of the entire manufacturing system, facilitating rapid design and optimization.
Flexible Manufacturing Systems (FMS)	Uses digitally-controlled tools to adapt to changes in production requirements quickly. Enables rapid reconfiguration of the manufacturing system to cater to different products or production volumes.
Integrated Product Life cycle Management (PLM)	Manages all stages of the product's life, from conception to disposal.

Digital Technology	Application
Integrates computer-based algorithms with physical processes.	Integrates with the manufacturing system to ensure alignment between product design, production and supply chain.
Advanced Analytics and Big Data	Analyzes vast amounts of data from various sources within the manufacturing system. Provides insights into performance metrics, KPIs, and areas for improvement.
Modular and Scalable System Design	Digital tools allow for the design of modular systems that can be easily scaled or reconfigured based on changing requirements.
Enables adaptability, autonomy, and real-time monitoring of manufacturing systems.	

Incorporating these digital technologies enables manufacturers to rapidly design adaptable and efficient manufacturing systems, respond swiftly to market changes, and reduce product time-to-market. Leveraging these digital technologies in manufacturing system design results in flexible, efficient, and responsive systems. These systems can quickly adapt to market demands, reduce lead times, minimize waste, and enhance overall operational efficiency.

Chapter 6 describes the Manufacturing System Development Entity, focusing on integrating LCE and 5S principles—Social, Sustainable, Sensing, Smart, and Safe—to enhance manufacturing system sustainability, productivity and efficiency. This chapter delves into each stage of manufacturing system development, from initial ideation, design to final deployment and decommissioning, illustrating how digital technologies can improve the facility's KPIs (quality, volume, costs, times and sustainability - reduce waste, resource efficiency).

1.3.1.5 Supply Chain Development Entity

Developing a supply chain involves a series of strategic choices and operational actions that form the backbone of product distribution and life cycle management. TABLE 1.10 depicts each stage of the LCE framework corresponding to this entity.

Here is a description of various components within the supply chain development entity. The process begins with identifying and prioritizing suppliers who meet the cost, quality, and delivery requirements and share a commitment to sustainability and ethical practices. This is significant because it reflects the company's product values and helps build a resilient and responsible brand image.

TABLE 1.10
Supply Chain Development Entity Description by Stage

Stage	Supply Chain Development Entity
Ideation—Imagination	***Supply Chain Strategy (Competitive Strategy).*** Plan the efficient flow of materials and information from sourcing to consumer, ensuring timely delivery and risk mitigation, with a focus on technology integration and adaptability to demand and market shifts. Define a competitive strategy of the supply chain: Operational Excellence, Customer Focus or Product Innovation.
Basic Development—Conceptualization	***Supply Chain Planning (Value Chain Strategy).*** Integrates forecasting, inventory, production, and distribution to meet customer demand. Uses data to predict sales, manages stock levels against costs and service goals. Ensures raw materials and components are secured from suppliers for uninterrupted production. Defines the Value Chain strategy: Vertical Integration, Horizontal Integration, Strategic Business Units and Collaboration Networks (Vertical or Horizontal).
Advanced Development—Design	***Supply Chain Operations (Production/Service Strategy).*** Addresses daily supply chain management, focusing on procurement, production, distribution, and logistics to streamline processes, manage inventory, ensure product delivery, and adapt to market changes. Defines production strategy: a) Production Strategy: MTS (Market to Stock), MTO (Make to Order), ATO (Assemble to Order), CTO (Configure to Order), Build to Order (BTO), ETO (Engineering to Order) b) Service Strategy: Services on Catalogues (SoC), Configuration of Services (CoS) and Design of Services (DoS)
Launching—Creation/Operation	***Supply Chain Execution (Drivers).*** The established supply chain is executed, including order fulfillment, logistics management, and the monitoring of supplier performance to ensure timely delivery and quality. It relies on real-time tracking and collaboration with all supply chain partners to ensure efficient operations and customer satisfaction, leveraging technology for automation and smooth information exchange. Key decisions are made regarding the core drivers of the supply chain: Inventory, Transport, Manufacturing facilities and Information Technology.

- **Supplier Selection:** Prioritize suppliers who adhere to sustainability and ethical practices.
- **Transportation:** Opt for transport methods with lower carbon footprints, like rail over road or electric vehicles.
- **Warehousing:** Implement energy-efficient warehousing practices.
- **End-of-Life Management:** Ensure that products at the end of their life are returned, recycled, or disposed of sustainably.

Digital technologies have transformed supply chain management and operations, enhancing visibility, agility, and resilience. Therefore, TABLE 1.11 shows some of the key digital technologies and their applications in supply chain management:

By leveraging these digital technologies, businesses can achieve a more integrated, transparent, and agile supply chain, better equipped to handle disruptions and cater to the dynamic needs of the market.

Chapter 7 explores the Supply Chain Development Entity, detailing stages from supply chain strategy to execution. This chapter emphasizes the integration of Life Cycle Engineering (LCE) with 5S principles (Social, Sustainable, Sensing, Smart, Safe) to enhance supply chain resilience, efficiency, and sustainability. Advanced technologies like IoT, AI, and blockchain are highlighted as essential tools for transforming supply chains into dynamic and transparent systems. Besides, a case study on Schneider Electric illustrates these principles in practice, showing how the company addresses Scope 3 emissions through strategic supplier collaboration, digital tool implementation, and sustainable operations.

1.3.1.6 Competitive Intelligence

Competitive intelligence in the industry and manufacturing is crucial because it enables organizations to gain market leadership and competitive advantage, especially as consumers increasingly value sustainability [82]. It involves the gathering and analyzing of information about competitors, market conditions, and any other external factors that can affect a company's competitive stance. By understanding the competitive landscape, companies can make more informed strategic decisions, anticipate market shifts, innovate, and maintain compliance with relevant standards like the ISO 14000 series for environmental management [73].

In the context of LCE, competitive intelligence can be used to enhance sustainability across the product life cycle. It helps assess the environmental impact of materials and processes and ensures that the products are designed and manufactured according to the highest environmental and quality standards. LCE emphasizes using advanced computational techniques, collaborative development, and feedback loops, all benefiting from the insights gathered through competitive intelligence. By applying competitive intelligence, organizations can identify opportunities for innovation, enhance

TABLE 1.11
Examples of Digital Technologies Employed during the Supply Chain Development Entity

Digital Technology	Application
Supply Chain Management (SCM) Software	Comprehensive platforms integrate various supply chain management functions, from procurement to distribution. Enables better planning, execution, and collaboration across the supply chain.
Industrial Internet of Things (IIoT)	Embeds sensors in products, vehicles, and warehouses to provide real-time data. Facilitates tracking, monitoring, and optimization of resources and processes.
Artificial Intelligence (AI) and Machine Learning (ML)	Enables demand forecasting, inventory optimization, and route planning. Identifies patterns and predicts disruptions or inefficiencies in the supply chain.
Blockchain	Provides a decentralized ledger for secure, transparent, and traceable transactions. Enhances authenticity, traceability, and trust in multi-stakeholder environments, especially beneficial for ensuring product authenticity and ethical sourcing.
Advanced Analytics and Big Data	Analyzes vast amounts of data to derive actionable insights. Helps in demand prediction, inventory management, and customer behavior analysis.
Cloud Computing	Facilitates data storage, sharing, and collaborative decision-making. Enables real-time communication and updates across global supply chains.
Digital Twins	Creates digital replicas of physical supply chain assets or processes. Enables real-time monitoring, simulation, and optimization.
Augmented Reality (AR) and Virtual Reality (VR)	AR assists in warehousing operations like picking and packing by overlaying digital information. VR provides immersive training and visualization of supply chain processes.
Autonomous Vehicles and Drones	Used for delivery and tracking of goods, especially in last-mile delivery. Reduces transportation costs and increases speed.
Robotics and Automation	Automates warehousing operations like sorting, picking, and packaging. Enhances storage and distribution centers' speed, efficiency, and accuracy.

Digital Technology	Application
Geolocation and GPS Tracking	Monitors the movement of goods in real-time. Provides timely updates, ensuring timely deliveries and reducing delays.
Radio Frequency Identification (RFID) and Near Field Communication (NFC)	Tracks products throughout their life cycle, from manufacturing to end-consumer. Enhances inventory management and reduces counterfeiting.
Smart Contracts	Automated contracts on platforms like blockchain.
	Ensures that agreements and transactions are executed when predefined conditions are met, reducing disputes and enhancing efficiency.
Integrated Business Planning (IBP) Platforms	Provides an integrated approach to business planning, linking strategic plans with operations and finance.
	Enhances alignment and collaboration across different functions of a supply chain.
e-Enterprise/intra-enterprise and inter-enterprise integration	e-Business (Business Intelligence, Data Analytics, AI - Artificial Intelligence); e-Customer (e.g. e-Marketing, CRM - Customer Relationship Management); e-Engineering (e.g. PLM, CAD, CAM, CAE); e-Supply (e.g. ERP - Enterprise Resource Planning, SCM- Supply Chain Management, SCI - Supply Chain Integration, LMS - Logistic Management Systems), e-Brokerage (e.g. ERP, SRM - Supply Relationship Management, RIS - Remote Inspection Systems), Total Integrated Factory (e.g. PLM, ERP, MES - Manufacturing Execution System, Shop Floor Control, Factory Automation).
	Customer-oriented company: helps in demand forecasting and enhancing customer satisfaction.

their processes, reduce costs, and ensure their products meet the evolving demands of consumers and regulations [74, 119].

Using competitive intelligence during LCE involves gathering data on competitors' practices and products, understanding regulatory trends, and monitoring technological advancements that could impact the product's life cycle from design to disposal. It is particularly important for informing decisions about eco-design, material selection, and manufacturing processes, which all contribute to a product's sustainability and marketability. However, while integrating competitive intelligence within LCE practices, companies may face challenges such as the need for significant investment, the complexity of life cycle assessments, the need for specialized skills to navigate digital technologies, rapid technological changes, and cybersecurity risks [35].

Chapter 8 explores the application of Competitive Intelligence (CI) across each entity within the Life Cycle Engineering (LCE) framework, emphasizing the strategic role CI plays in sustaining a competitive advantage. This chapter introduces CI's systematic approach to gathering, analyzing, and utilizing information about competitors, market trends, technologies, and regulatory frameworks to support strategic decision-making and innovation.

Through a case study on nearshoring to Mexico, the chapter illustrates CI's importance in addressing challenges like resource supply, infrastructure, and regulatory issues. This case highlights how CI informs strategic planning, enabling Mexican companies to leverage proximity to the U.S. market while optimizing their supply chains. Integrating CI within the 5S principles—Social, Sustainable, Sensing, Smart, and Safe—enables companies to make data-driven decisions that enhance competitiveness, sustainability, and resilience across all stages of product life cycles.

Another way to review the Life cycle Engineering Framework from TABLE 1.1 is by Stages.

1.3.2 Stages

The LCE is segmented into four main stages: Ideation/Imagination, Basic Development/Conceptualization, Advanced Development/Design, and Launching/Creation/Operation. This spans various entities: Material, Product, Process, Manufacturing System, and Supply Chain. Each entity has specific activities and outputs for every stage.

1.3.2.1 Idea—Imagination

This stage is crucial for setting the foundation of the development process. It involves understanding future trends (megatrends and microtrends), exponential and disruptive technologies, future society needs, and envisioning how materials, products, processes, manufacturing systems and supply chain can cope with the performance requirements, sustainability goals and society behavior. The significance lies in establishing a clear direction and goal for the project.

- **Material Development Entity:** At this stage, materials are conceptualized based on user needs, functionality, performance, trends and disruptive technologies. This is where the inherent properties required of materials are determined, considering what the end user will expect in terms of structural strength, conductivity, or any other relevant property.

- **Product Development Entity:** Product ideas are generated and screened for their technical and economic viability. Using trends, patents, benchmarking and user behaviors. It is a brainstorming phase to conceptualize what products could be developed and to outline the project's scope and plan.
- **Process Development Entity:** Product design and its components are analyzed in detail to specify the components that must be manufactured, including their geometry, material, and production volumes. This is significant for understanding the customer requirements and aligning the product design accordingly.
- **Manufacturing System Development Entity:** This stage focuses on preparing the manufacturing system through one of the three potential pathways. Each pathway ensures the manufacturing system is optimized to align with the project's requirements:
  - ***Product Transfer (Outsourcing):*** Individual component specifications are defined, and product information is collected. A Bill of Materials (BOM) is created to identify required materials, standard components, quality requirements, and delivery times. The product and its components are outsourced to be produced with strategic suppliers.
  - ***Technology Transfer:*** Individual component information is captured and organized into three key aspects: geometry (drawings), materials (specifications), and production volume (batch size). The manufacturing technology and process to produce such components is analyzed and specified to be transferred to suppliers, so they are able to manufacture the product or component.
  - ***Facility Development:*** Technical information about the product and manufacturing processes is used to design and establish a new facility tailored to production needs.
- **Supply Chain Development Entity:** The focus here is on developing a strategic plan for the supply chain. This encompasses everything from sourcing raw materials to delivering the final product to the consumer, highlighting the importance of technology integration and flexibility in response to market changes.

During the ideation phase, the goal is to brainstorm and gather various product ideas and component specifications based on analyzing user needs and technology requirements. Material functionality and performance are critical factors considered at this stage. Thus, this stage is crucial for setting the foundation of product development. By understanding user needs and potential product ideas, developers can create innovative solutions that are both relevant and feasible.

1.3.2.2 Basic Development—Conceptualization

The initial ideas are transformed into tangible concepts with specified characteristics. Materials are engineered at a molecular level, and conceptual designs are drafted. This stage is significant for ensuring the feasibility of the ideas and planning for their real-world application.

- **Material Development Entity:** Designers and engineers determine the molecular structure and synthesis processes needed to create the desired material properties. This is vital for establishing the feasibility and sustainability of the material in the context of the product's needs.
- **Product Development Entity:** A conceptual design is formulated during this phase, considering constraints and customer requirements. The significance lies in translating customer needs into a tangible design that can be further developed.
- **Process Development Entity:** Selecting the appropriate manufacturing processes for different components becomes the focus, deciding whether parts should be purchased as standard or manufactured in-house.
- **Manufacturing System Development Entity:** Detailed manufacturing and quality control requirements are established, which include defining product attributes and control variables. It's a critical step for ensuring the product can be made to the required standards and within budget.
- **Supply Chain Development Entity:** Supply chain planning is developed to efficiently balance inventory levels, production schedules, and distribution to meet customer demand. It's significant for aligning production with market demand and managing costs.

In this phase, the focus shifts to creating molecular structures and conceptual designs that incorporate design constraints and customer requirements. The classification of the manufacturing process is also determined, setting the stage for further development. Hence, conceptualization is significant because it translates theoretical ideas into tangible designs. It also ensures that the developed products align with customers' needs and can be manufactured within quality control standards.

1.3.2.3 Advanced Development—Design

This is where the practical design occurs, creating detailed blueprints and specifications. For manufacturing systems, it involves selecting suppliers and finalizing the process plan. The significance of this stage is in bridging the gap between concept and reality, setting the stage for actual production.

- **Material Development Entity:** The detailed engineering design of the materials is completed, ensuring they meet the specified manufacturing requirements.
- **Product Development Entity:** Detailed and engineering designs of all parts are completed. This includes exact arrangements, dimensions, and surface properties, all necessary for manufacturing.
- **Process Development Entity:** A detailed manufacturing process plan for each component is created, specifying the time and cost implications and ensuring that production volumes and quality standards can be met.
- **Manufacturing System Development Entity:** This includes the selection of manufacturing partners and suppliers and the detailed manufacturing process planning, including equipment and layout of the production facility.
- **Supply Chain Development Entity:** The focus shifts to supply chain operations, managing the daily activities needed to produce and deliver products. It is significant for maintaining a smooth, uninterrupted flow of goods to the market.

Detailed engineering and design activities are performed, resulting in engineering drawings and detailed manufacturing plans. The selection of suppliers and the design of manufacturing systems are key activities at this stage. This phase is crucial for finalizing product design and preparing for manufacturing. The activities ensure that the product can be produced with the available technology and that the manufacturing systems are strategically planned.

1.3.2.4 Launching—Creation/Operation

The final stage involves bringing the product to life through manufacturing, assembly, and launching the supply chain operations. It is significant because it tests the designs in real-world conditions, ensuring they meet the required standards before full-scale production and distribution.

- **Material Development Entity:** Synthesis and prototyping of materials take place, with thorough analysis and characterization to ensure they meet design requirements. It is the stage where the material is tested in a real-world scenario to verify it performs as expected.
- **Product Development Entity:** Prototypes of the product are created and tested, which is a crucial phase for identifying and rectifying any issues before mass production.
- **Process Development Entity:** The manufacturing process is fine-tuned with the help of simulation tools, ensuring that the final process can produce the product at the desired volumes and standards.

- **Manufacturing System Development Entity:** The actual manufacturing and assembly of the product commence, alongside the setting up of the production facilities. This stage is critical for transitioning from planning to actual operation.

- **Supply Chain Development Entity:** The execution of the supply chain plan begins, which includes fulfilling orders, managing logistics, and monitoring supplier performance to ensure timely delivery and product quality. It's significant for meeting customer expectations and achieving business success.

The final stage involves prototyping, assembly, and evaluating the manufacturing process. The facility design is finalized, and the supply chain operations are executed. Therefore, launching is when the product comes to life. Testing prototypes, evaluating processes, and implementing supply chain operations are critical to ensuring the product meets design specifications and market demands.

The end-of-life stage of products is critical within the LCE framework, especially when aligned with the 5S principles. This stage involves strategies to ensure products are disposed of in a manner that minimizes environmental impact and maximizes resource recovery. Additionally, the adaptation of the 10R framework—Refuse, Rethink, Reduce, Reuse, Repair, Refurbish, Remanufacture, Repurpose, Recycle, and Recover—within the LCE framework further enhances this approach by focusing on sustainable end-of-life management. Therefore, the integration of the 10R framework within the LCE approach involves systematic planning and execution across all stages of the product life cycle:

- **Design Phase:** Incorporating the 5S and 10R principles at the beginning of product design ensures that every aspect, from material selection to product architecture, is aligned with sustainability and efficiency goals.

- **Manufacturing Process:** Adapting manufacturing operations to accommodate refurbished or remanufactured components and designing systems that facilitate easy disassembly and recycling at the end of the product's life.

- **End-of-Life Management:** Developing robust systems for collecting and processing end-of-life products, which support the efficient implementation of the 10R strategies through smart and sensing technologies.

Organizations can create a sustainable and efficient life cycle management strategy by embedding the 10R framework into the 5S-enhanced LCE framework. This strategy maximizes resource utilization and minimizes environmental impacts, supporting the creation of a circular economy and contributing to sustainable development goals.

1.3.3 Techniques and Methods Required During the Development

Following are listed some of the strategies and tools for LCE development:

- **Advanced Computational Techniques:** Predictive Modeling: Use computational modeling to predict material properties and behaviors before physical testing.
- **Machine Learning and AI:** Using data-driven methods to analyze and predict optimal material compositions quickly.
- **High-Throughput Experiments:** Design experiments to simultaneously test multiple material compositions. Use automated systems to screen large numbers of material samples rapidly.
- **Feedback Loops:** Quickly integrate feedback from testing and real-world applications to refine the material properties. Adapt and iterate based on performance data.
- **Collaborative Development:** Foster collaborations between researchers, industries, and institutions to share knowledge, resources, and best practices.
- **Open Source Platforms:** Sharing material data and research findings openly can accelerate material development by avoiding redundant efforts and promoting collaborative improvements.
- **Life Cycle Assessment (LCA):** A technique to assess the environmental aspects and potential impacts of a product, process, or service.
- **Eco-design:** Designing products with consideration for their environmental impacts throughout their life cycle.
- **Circular Economy Principles:** Designing systems where resources are reused and recycled, minimizing waste.
- **Green Supply Chain Management:** Incorporating eco-friendly practices throughout the supply chain.
- **Regulations and Standards:** Complying with, or surpassing, international standards like ISO 14000 (Environmental Management).

By incorporating LCE into decision-making processes, companies can ensure their operations are more sustainable, resilient, and future-proof. It helps reduce environmental impacts and often results in cost savings in the long run.

1.3.4 Core Technologies to Employ for the Accelerated Development of the Entity

The accelerated development of manufacturing entities sustainably and efficiently requires the integration of advanced digital technologies. Therefore, ten core digital technologies form the foundation for innovation across all entities and stages of the manufacturing life cycle.

The following technologies are identified as core due to their widespread applicability and transformative potential:

1. **Artificial Intelligence and Machine Learning (AI/ML):** Enables intelligent decision-making, predictive analytics, and process optimization across entities and stages.
2. **Simulation Tools:** Provide virtual testing environments to model behaviors, evaluate feasibility, and refine designs before physical implementation.
3. **Internet of Things (IoT):** Facilitates real-time monitoring and interconnected systems to enhance operational efficiency.
4. **Digital Twins:** Create virtual replicas of physical systems to simulate and optimize performance dynamically.
5. **Additive Manufacturing:** Supports rapid prototyping and efficient production of components and materials.
6. **Computer-Aided Tools (CAD, CAM, CAE, and CAPP):** Simplify design, engineering analysis, manufacturing, and process planning, integrating AI to enhance capabilities.
7. **Big Data Analytics and Blockchain:** Offer insights through data analysis while ensuring traceability and secure data handling across processes.
8. **Cloud Computing:** Provides scalable infrastructure for data storage, processing, and collaboration across multiple entities.
9. **Augmented and Virtual Reality (AR/VR):** Enable immersive design, prototyping, training, and operational planning.
10. **Robotics and Automation:** Streamline production processes, reduce human error, and enhance scalability.

Thus, TABLES 1.12 and 1.13 exemplify how these core technologies can be applied in detail across various entities and stages.

While these TABLES showed the ten core digital technologies, other relevant technologies will be described in each chapter. These supplementary technologies build upon the foundation established here, providing additional tools and methods for specific challenges and applications. However, the ten core technologies outlined above remain pivotal, offering the versatility and innovation necessary for the accelerated development of all manufacturing entities.

TABLE 1.12
Core Digital Technologies for Each Entity for Stages 1 and 2.

	ENTITY						
				Manufacturing System			
STAGES	Material	Product	Process	*Product Transfer*	*Technology Transfer*	*Facility Design*	Supply Chain
Ideation — Imagination	• **AI/ML** for identifying promising material combinations based on desired properties. • **Simulation Tools** to model material behavior under different conditions. • **Big Data Analytics** can analyze historical data for insights on material performance trends.	• **CAD** to draft initial designs and **AI/ML** to optimize product specifications. • Implement **AR/VR** to visualize design concepts.	• **AI/ML** to identify optimal process parameters. • **Simulation Tools** evaluate process feasibility.	• **AI/ML** for product compatibility analysis across different manufacturing setups. • **Simulation Tools** to model production lines for the transferred product.	• **Simulation Tools** and **AI/ML** to assess technology adaptability.	• **CAD/CAE** tools to design energy-efficient layouts. • **Simulation Tools** to evaluate different facility configurations.	• **Big Data Analytics** and **AI/ML** for demand forecasting and supply chain optimization.
Basic Development — Conceptualization	• **AI/ML** to optimize material properties through predictive algorithms. • **CAE** tools simulate stress, heat transfer, and other engineering properties. • **Additive Manufacturing** prototypes materials for initial testing. • **Cloud Computing** stores and processes large datasets of material characteristics.	• **AI/ML** for predicting product behavior under various conditions. • **CAE** supports detailed analysis of mechanical, thermal, and structural properties. • **Simulation Tools** ensure functionality and identify potential design flaws.	• **AI/ML** to refine workflows for efficiency. • **Variant CAPP** for standardizing processes. • **CAE** to test manufacturing scenarios virtually.	• **AI/ML** to optimize production workflows for the transferred product. • **Digital Twins** to simulate the production environment before actual transfer. • **Blockchain** for secure and traceable product data transfer.	• **Simulation Tools** and **AI/ML** to assess technology adaptability. • **Blockhain** for secure technology IP transfer.	• **AI/ML** to optimize energy consumption patterns for preliminary designs. • **Digital Twins** of the facility to test operations virtually.	• **AI/ML** to optimize logistics pathways. • **Digital Twins** for simulating supply chain operations. • **Blockchain** for secure and transparent supply chain tracking.

TABLE 1.13
Core Digital Technologies for Each Entity for Stages 3 and 4.

	ENTITY						
				Manufacturing System			
STAGES	**Material**	**Product**	**Process**	*Product Transfer*	*Technology Transfer*	*Facility Design*	**Supply Chain**
Advanced Development — Design	• **AI/ML** to refine material production processes using real-time analytics. • **Digital Twins** for testing material performance virtually in different environments. • **Robotics** and **Automation** streamline material production processes.	• **AI/ML** to personalize product features for specific market segments. • **Digital Twins** of the product to test performance in real-world conditions. • **Additive Manufacturing** aids in finalizing prototypes.	• **AI/ML** to implement adaptive control systems for variability. • **Hybrid CAPP** with **AI/ML** for adaptive process planning. • **IoT** for monitoring and improving process efficiency.	• **Robotics** and **Automation** to optimize production lines.	• **AI/ML** to adapt control algorithms for seamless integration. • **Digital Twins** for testing new technology integrations.	• **Robotics** and **Automation** to optimize facility operations. • **IoT** to monitor real-time production line adjustments.	• **AI/ML** to predict risks and disruptions. • **Robotics** and **Automation** to automate logistics. • **IoT** for real-time tracking and inventory management.
Launching— Creation/ Operation	• **IoT** for real-time monitoring of material quality in production.	• **IoT**-enabled devices to monitor product performance post-launch. • **Blockchain** for product traceability and authenticity.	• **Robotics** and **Automation** to automate processes. • **Cloud Computing** to store and analyze process performance data. • **AR/VR** to identify potential challenges for a more grounded view of process stages.	• **IoT** to monitor the real-time product transfer process.	• **IoT** and **Robotics** to automate deployment and monitor performance.	• **IoT** systems to monitor energy consumption and operational efficiency. • **AR/VR** to conduct virtual walkthroughs for stakeholders.	• **AI/ML** to enhance real-time decision-making. • **Cloud Computing** for scalable supply chain data storage and processing.

1.4 The Micromachine: A Case Study in Life Cycle Engineering

The Micromachine is a versatile and modular CNC (Computer Numerical Control) machine tool (FIGURE 1.2). It is designed for adaptability and flexibility, catering to various machining operations. Key features include a modular design for easy upgrades, an open software system for customization, and a range of modules for diverse functionality. The Micromachine is suitable for educational settings and small to medium-sized businesses, emphasizing innovation and skill development. Technological megatrends and patent analyses have influenced its design and development. The machine's versatility is underpinned by its reconfigurability, high configurability, and a workspace defined by X, Y, and Z axes. A CNC controller manages its operation and works with materials like metal, plastic, and wood [2].

FIGURE 1.2
Key features of the micromachine.

1. **Reconfigurable:** It is easy to assemble, set up and ramp up to begin production once it has been acquired.
2. **Customization:** It is tailored to carry out multiple processes on the same machine tool.
3. **Workspace:** The workspace for the development of functional prototype parts must offer a minimum of 10 cm of displacement per installed axis.
4. **System Anatomy:** The Micromachine works with standard tools, offers precision in product creation, and its components are easily replaceable.

5. **Working Materials:** The Micromachine processes wood, aluminum, plastics, and low-impact environmental sources.

1.4.1 Walkthrough of the Entire Life Cycle Process, from Idea to Launching

The Micromachine's advanced features and modular design make it highly applicable for engineering education and specialized manufacturing products. The open software system allows seamless integration with digital twin technology, facilitating real-time monitoring and simulation. This enables predictive maintenance, process optimization, and efficient design alterations, enhancing the machine's performance and applications in various industries. The combination of physical versatility and digital connectivity positions the Micromachine as an ideal candidate for digital twin applications.

Regarding the stages of the LCE framework, the Micromachine's journey begins with the *Idea—Imagination* stage, where initial concepts are generated, considering both customer needs and technological possibilities. As it transitions to the *Basic Development—Conceptualization* stage, these ideas are transformed into tangible designs with careful consideration of material choices and manufacturing processes.

The *Advanced Development—Design* stage marks a critical point where detailed engineering designs are finalized, ensuring the Micromachine's components are optimized for manufacturing. Here, the focus is on precision, cost-effectiveness, and sustainability.

The Micromachine moves from the drawing board to reality in the *Launching—Creation/Operation* stage. Production begins, emphasizing quality control and efficiency. The supply chain is managed to ensure timely delivery, and the Micromachine is introduced to the market.

Finally, it is important to mention that all the LCE stages for each entity must be done concurrently so the Simultaneous Engineering teams can exchange continuously ideas, concepts, restrictions and decisions on each entity.

1.4.1.1 Micromachine: Material Development

Material development for the micromachine is the selection of the materials that will satisfy the performance of the micromachine based on the manufacturing process it will perform like metal cutting, 3D printing, laser engraving, and plastic extrusion. This is a crucial stage where the feasibility and sustainability of the materials are established in the context of the Micromachine's requirements (FIGURE 1.3):

1. **User Needs, Functionality and Performance:** the Micromachine requires durable, lightweight materials and possibly has specific electrical or thermal properties, this stage would involve

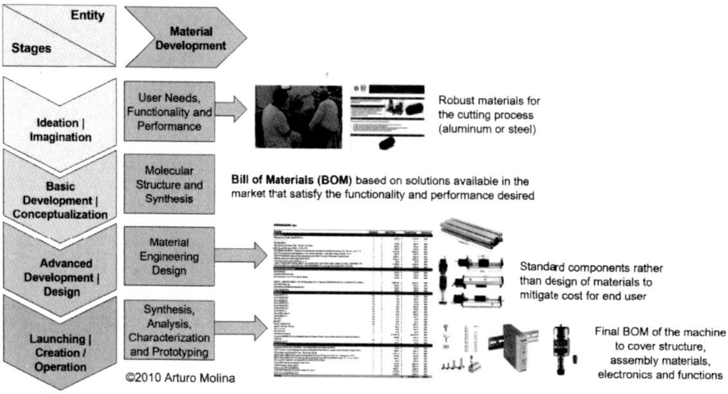

FIGURE 1.3
Material development and selection for micromachine.

extensive research and development to identify and select such materials. A key variable is low cost and sustainability materials.

2. **Molecular Structure and Synthesis:** the analysis of the structure and mechanical forces of the micromachine determines the materials to be used: steel and aluminum. Two different prototypes were developed using steel and aluminum [114].
3. **Material Engineering Design:** the materials selected for the micromachine is the Bill of Materials (BOM) based on already available materials in the market that satisfy functionality and performance levels. In this development, there was no need to develop and engineer new materials.
4. **Synthesis, analysis, characterization and prototyping:** standard materials and components were selected rather than materials design to mitigate cost for end-user. Final BOM of the machine to cover structure, assembly materials, electronics, and functions.

The 3D Food Printer: A Specific Case for Materials Development for Printing Food

The specific case presented here is an adaptation from the material developed by [97] and for the case of a 3D food printer [91] and plastic extruder [28]. In the Idea Stage of material development for the 3D Food printer, the primary goal is to conceive and evaluate suitable materials for printing food. The primary focus is on conceptualizing materials that align with the overarching goals of sustainability and functionality. The team sets out to identify food-safe, environmentally friendly materials, and compatible with 3D printing technologies.

- **Environmental and Performance Requirements Identification:** The team identifies the key environmental and performance requirements the materials need to meet. This includes biodegradability, recyclability, non-toxicity, and the ability to handle different types of food substances without degradation.
- **User Needs and Expectations:** Understanding user needs is crucial in this phase. Feedback from potential users, such as culinary professionals, educators, and food enthusiasts, is gathered to understand food safety and material durability expectations.

During the *Basic Development Stage*, the team delves deeper into the engineering aspects of the materials to be used in the Micromachine. The focus is on designing materials that can withstand the operational demands of 3D food printing while ensuring minimal environmental impact.

- **Material Selection:** Various candidate materials are evaluated for their mechanical properties, safety, and environmental impact. Materials such as food-grade bioplastics or composite materials that can decompose naturally or be easily recycled are prioritized.
- **Prototyping and Testing:** Small-scale prototypes of material samples are created to test their performance in simulated operational conditions. These tests help assess the materials' feasibility for effective food printing and handling.

The *Advanced Development Stage* involves finalizing the material choices by conducting extensive testing and refinement to ensure that the materials meet the functional requirements and uphold the sustainability standards set in the earlier stages.

- **Engineering Design:** Detailed engineering designs of the materials are developed. This includes the formulation of composites or blends that enhance the printability and structural integrity of the materials during and after printing.
- **Sustainability Optimization:** The sustainability of each material is evaluated in depth. This includes life cycle assessments to measure the carbon footprint and environmental impact of the materials throughout their life cycle from production to disposal.
- **Regulatory Compliance and Safety Testing:** Ensuring compliance with food safety and environmental regulations is critical. Materials undergo rigorous safety testing to meet global standards, ensuring they are safe for food contact and do not leach harmful substances.

Finally, during the *Launching Stage* and with the materials fully developed and tested, the final stage focuses on integrating these materials into the production processes of the Micromachine and preparing them for full-scale use and distribution.

- **Scale-up and Quality Control:** Production processes for the materials are scaled up, with a strong emphasis on maintaining quality control during mass production. The processes are optimized to minimize waste and enhance efficiency.
- **Supply Chain Integration:** The supply chain for sourcing and delivering these materials is established. Suppliers who adhere to the same environmental and ethical standards are selected to ensure consistency and sustainability across the board.
- **Market Introduction and Feedback Loop:** As the Micromachine enters the market, the initial batches of materials are closely monitored. Customer feedback is solicited to understand real-world performance and satisfaction, allowing for ongoing improvements and refinements.

Throughout the material development process for the 3D Food Printer, there is a continuous emphasis on innovation, sustainability, and user-centric design. By rigorously testing and refining the materials used, the project not only meets its functional goals but also contributes positively to environmental sustainability, aligning with the broader objectives of life cycle engineering.

1.4.1.2 Micromachine: Product Development

During the product development, ideas for the Micromachine are conceived and screened for their technical and economic viability (product idea). This phase is crucial for conceptualizing the products needed to develop the Micromachine. It also involves outlining the project's scope and plan. Besides, the Micromachine's product design is analyzed in detail. This analysis specifies the components that must be manufactured, including their geometry, material, and production rates. FIGURE 1.4 depicts the Micromachine Product Development Entity for each stage of the LCE framework.

- **Idea—Imagination Stage:** Micromachine product development begins with the ideation phase, where brainstorming and creativity are key to defining the product's concept and scope.
  - ***Product Conceptualization:*** The team generates various ideas for the Micromachine, envisioning it as a multifunctional manufacturing process machine that can also serve educational purposes in technical and engineering schools. The primary aim is to integrate advanced technology with user-friendly interfaces to enhance learning and food preparation.
  - ***Feasibility Studies:*** Preliminary feasibility studies assess the technical and economic viability of the proposed ideas. This includes considering potential market demand, technological requirements, and the product's educational impact.

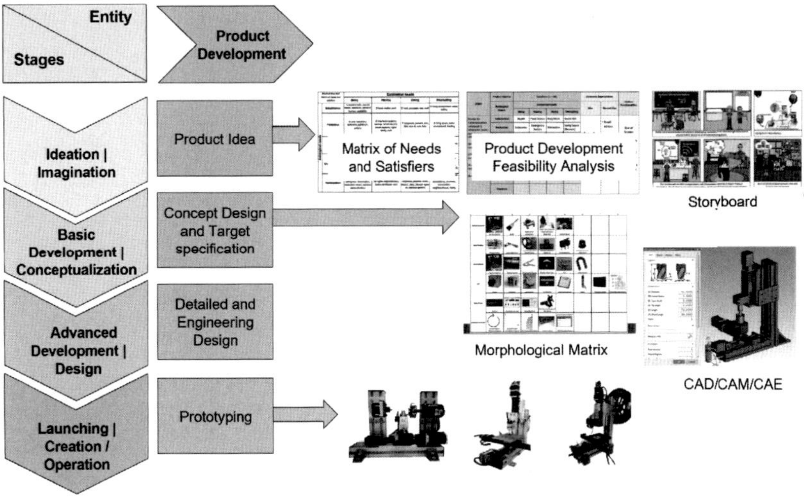

FIGURE 1.4
Product development for micromachine.

- **Basic Development—Conceptualization Stage:** With a clear product concept, the team moves to the conceptual design phase, where more detailed planning and specification development occur.
  - ***Design Specifications:*** Detailed design specifications are established, which include the size, functionality, user interface, and compatibility with various manufacturing process. The specifications are designed to ensure the product can be easily used in educational and manufacturing facilities.
  - ***Prototyping:*** Early prototypes of the Micromachine are developed to test the concept in practical settings. These prototypes help identify any design flaws and gather initial user feedback, which is crucial for further refinements.
- **Advanced Development—Design Stage:** During the advanced development stage, the design of the Micromachine is finalized. This phase involves detailed engineering, ensuring every component functions optimally and meets all regulatory standards.
  - ***Engineering Design:*** This involves finalizing the product's mechanical, electrical, and software components. CAD, CAM and CAE tools are extensively used to model and simulate the product's operations to ensure precision and efficiency.
  - ***User Interface Design:*** Special attention is given to the user interface design to ensure it is intuitive and accessible for all users, including students and professional engineers. This includes

touchscreen controls, customizable settings, and possibly AI-driven features to aid in recipe creation and modification.
  - ***Testing and Compliance:*** Rigorous testing ensures the product meets all safety and quality standards. This includes functional testing of the manufacturing process mechanisms and safety testing. Compliance with relevant local and international standards is verified to prepare the product for widespread distribution.

- **Launching—Creation/Operation Stage:** The Micromachine is ready for full-scale production and market introduction in the final stage. This stage ensures that the manufacturing and launch processes are aligned with the product's quality standards and market expectations.
  - ***Manufacturing Setup:*** The manufacturing processes are set up with an emphasis on quality control and efficiency. Supply chains for components are secured, focusing on sustainability and reliability.
  - ***Market Launch and Promotion:*** The product is launched with targeted marketing campaigns for educational institutions, professionals, and tech enthusiasts. Promotional activities focus on the innovative aspects of the Micromachine and its benefits in engineering education and industry manufacturing process.
  - ***Customer Support and Feedback:*** Post-launch, the focus shifts to customer support and gathering feedback. This feedback loop is crucial for identifying any issues in real-world usage and for continuous improvement of the product. Customer service teams are trained to provide effective support and gather user insights that can inform future updates or new product developments.

The product development process for the Micromachine is meticulously planned to ensure that the final product not only meets the functional requirements for different manufacturing processes but also adds value in educational contexts, thereby enhancing learning experiences and engineering innovation.

1.4.1.3 Micromachine: Process Development Entity

In the Processes Development the focus in on selecting the appropriate manufacturing processes for different Micromachine components and decides whether parts should be purchased as standard components or outsourced or manufactured in-house. Factors like cost, availability and the specific requirements of the Micromachine influence this decision. FIGURE 1.5 depicts the Micromachine Process Development Entity for each stage of the LCE framework.

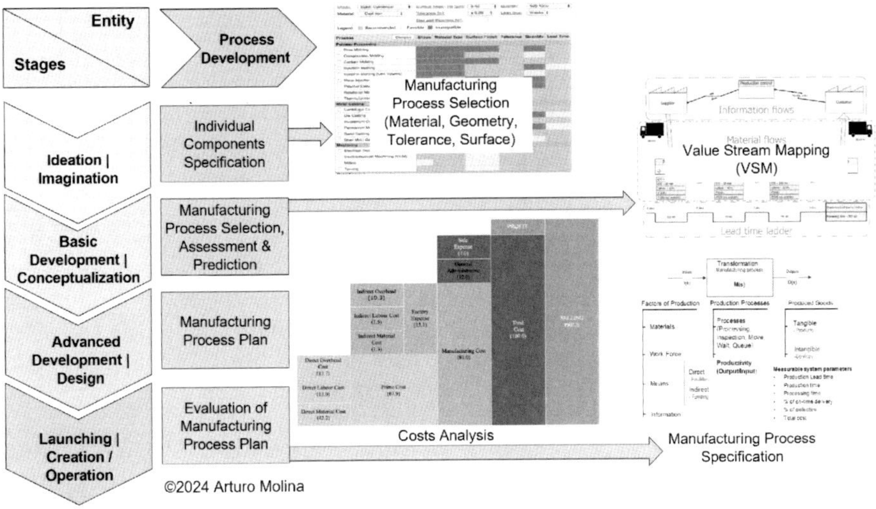

FIGURE 1.5
Process development for micromachine.

- **Idea—Imagination Stage:** The process development for the Micromachine starts with a thorough analysis of the product design received from the product development team. This stage focuses on decomposing the product into individual components and identifying the specific requirements for each.
  - ***Component Specification:*** The detailed product design is reviewed to identify all individual components that need to be manufactured. This includes understanding the geometry, material, and expected production rates of each part, which are crucial for defining the scope of the manufacturing processes.
  - ***Project Planning:*** A comprehensive project plan is created, outlining the steps required to develop the manufacturing processes. This includes timelines, resource allocation, and preliminary risk assessment to ensure that the manufacturing phase can proceed smoothly without delays.
- **Basic Development—Conceptualization Stage:** The next step is to select the appropriate manufacturing processes once the components are specified. This stage is critical for determining how each part will be produced, whether in-house or outsourced.
  - ***Manufacturing Process Selection:*** The appropriate manufacturing processes are selected for each component identified in the ideation stage. This decision is based on the material and production rate requirements, as well as on the complexity of the part. Some

components may be standard and purchased from suppliers, while others may require custom manufacturing processes.

- *Supplier and Partner Engagement:* Potential suppliers and manufacturing partners are evaluated and engaged based on their capabilities to meet the quality, cost, and timeline requirements. This early engagement is essential for ensuring that all parts meet the high standards required for the Micromachine.

- **Advanced Development—Design Stage:** Once the manufacturing processes and suppliers are in place, the next focus is on developing detailed manufacturing plans for each component. This stage involves finalizing the processes and ensuring they are optimized for efficiency and quality.
  - *Manufacturing Process Planning:* Detailed plans for each manufacturing process are developed, including the required equipment, materials, and labor. Time and cost estimates are refined to ensure that the manufacturing phase can be executed within the allocated budget and schedule.
  - *Process Optimization and Simulation:* Advanced simulation tools are used to model the manufacturing processes, allowing for optimization before physical production begins. This includes using software like Finite Element Analysis (FEA), CAM (Computer-Aided Manufacturing), and CAPP (Computer-Aided Process Planning) to ensure that all manufacturing processes are robust and capable of producing components to the required specifications.

- **Launching—Creation/Operation Stage:** The final stage of the process development involves implementing the manufacturing processes and ramping up production. This stage is critical for ensuring that all components are produced to the highest standards and are ready for assembly and final product integration.
  - *Production Implementation:* The manufacturing processes are implemented on the production floor. This includes setting up machinery, calibrating equipment, and training staff. Each process is closely monitored to ensure it performs as expected and produces components that meet all specifications.
  - *Quality Assurance and Control:* Rigorous quality control measures are implemented to ensure that every component meets the required standards. This includes regular inspections and parts testing as they come off the production line. Any deviations from the specifications are addressed immediately to prevent defects from affecting the final product.
  - *Production Ramp-Up:* Once the manufacturing processes are stabilized, production is gradually ramped up to meet the anticipated

market demand. This phase is closely monitored to ensure that scaling up production does not compromise the quality of the components or the final product.

The process development entity for the Micromachine ensures that every component is manufactured efficiently, cost-effectively, and to the highest quality standards. This meticulous planning and execution of manufacturing processes are essential for the successful production of the Micromachine, ensuring that it meets the functional and educational expectations of its users.

1.4.1.4 Micromachine: Manufacturing System Development

This entity is critical for planning the manufacturing phase of the Micromachine, ensuring quality, production costs, volumes, and delivery times. It encompasses the decision of how the Micromachine, and all its components, materials, and parts are going to be fabricated (FIGURE 1.6).

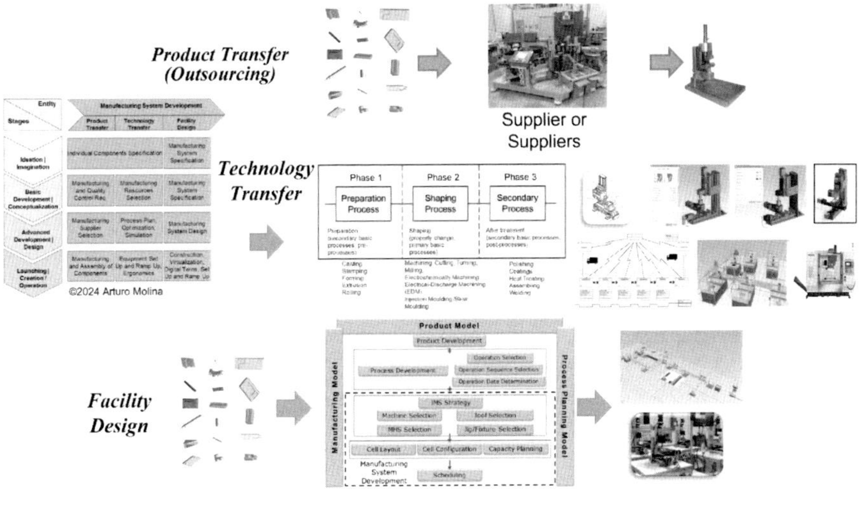

FIGURE 1.6
Manufacturing system development for micromachine.

- **Idea—Imagination Stage:** This stage lays the groundwork for establishing a comprehensive manufacturing system tailored to the micromachine's specifications. There are three alternatives and also a combination of these three alternatives is possible:
    - ***Product Transfer (Outsourcing):*** Outsourcing all the fabrication of the Micromachine to a supplier or suppliers.

  - ***Technology Transfer:*** Select some components of the Micromachine to be manufactured by specific suppliers and outsource these components and its technology to be manufactured. The assembly and remaining components are manufactured in-house.
  - ***Facility Design (New Manufacturing Facility):*** Design and create a new manufacturing facility to fabricate the micromachine in-house.

- **Basic Development—Conceptualization Stage:** In this stage, detailed planning and conceptualization of the manufacturing system are developed, focusing on scalability and efficiency.

  - ***Product Transfer:*** Specifications for manufacturing and quality control are specified to be outsourced. This includes establishing critical product attributes such as material, dimension, tolerances, and surface finish, ensuring the system meets the Micromachine's quality requirements.
  - ***Technology Transfer:*** Selection of manufacturing resources occurs here, including the evaluation and selection of the best equipment to manufacture components of the Micromachine, such as machines, material handling equipment, tooling, and fixtures. This is important to outsource the specific components defining precisely what manufacturing resources are needed to manufacture such component.
  - ***Facility Design:*** The manufacturing system specification continues, involving a detailed evaluation of available technology and equipment selection tailored to the product's specific manufacturing needs. A new facility is conceptualized that will meet the production requirements in terms of cost, volume, quality and times.

- **Advanced Development—Design Stage:** This stage involves finalizing the detailed designs and setups for the manufacturing process, ensuring operational efficiency and quality control.

  - ***Product Transfer:*** Supplier selection is carried out, evaluating and choosing partners based on their capacity to meet demand and quality standards. This ensures a seamless integration of the supply chain.
  - ***Technology Transfer:*** Detailed process planning is developed, including elaborating process charts to define raw materials, tools, fixtures, and gauges necessary for manufacturing components. The supplier's selection is carried out as well.
  - ***Facility Design:*** The complete manufacturing system design is finalized here. This includes defining manufacturing strategies and configuring stations, cells, shopfloor, and factory layouts. Capacity plans are assessed using simulation tools to ensure the facility can handle projected production volumes.

- **Launching—Creation/Operation Stage:** This stage focuses on implementing the manufacturing system, testing its effectiveness, and ensuring it can scale up to meet production demands.
  - ***Product Transfer:*** The final manufacturing and assembly of components occur by the suppliers. Components are delivered, assembled, and subjected to quality control checks to meet all specified requirements. Usually, a supplier integrates the final manufacturing of a Micromachine.
  - ***Technology Transfer:*** Equipment setup and the production ramp-up are critical here. Equipment is installed, tested, and calibrated to begin production at the supplier's manufacturing facility. This phase ensures that all technology transfers have been successful, and the manufacturing process can operate at full capacity.
  - ***Facility Design:*** The construction and setup of the manufacturing facility are completed. Production begins with a ramp-up phase where the facility's performance is closely monitored and adjusted to optimize output and efficiency.

By structuring the manufacturing system development into these detailed phases—Product Transfer, Technology Transfer, and Facility Design—the approach ensures a holistic development that not only meets the technical and production requirements of the micromachine but also integrates seamlessly with the broader operational goals of the organization. This methodical approach facilitates a smooth transition from concept to full-scale production, optimizing resource use and ensuring product quality and compliance. It is important to point out that a combination of the three approaches is feasible, or using all the three alternatives is also a possibility. The decision of how to design the manufacturing system is generally based on costs, volume, quality, and time constraints. For example, it could be necessary to design an assembly facility for the Micromachine, outsource different components and transfer some manufacturing process to a supplier to ensure the quality and cost for those specific components.

1.4.1.5 Micromachine: Supply Chain Development

This entity involves strategically planning and managing materials, components, and finished products. It includes sourcing raw materials, coordinating with suppliers, managing inventory, and overseeing the distribution of the Micromachine to customers. Effective supply chain management ensures timely production, maintains cost efficiency, and meets market demands. This entity is key in aligning the production process with market needs and sustainability goals, when outsourcing and technology transfer to fabricate the Micromachine has been decided (FIGURE 1.7)

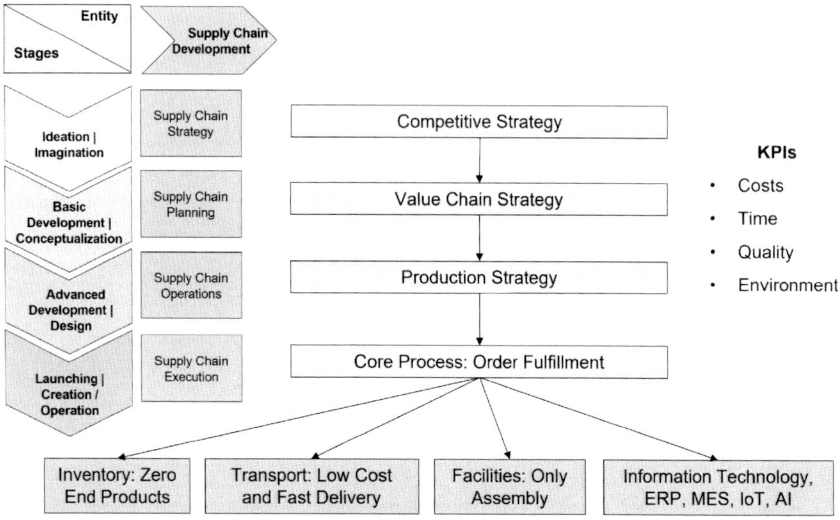

FIGURE 1.7
Supply chain development for micromachine.

- **Idea—Imagination Stage:** This initial stage involves strategizing and conceptualizing the supply chain to ensure efficient and sustainable operations from the outset.
  - ***Supply Chain Strategy (Competitive Strategy):*** The strategic planning for the supply chain begins with defining the efficient flow of materials and information from sourcing to consumer delivery. Key considerations include ensuring timely delivery, mitigating risks, and integrating technology to adapt to market demands and shifts. The goal is to establish a robust framework that supports the Micromachine's life cycle from raw material sourcing to end-user delivery. The Competitive Strategy defined for the Micromachine is Operational Excellence focusing on costs and time to market.
- **Basic Development—Conceptualization Stage:** During this stage, detailed supply chain planning is developed, focusing on optimization and responsiveness to meet anticipated customer demands.
  - ***Supply Chain Planning (Value Chain Strategy):*** This involves integrating forecasting, inventory, production, and distribution plans to manage the flow of goods efficiently. The process utilizes data to predict sales trends, manage stock levels against cost and service goals, and ensure raw materials and components are secured from suppliers for uninterrupted production. This stage is crucial for aligning the supply chain with the production requirements of the Micromachine, ensuring

that all components are available when needed by using a Vertical Network of suppliers selected by their time response and costs.

- **Advanced Development—Design Stage:** Here, the operational aspects of the supply chain are fine-tuned to ensure seamless day-to-day management, focusing on procurement, production, distribution, and logistics.
  - ***Supply Chain Operations (Production/Service Strategy and Core Process Selection/Definition and KPIs—Key Performance Indicators):*** This phase addresses the management of daily supply chain activities. It focuses on optimizing procurement, overseeing production processes, managing logistics, and ensuring that distribution channels are efficient and capable of adapting to market changes. The operations phase is designed to streamline processes, manage inventory effectively, and ensure consistent product delivery, which is critical for the success of the Micromachine. The production strategy selected for the Micromachines is MTO (Maker To Order) to avoid the inventory of end-products, also the most critical KPIs are costs, time, quality and environmental impact (carbon footprint). The core process is Order Fulfillment process.
- **Launching—Creation/Operation Stage:** The final stage involves fully executing the supply chain, implementing the plans and strategies developed in earlier stages.
  - ***Supply Chain Execution (Drivers):*** In this phase, the established supply chain plan is put into action. This includes fulfilling orders, managing logistics, and monitoring the performance of suppliers to ensure timely delivery and maintain quality standards. Real-time tracking and collaboration with all supply chain partners are essential to ensure efficient operations and customer satisfaction. Leveraging technology for automation and smooth information exchange is critical at this stage to handle the complexities of the Micromachine's supply chain. The decisions about the drivers are Inventory (Zero Inventory of End Products), Transport (Low Cost and Fast delivery), Facilities (Only assembly, all components will be outsourced), and Information technology to be used (ERP, MES, IoT and AI for operations and manufacturing execution).

The approach ensures a comprehensive and integrated system that supports the micromachine's production and market distribution by structuring the supply chain development into these detailed phases. From strategic planning through to execution, each phase is designed to build upon the previous, enhancing efficiency, sustainability, and responsiveness to market needs. This structured approach ensures that the supply chain is robust, agile, and capable of supporting the Micromachine's success in the competitive market.

1.4.2 Application of the 5S Principles

The Micromachine considers the 5S principles—Social, Sustainable, Sensing, Smart, and Safe. It exemplifies how integrating these principles into product development and life cycle management can lead to innovations that are not only technologically advanced but also environmentally sustainable and socially responsible.

- **Social:** The development of the Micromachine (for example Food Printer) considers the social impact of technology. It supports community engagement and promotes nutritional awareness by enabling local food production and customization. It also opens educational opportunities, as it can be used in schools to teach students about sustainable practices and the technology behind 3D printing.
  - ***Engagement and Ethics:*** Throughout the life cycle of the Micromachine, engaging with stakeholders ethically is a priority. This includes ensuring fair labor practices within the supply chain, engaging with local communities where materials are sourced, and maintaining transparency in operations. Social responsibility initiatives are integrated into marketing strategies to enhance brand loyalty and consumer trust.
- **Sustainable:** Sustainability is at the core of the Micromachine's life cycle. From the selection of biodegradable materials used in the 3D printing process to the energy-efficient design of the machine itself, every aspect is designed to minimize environmental impact. Additionally, the machine promotes local food ingredients, reducing the carbon footprint associated with food transport.
  - ***Eco-friendly Practices:*** This starts with the selection of materials that are both efficient and environmentally friendly. Strategies such as waste reduction, energy efficiency, and minimizing carbon emissions are emphasized during the manufacturing process. The product design incorporates features that enhance recyclability and reduce environmental impact during the end-of-life phase.
- **Sensing:** Integrated sensors in the Micromachine provide real-time data on its operation, which is essential for ensuring optimal performance and minimizing waste. These sensors detect potential issues before they become problematic, thus conserving materials and energy. In supply chain management, sensing technologies ensure accurate tracking of goods from production to delivery, optimizing logistics and reducing delays.
- **Smart:** The Micromachine incorporates smart technologies, including AI and machine learning, to optimize, for example, food printing processes and adapt to user preferences. This capability enhances the user experience and improves the machine's efficiency and resource use.

  - ***Advanced Technologies and AI:*** The Micromachine leverages smart technologies, including AI and machine learning, to predict maintenance needs, optimize production processes, and enhance product functionality. These technologies also contribute to developing a digital twin of the Micromachine, which simulates the product's performance and helps refine design and operational strategies before they are implemented.

- **Safe:** Safety is paramount in the design and operation of the Micromachine. It includes food safety controls to ensure that the materials used are safe for consumption and that the printing process maintains food integrity. Additionally, the machine features user-friendly interfaces to prevent operational errors.

By applying these 5S principles throughout the Micromachine's life cycle, the project adheres to best practices in terms of efficiency and sustainability and ensures that the product remains competitive and in line with consumer and regulatory expectations. This holistic approach facilitates an operationally effective, environmentally responsible, and socially conscious life cycle.

1.4.3 Digital Technologies Applied to the Micromachine

Integrating digital technologies into the Micromachine life cycle exemplifies its alignment with the 5S principle. Each life cycle stage leverages advanced tools and methods to enhance efficiency, functionality, and sustainability, as depicted in FIGURE 1.8.

1. **Material Development:** The material development stage leverages digital technologies to ensure the selection and characterization of materials that meet the Micromachine's performance, cost, and sustainability criteria.
    - ***Computer-Aided Ideation (CAI):*** Patent analysis and benchmarking tools enable innovation in material selection by identifying trends, potential patents, and best practices.
    - ***Intelligent CAD and Knowledge-Based Engineering Systems (KBES):*** These tools provide data-driven insights to model and analyze materials, ensuring suitability for the Micromachine's functional requirements.
    - ***Material Databases:*** Comprehensive databases are used to evaluate materials based on environmental impact, mechanical properties, and cost.
    - ***Material Engineering Design Software:*** Advanced design tools simulate and optimize material performance under various conditions, aiding in prototype creation.

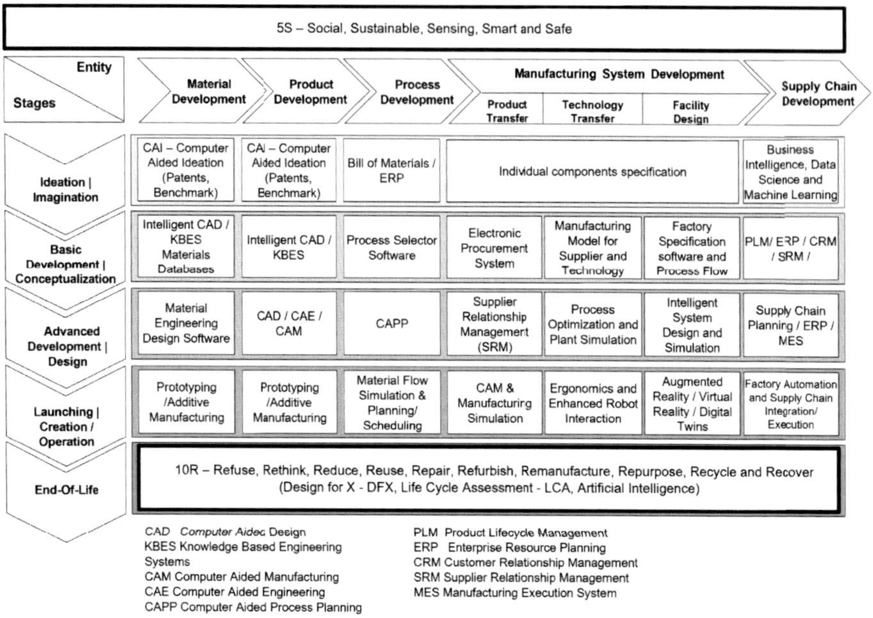

FIGURE 1.8
Digital technologies to support life cycle engineering of micromachine development.

- ***Prototyping and Additive Manufacturing:*** Rapid prototyping using 3D printing allows testing of materials and designs iteratively, reducing waste and development time.

2. **Product Development:** The product development stage focuses on transforming the conceptual design into a functional prototype, incorporating multiple digital solutions.
   - ***Computer-Aided Ideation (CAI):*** Patents and benchmarks are revisited to refine product concepts, ensuring alignment with market demands and technological trends.
   - ***Intelligent CAD/KBES:*** CAD and KBES tools facilitate the creation of precise product designs tailored to the Micromachine's manufacturing processes.
   - ***CAD/CAE/CAM Integration:*** Engineering tools like Computer-Aided Design (CAD), Engineering (CAE), and Manufacturing (CAM) ensure detailed simulations, structural analysis, and manufacturability.
   - ***Prototyping and Additive Manufacturing:*** Additive manufacturing enables physical prototypes to be built quickly, facilitating design validation and iterative improvements.

3. **Process Development:** Process development optimizes the manufacturing and assembly processes of the Micromachine using advanced planning tools.
   - ***Bill of Materials (BOM) and ERP Systems:*** Digital tools integrate the BOM with enterprise resource planning (ERP) systems for seamless material management.
   - ***Process Selector Software:*** Software solutions analyze manufacturing requirements to recommend the most efficient processes for different components.
   - ***Computer-Aided Process Planning (CAPP):*** Digital process planning optimizes machining sequences, tooling, and labor requirements.
   - ***Material Flow Simulation and Planning/Scheduling:*** Simulation tools model production layouts and workflows, enabling efficient resource allocation and minimizing bottlenecks.
4. **Manufacturing System Development:** This stage focuses on implementing manufacturing systems that are efficient, flexible, and sustainable, supported by advanced digital technologies.
   - ***Product Transfer, Technology Transfer, and Facility Design:*** Tools support the integration of in-house and outsourced manufacturing processes with precise component specifications.
   - ***Process Optimization and Plant Simulation:*** Advanced simulation tools refine production layouts, resource use, and overall efficiency.
   - ***Intelligent System Design and Simulation:*** AI-driven systems enhance manufacturing system adaptability and operational performance.
   - ***CAM and Manufacturing Simulation:*** CAM software ensures precision in machining, while simulations validate manufacturing workflows.
   - ***Augmented Reality (AR), Virtual Reality (VR), and Digital Twins:*** These technologies simulate manufacturing environments, enabling predictive maintenance, real-time monitoring, and system optimization.
5. **Supply Chain Development:** The supply chain is managed with intelligent tools to ensure agility, responsiveness, and sustainability.
   - ***Business Intelligence, Data Science, and Machine Learning:*** Predictive analytics optimize demand forecasting, inventory management, and logistics.

- ***PLM, ERP, CRM, and SRM Systems:*** Integrated platforms manage product life cycle, enterprise resources, customer relationships, and supplier coordination seamlessly.
- ***Supply Chain Planning (ERP/MES):*** Advanced planning tools ensure efficient material and information flow across the supply chain.
- ***Factory Automation and Supply Chain Integration/ Execution:*** Automated systems and IoT technologies enable real-time tracking and streamlined production-to-delivery processes.

6. **End-of-Life:** The Micromachine's life cycle closes with a focus on sustainability, leveraging digital tools to maximize resource efficiency and environmental stewardship.
   - ***10R Framework Implementation:*** Digital tools support strategies for Refusing, Rethinking, Reducing, Reusing, Repairing, Refurbishing, Remanufacturing, Repurposing, Recycling, and Recovering components and materials.
   - ***Design for X (DFX):*** DFX tools ensure that the Micromachine is designed for disassembly, recyclability, and reusability.
   - ***Life Cycle Assessment (LCA):*** LCA tools evaluate the environmental impact of the Micromachine across its life cycle, identifying areas for improvement.
   - ***Artificial Intelligence (AI):*** AI enhances predictive analytics and optimizes resource recovery processes.

These tools not only enhance the machine's efficiency and functionality but also support broader goals of environmental responsibility and social impact, ensuring the Micromachine's relevance in a rapidly evolving technological landscape.

1.5 The Future of Sustainable Manufacturing

Sustainable manufacturing is a multidimensional approach that integrates environmental, economic, and social dimensions to minimize resource consumption while maintaining productivity and profitability. Recent advances have increasingly focused on leveraging digital and clean technologies to create environmentally sound processes and systems, which have become essential for achieving global sustainability goals [13, 41].

The integration of digital technologies into manufacturing is a prominent trend, providing tools like AI, IoT, and predictive analytics to enhance sustainability by optimizing energy use, reducing waste, and improving supply chain transparency [72, 122]. These advancements align closely with circular economy principles, emphasizing the 10R framework.

Moreover, there is an increasing push towards aligning sustainable manufacturing practices with the United Nations' Sustainable Development Goals (SDGs), particularly those promoting sustainable industrialization. Regulatory frameworks and incentives from governments further bolster this shift, encouraging industries to adopt greener methods and technologies [147].

This evolving landscape of sustainable manufacturing reflects the critical interplay between technological advancements, regulatory frameworks, and societal demands, shaping the future of the manufacturing industry. The future of sustainable manufacturing, as illuminated by the LCE framework, is intrinsically linked to the integration of digital technologies. This emerging landscape is characterized by a shift towards holistic product life cycle management, where every aspect, from raw material extraction to end-of-life disposal, is optimized for minimal environmental impact [56]. The LCE framework incorporates advanced computational techniques, high-throughput experiments, feedback loops, collaborative development, and eco-design principles. It aligns with circular economy principles and green supply chain management, complying with international environmental standards.

The evolution of digital transformation in manufacturing is segmented into several phases. Initially focusing on constructing an information society, it progresses towards integrating decision-making tools and enhancing the overall performance of digital transformation [127]. This evolution signifies a deeper integration of digital technologies in manufacturing processes, contributing to sustainable practices.

The technological trajectory in manufacturing, guided by the LCE framework, identifies the developmental stages of digital transformation. This trajectory is crucial for recognizing potential future developments in sustainable manufacturing [95].

The introduction of digital technologies in manufacturing aligns with sustainable practices. Technologies such as AI, machine learning, and digital twins are becoming increasingly important in rapidly developing new materials and innovation in material science [12, 69]. Besides, digital tools are pivotal for designing modular systems that can be easily scaled or reconfigured to meet changing requirements. This flexibility allows manufacturers to adapt to market changes swiftly, minimizing waste and enhancing operational efficiency.

Digital technologies like AI, ML, CAD/CAM/CAPP, 3D printing, Digital Twins, IIoT, and robotics play crucial roles in LCE. They are vital in the material development and supply chain processes. They promote the use of predictive modeling and real-time feedback loops, enhancing sustainability in manufacturing.

Moreover, the integration of digital technologies in sustainable manufacturing aligns with the United Nations' Sustainable Development Goals, particularly the goal of promoting sustainable industrialization through new technologies [51]. This underlines the importance of developing sustainable manufacturing technologies as a vital component of engineering, which not only consumes energy and manpower but also systematically converts raw materials into products beneficial for society.

The future of sustainable manufacturing lies in integrating LCE with the 5S principles. This approach promotes not only environmental stewardship but also operational. FIGURE 1.9 exemplifies an approach of this in the future:

1. **Holistic Product Development:** Future product development will increasingly incorporate eco-design from the ideation phase, ensuring that every product is designed with its entire life cycle in mind. This includes using recyclable or biodegradable materials and designing for easier disassembly.
   - ***Social Impact:*** A stronger focus will be on creating positive social impacts, such as supporting fair trade and labor practices and enhancing community engagement.
2. **Advanced Manufacturing Processes:** Leveraging AI and IoT, smart factories will become the norm, with systems that can predict maintenance needs, optimize energy use, and minimize waste in real-time. These factories will use sensing technologies to continuously improve the manufacturing process and ensure it remains safe for workers and the environment.
   - ***Adaptive Manufacturing Systems:*** Facilities will be designed to be flexible and modular, able to quickly adapt to new manufacturing methods or changes in product design without significant downtime or waste.
3. **Sustainable Supply Chains:** Blockchain and IoT will enable more transparent and traceable supply chains, ensuring that all materials sourced meet sustainability standards. This will also help companies monitor their supply chain in real-time, reducing risks and improving efficiency.
   - ***Green Logistics:*** Emphasis will be placed on optimizing logistics to reduce carbon footprints, utilizing electric and autonomous vehicles, and improving route planning to decrease energy consumption.
4. **Enhanced End-of-Life Management:** The focus will shift towards circular economy models, where the end-of-life of products is considered at the design stage. Products will be designed for reuse, refurbishment, and easy recycling, supporting a loop of continuous material use.

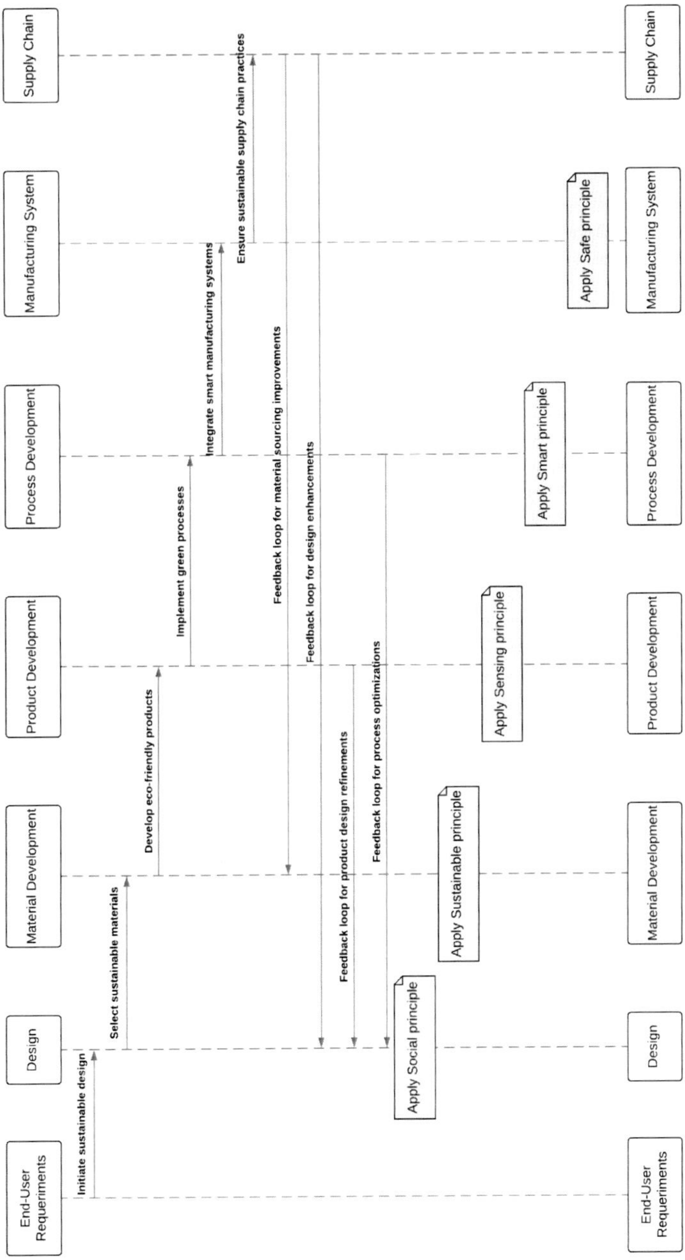

FIGURE 1.9
Future of sustainable manufacturing considering the 5S principles.

- ***Consumer Participation:*** Consumers will be more involved in sustainability through programs encouraging the return of end-of-life products for recycling or refurbishing.

5. **Regulatory and Market Adaptation:** Compliance with global sustainability standards will become a competitive edge rather than a regulatory requirement. Manufacturers that can prove their commitment to sustainability will have a market advantage.
   - ***Incentivization:*** Governments and international bodies will likely introduce more incentives for companies practicing sustainable manufacturing, such as tax breaks, grants, and premium market positions.
6. **Innovation and Collaboration:** The future will see an increase in open innovation where companies share knowledge and resources to solve common sustainability challenges. This collaborative approach will speed up innovation and lead to more sustainable solutions across industries.
   - ***Cross-Sector Partnerships:*** Partnerships across different sectors will become crucial in solving complex sustainability challenges. For example, tech companies might partner with manufacturing firms to develop new materials that are both high-performing and sustainable.

Integrating LCE with the 5S principles presents a forward-thinking approach to manufacturing that is environmentally friendly, economically viable, and socially responsible. This integrated approach will likely become essential as global pressures on resources intensify and consumer demand for sustainable products increases. The future of sustainable manufacturing will rely heavily on innovation, smart technologies, and a commitment to improving production's social and environmental impact.

1.6 Conclusion

This chapter highlights the essential nature of digital technologies in advancing sustainable manufacturing practices. These technologies contribute to creating materials and products that are innovative and efficient and align with the Sustainable Development Goals of promoting sustainable industrialization. The LCE framework, thus, represents a strategic approach to sustainable manufacturing, ensuring that engineering processes systematically convert raw materials into products that benefit society while conserving resources and reducing environmental impact.

Hence, the LCE framework, when integrated throughout the product development life cycle—from material selection to product design, process development, manufacturing system development, and supply chain management—can significantly improve the sustainability and resilience of manufacturing operations.

The Micromachine case study is a practical example of the LCE framework, demonstrating how digital twins and advanced manufacturing techniques such as CNC can be leveraged to create adaptable, efficient, and sustainable manufacturing solutions. The case study walks through the entire life cycle process of the Micromachine, from idea generation and conceptual design to its launching.

The future of sustainable manufacturing is envisioned to be highly dependent on integrating digital technologies. This integration will lead to a manufacturing landscape where holistic product life cycle management minimizes environmental impact. The LCE framework plays a crucial role in this future, incorporating advanced computational techniques, high–throughput experiments, collaborative development, and adherence to international environmental standards.

2

Introducing 5S Systems: Principles, Concepts, and Definitions

This chapter provides a holistic approach and a comprehensive view of integrating 5S principles (Social, Sustainable, Sensing, Smart, and Safe) in the Life Cycle Engineering framework. It explores how these principles can be applied throughout the life cycle of a system, from ideation to end-of-life management, to optimize its sustainability, efficiency, and cost-effectiveness. This led into the deployment of systems that not only meet current technological and operational standards but also adapt to future changes, thus ensuring sustainability and resilience.

2.1 Introduction

The 5S systems, integrating Social, Sustainable, Sensing, Smart, and Safe principles, are crucial in addressing the multifaceted demands of modern and future environments. Unlike traditional systems, which often focus narrowly on efficiency and cost-effectiveness, 5S systems adopt a holistic approach (see FIGURE 2.1). They ensure operations are economically viable, socially responsible, environmentally sustainable, and safe for all stakeholders. These systems are characterized by their adaptability and intelligence, leveraging smart technologies and sensing capabilities to dynamically respond to varying conditions, a feature that conventional systems typically lack.

Ethical and inclusive considerations are at the core of 5S systems. These type of systems strongly emphasize ethical and inclusive practices, ensuring that the benefits of technological advancements are equitably shared. This promotes inclusivity, fair labor practices, and positive community engagement, setting a higher ethical standard. This is a stark difference from conventional methodologies, where the broader social impact may be overlooked in favor of narrow, short-term objectives.

The sustainability focus of 5S systems is another distinguishing factor, as they strive to contribute to long-term ecological balance, using renewable resources and minimizing environmental footprints, thus addressing a critical shortcoming of traditional systems often criticized for their unsustainable

DOI: 10.1201/9781003501619-2

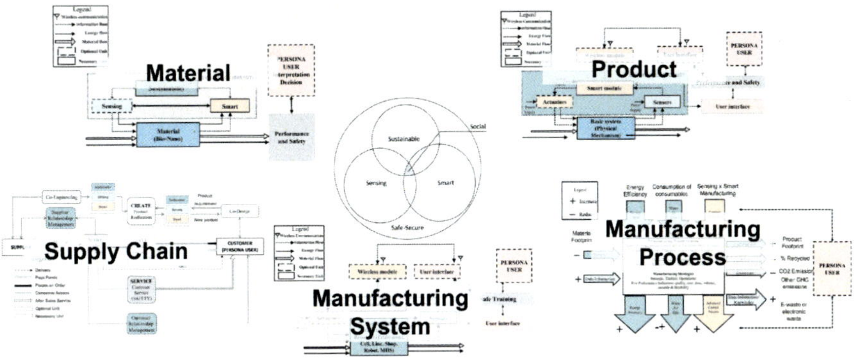

FIGURE 2.1
5S systems: material, product, manufacturing process, manufacturing system and supply chain.

practices. 5S systems are designed to minimize environmental footprints and promote the conservation of resources, contrasting with the less environmentally focused practices of conventional systems that can lead to unsustainable exploitation of resources.

Additionally, safety and security are paramount in 5S systems. They incorporate robust safety protocols, data security, and risk mitigation strategies to ensure the well-being of individuals and communities. For instance, the concept of Industry 5.0 integrates these principles to create systems that democratize knowledge co-production from Big Data, enhancing both innovation and safety in manufacturing [108]. In contrast, conventional setups may overlook these aspects in pursuit of other objectives. 5S systems represent a forward-thinking, comprehensive approach, designed to be resilient, adaptable, and positively impactful on society and the environment, setting them apart from conventional systems.

5S systems represent a significant evolution from conventional methodologies, marked by a shift towards a more integrated and forward-thinking approach. Unlike traditional methods that prioritize financial gains and operational efficiency, 5S systems encompass a holistic perspective, addressing not only economic objectives but also social, environmental, technological, and safety concerns. This broader focus ensures that actions and decisions within 5S frameworks are both ethically responsible and sustainable over the long term. In contrast to the often static and reactive nature of conventional systems, 5S systems are adaptive and intelligent, leveraging advanced technologies and sensing capabilities to proactively respond to changing conditions and needs. This allows for a dynamic and flexible approach to challenges, as opposed to the rigid and sometimes outdated processes prevalent in traditional methodologies.

Moreover, 5S systems place a strong emphasis on safety and security, incorporating robust measures to protect individuals and communities, as well as ensuring data integrity and security. This comprehensive approach to safety is often more advanced than in conventional methodologies, where safety and security measures may be implemented more reactively and with varying degrees of thoroughness. For instance, in [113], the authors discussed the integration of Social, Sustainable, Sensing, and Smart principles in assessing solar companies for supplying electrical energy in manufacturing processes.

Overall, 5S systems offer a more comprehensive, responsive, and responsible approach compared to conventional methodologies, aligning with modern demands for sustainability, ethical practices, and innovative technological integration, thereby setting a new standard for operational and strategic excellence. A 5S systems approach that integrates Social, Sustainable, Sensing, Smart, and Safe principles represent a comprehensive framework designed to enhance operational and development of materials, products, manufacturing processes, manufacturing system and supply chains across various industries. This approach aims to create systems that are not only efficient and technologically advanced but also responsible and mindful of broader societal and environmental impacts. Each principle within this framework is described as:

- **Social:** Focuses on creating systems that positively impact society, promoting ethical practices, inclusivity, and community engagement. It emphasizes the importance of fair labor practices, diversity, and contributing to the social well-being of communities affected by the system's operations.
- **Sustainable:** Ensures that systems are designed and operated in an environmentally friendly manner, minimizing ecological footprints, and promoting the conservation of resources. Sustainability in this context involves utilizing renewable resources, reducing waste, and implementing practices that support long-term ecological balance.
- **Sensing:** Involves the integration of sensors and data collection technologies to gather insights about the system's performance, environmental conditions, and other relevant metrics. Sensing capabilities enable real-time monitoring and data-driven decision-making, enhancing the efficiency and responsiveness of systems.
- **Smart:** Refers to the incorporation of advanced technologies, such as artificial intelligence (AI), machine learning, and automation, to make systems more intelligent and adaptable. Smart systems can analyze data, learn from it, and make autonomous decisions to improve operations, user experiences, and adapt to changing conditions.
- **Safe:** Prioritizes the safety and well-being of individuals and communities interacting with or affected by the system. This includes implementing robust safety protocols, ensuring data security, and designing systems to prevent accidents and mitigate risks.

By embodying these 5S principles, systems are developed with a holistic view that transcends traditional performance metrics, incorporating ethical, environmental, and technological dimensions. This approach fosters the creation of innovative solutions that are prepared to meet current challenges while being adaptable to future changes, ensuring they are beneficial for both people and the planet.

2.2 5S Systems

This section describes each system, starting with materials, then products, followed by manufacturing process, manufacturing system, and ending with the supply chain.

2.2.1 5S Materials

Materials defined within a 5S system—incorporating the principles of being Social, Sustainable, Sensing, Smart, and Safe—represent an advanced approach to material science and engineering that goes beyond traditional functionalities. This framework ensures that materials are not only innovative and efficient but also ethically produced, environmentally friendly, responsive to external conditions, integrated with technology, and safe for all users. Therefore, 5S materials can be described as:

- **Social:** Materials are developed and produced with a strong emphasis on positive social impact, including ethical labor practices—how a material is produced, community engagement—how the material is used, society impact—how the material is re-used/recycled/disposed, and ensuring that the benefits of technological advancements are accessible and equitable. This principle also encompasses efforts to enhance the quality of life and support the development of the communities involved in the production process.
- **Sustainable:** Sustainability in materials focuses on minimizing environmental impact throughout the life cycle—from extraction or synthesis of raw materials to production, usage, and eventual disposal. Sustainable materials are those that are recyclable, biodegradable, or made from renewable resources, designed to reduce waste and conserve natural resources.
- **Sensing:** Sensing materials have the capability to collect data about their environment or condition. This could include materials that change color in response to chemical exposure, textiles that monitor physiological data, or building materials that sense structural stress. Sensing capabilities enable

real-time monitoring and responsiveness to changes, enhancing functionality and safety.

- **Smart:** Smart materials exhibit adaptive behaviors or functionalities in response to external stimuli, such as temperature, pressure, moisture, or electrical inputs. These materials can change their properties (e.g., shape, color, conductivity) intelligently, offering innovative applications in various fields like healthcare, construction, and wearable technology, making systems more interactive and efficient.
- **Safe:** Safety is paramount, ensuring that materials are non-toxic, do not pose hazards during their life cycle, and are designed with end-user and environmental safety in mind. This includes considering the health impacts of materials during production, use, and disposal, and designing materials that contribute to safer products and environments.

FIGURE 2.2 denotes the principles of 5S materials. By integrating these 5S principles, materials become more than just substances used in production; they embody a holistic approach to innovation that respects people and the planet, while pushing the boundaries of what materials can do through technology. Energy is a very important concept to address in the development of 5S materials as they allow the materials to have sensing and smart functionalities. Sustainability must consider the development of materials with Net Zero Footprint during the life cycle of the material fabrication, usage and disposal. This approach not only addresses current demands for more responsible and intelligent materials but also sets a foundation for future advancements in material science and engineering.

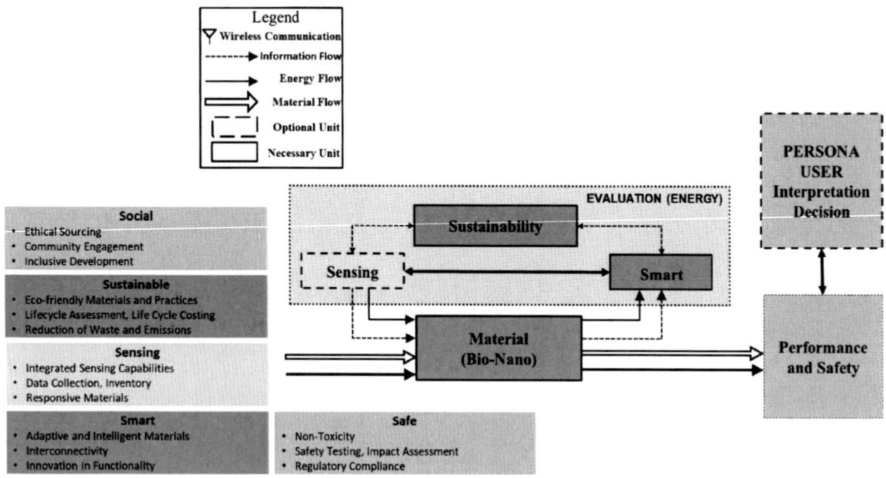

FIGURE 2.2
5S materials (social, sustainable, sensing, smart and safe).

2.2.2 5S Products

Defining products as a 5S system represents a holistic and forward-thinking approach to product development and management. This framework ensures products are not only efficient and technologically advanced but also ethically produced, environmentally responsible, and designed with user safety and societal well-being in mind. Here's a detailed breakdown of each principle in the context of 5S products:

- **Social:** Products designed with social considerations prioritize ethical practices, positive community impact and user center design. They are accessible and inclusive, meeting diverse user needs and contributing to societal well-being. Socially focused products also aim to enhance connectivity and interaction within communities, fostering engagement and understanding (i.e., connected products).
- **Sustainable:** Sustainable products are developed with an emphasis on environmental stewardship. They utilize eco-friendly materials, minimize waste through efficient design, and are recyclable or biodegradable at the end of their life cycle (zero footprint). Sustainable products aim to reduce the carbon footprint and resource consumption, encouraging a shift towards a more circular economy.
- **Sensing:** Sensing products are equipped with technologies that enable them to collect and respond to data about their environment or user interaction. This capability can enhance user experience, optimize product performance, and contribute to safety and efficiency. Sensing products adapt and react to changes, providing insights and data that can inform users and manufacturers alike (i.e., connected products).
- **Smart:** Smart products incorporate advanced technologies like AI, IoT, and machine learning to offer adaptive, intelligent functionalities. They can learn from user interactions, integrate seamlessly with other devices and systems, and provide personalized experiences. Smart products are characterized by their ability to improve over time, offering innovative solutions to everyday challenges.
- **Safe:** Safety is a paramount consideration, ensuring products are designed and manufactured to eliminate or minimize risks to users and the environment. This includes rigorous testing for toxic substances, ensuring data privacy and security in smart products, and adhering to high safety standards in design and function. Safe products inspire trust and confidence among users.

Integrating these 5S principles into product development and management leads to products that are not only functional and innovative but also responsible, sustainable, and beneficial for society and the environment. FIGURE 2.3 depicts all the principles and concepts required to be addressed while developing a 5S product. This approach reflects a comprehensive understanding of the role products play in shaping a sustainable future, emphasizing the need for thoughtful design, ethical production, and continuous improvement.

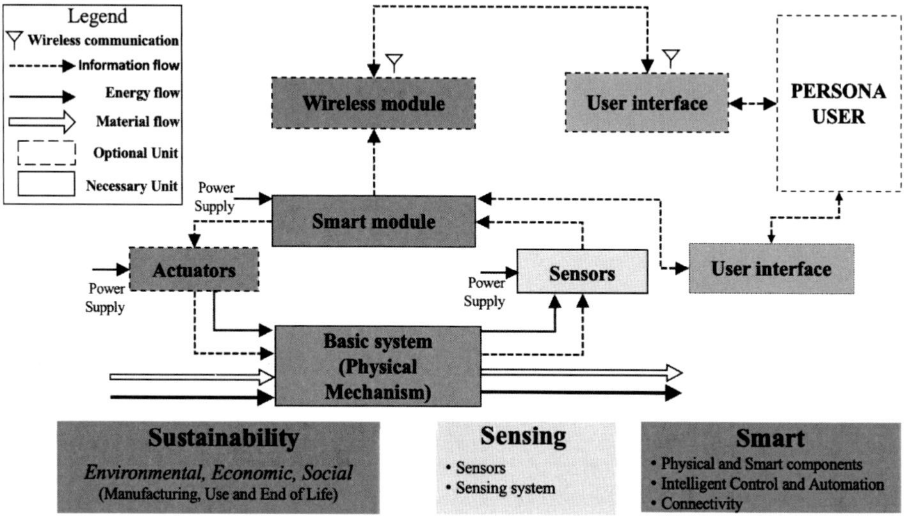

FIGURE 2.3
5S products (social, sustainable, sensing, smart and safe).

2.2.3 5S Manufacturing Process

Defining a manufacturing process within a 5S system envisions a comprehensive approach to manufacturing that prioritizes not only costs, efficiency, and productivity but also ethical practices, environmental responsibility, advanced technology integration, and safety. Each principle can be applied to a 5S manufacturing process in the following manner:

- **Social:**
  - ***Ethical Practices:*** Ensuring fair labor practices, equitable working conditions, and positive community engagement throughout the manufacturing process. This includes supporting workforce development, diversity, and inclusion.
  - ***Community Impact:*** The manufacturing process contributes positively to local communities through economic support, community programs, or minimizing disruptive impacts.
- **Sustainable:**
  - ***Resource Efficiency:*** Optimizing the use of resources, including materials, water, and energy, to minimize waste and reduce environmental footprint.

  - ***Eco-friendly Materials and Practices:*** Using materials and implementing practices that have a lower environmental impact, such as recycling, reducing emissions, and utilizing renewable energy sources.

- **Sensing:**
  - ***Data Collection and Analysis:*** Incorporating sensors and IoT devices to collect real-time data on various aspects of the manufacturing process, from machine performance to environmental conditions, enabling data-driven decision-making.
  - ***Predictive Maintenance and Optimization:*** Using sensing technologies to predict machinery maintenance needs and optimize production processes, enhancing efficiency and reducing downtime.

- **Smart:**
  - ***Automation and Robotics:*** Implementing automation and robotics to improve precision, efficiency, and productivity, while reducing human error and operational costs.
  - ***Adaptive Manufacturing Systems:*** Utilizing AI and machine learning to create manufacturing systems that can adapt to changes in production demand, material supply, or design specifications in real time.

- **Safe:**
  - ***Worker Safety:*** Prioritizing the health and safety of the workforce by implementing robust safety protocols, emergency preparedness plans, and continuous safety training.
  - ***Product and Environmental Safety:*** Ensuring that manufacturing processes do not compromise the safety of the end products or negatively impact the environment, adhering to stringent safety and environmental regulations.

Incorporating the 5S principles into a manufacturing process means building a system that is ethically grounded, environmentally conscious, technologically advanced, and inherently safe. The manufacturing process described in FIGURE 2.4 describes the most important concepts of a process that must achieve the principles of sustainability (reduction of waste, CO_2 emissions, usage efficient of energy, limiting use of consumable—water, air, oils). Also, the usage of sensors and advanced control to achieve intelligent manufacturing. This approach not only enhances the sustainability and efficiency of the manufacturing process but also supports the well-being of workers (Persona-user), communities, and the planet, fostering a positive impact that extends beyond the factory floor.

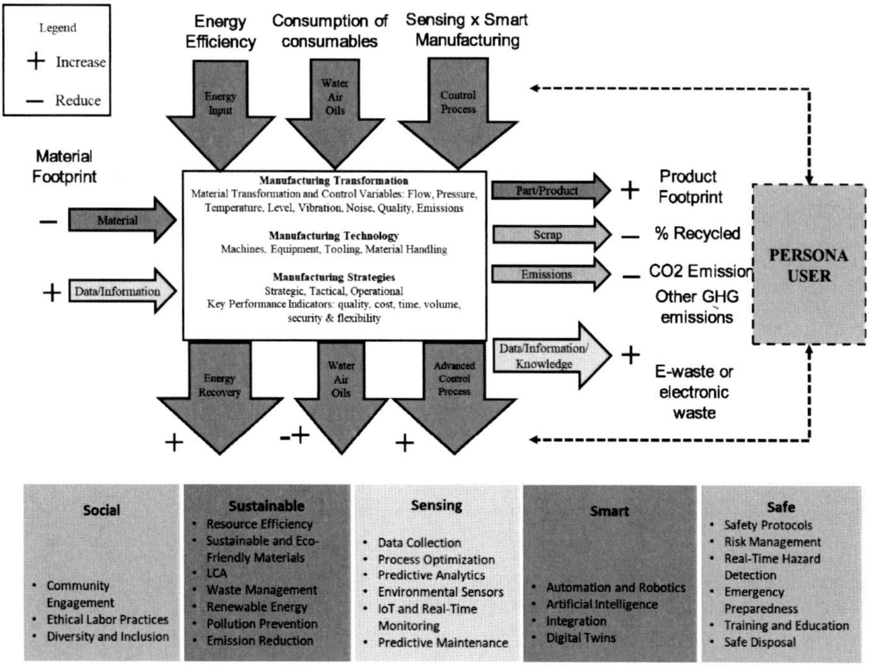

FIGURE 2.4
5S manufacturing process (social, sustainable, sensing, smart and safe).

2.2.4 5S Manufacturing System

A Manufacturing System defined within a 5S framework integrates a holistic and systemic approach to manufacturing that aligns operational efficiency with ethical, environmental, and technological imperatives.

- **Social:** Focuses on the societal impact of the manufacturing system, emphasizing ethical labor practices, community engagement, and social equity. It involves creating value for both the workforce and the broader community by ensuring fair employment practices, supporting community development projects, and ensuring the manufacturing system is transparent in its operations, is accessible and beneficial to its stakeholders.
- **Sustainable:** Pertains to the environmental responsibility of the manufacturing system. This involves minimizing the environmental footprint through efficient resource use, reducing waste and emissions, and implementing practices that promote the longevity and recyclability of products and its production system. Sustainable practices also encompass using renewable energy sources and pursuing innovations that contribute to a circular economy.
- **Sensing:** Involves the integration of sensors and data analytics to monitor and optimize the manufacturing process and manufacturing system

continuously. Sensing capabilities allow for real-time tracking of production efficiency, energy consumption, and environmental impacts, enabling proactive adjustments and improvements. This data-driven approach enhances process efficiency, product quality, and sustainability outcomes.

- **Smart:** Refers to incorporating advanced technologies, such as AI, IoT, robotics, and automation, to create an intelligent manufacturing system capable of adaptive, efficient, and high-quality production. Smart systems can learn from data, predict maintenance needs, automate routine tasks, and adapt to changes in demand or operating conditions, significantly improving operational agility and innovation.
- **Safe:** Ensures the health and safety of all individuals involved in or affected by the manufacturing process and manufacturing system, as well as the safety of the produced goods and the surrounding environment. This encompasses implementing rigorous safety standards, emergency preparedness, and risk management practices to prevent accidents, injuries, and environmental harm. It also includes safeguarding data and systems against cyberthreats in increasingly connected manufacturing environments.

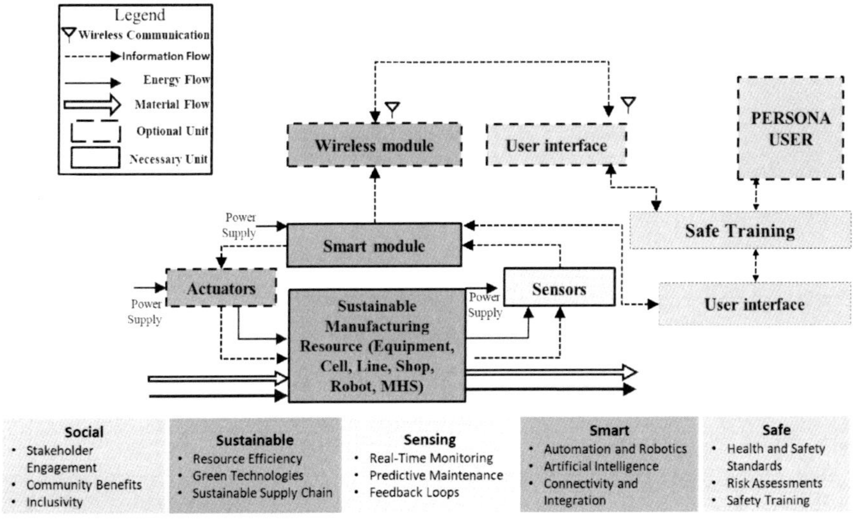

FIGURE 2.5
5S manufacturing system (social, sustainable, sensing, smart and safe).

By embodying these 5S principles, a manufacturing system transcends traditional focus areas to embrace a more comprehensive and forward-looking perspective. FIGURE 2.5 describes such a system that is designed not only to achieve operational excellence and innovation but also to uphold social responsibility, environmental sustainability, and safety, ensuring resilience and positive impact in a rapidly evolving global landscape.

2.2.5 5S Supply Chain

Defining a supply chain as a 5S system transforms it into a comprehensive and forward-thinking network that transcends traditional logistics and procurement practices. This approach integrates advanced operational and ethical considerations, focusing on efficiency, innovation, responsibility, and adaptability. The following descriptions relate to the application of the framework to a 5S supply chain:

- **Social:**
  - ***Ethical Practices and Equity:*** Prioritizes fair labor practices, equitable working conditions, and positive community impact throughout the supply chain. This includes ensuring all participants in the supply chain adhere to ethical standards and contribute to social well-being.
  - ***Community Engagement:*** Engages with communities affected by supply chain operations to ensure their needs and well-being are considered, fostering positive relationships and support for local development.
- **Sustainable:**
  - ***Environmental Stewardship:*** Implements eco-friendly practices across the supply chain, from sourcing raw materials to delivering the final product. This involves reducing carbon emissions, minimizing waste, and using sustainable materials to lessen the environmental impact.
  - ***Resource Efficiency:*** Focuses on optimizing the use of resources throughout the supply chain, promoting recycling, and encouraging the use of renewable resources to achieve a more sustainable operation.
- **Sensing:**
  - ***Data-Driven Insights:*** Utilizes sensors, IoT devices, and other technologies to collect real-time data across the supply chain. This data informs decision-making, improves transparency, and enhances the ability to respond to changes and challenges.
  - ***Monitoring and Tracking:*** Employs sensing technologies to monitor the condition and location of goods, ensuring integrity and quality from production to delivery, while also enabling predictive analytics for demand and supply planning.
- **Smart:**
  - ***Technological Integration:*** Incorporates advanced technologies such as AI, data analytics, machine learning, blockchain, and automation to streamline operations, improve efficiency, and foster innovation. Smart supply chains are adaptive, responsive, and capable of optimizing themselves based on predictive analytics and real-time data.

  - ***Connectivity and Collaboration:*** Enhances connectivity between supply chain participants through digital platforms, facilitating seamless communication, collaboration, and exchange of information, which supports more coordinated and agile operations.

- **Safe:**
  - ***Risk Management:*** Implements comprehensive safety and risk management practices to protect workers, consumers, and communities from potential harm. This includes ensuring product safety, securing data and information systems, and preparing for disruptions or emergencies.
  - ***Health and Safety Standards:*** Adheres to strict health and safety standards throughout the supply chain, minimizing the risk of accidents, ensuring the safe handling of materials, and promoting a culture of safety and well-being.

By adopting a 5S approach, a supply chain becomes not just a conduit for moving goods, but a dynamic, responsible network that supports sustainable growth, innovation, and resilience. This framework ensures that supply chain operations contribute positively to society and the environment, embrace cutting-edge technologies, and prioritize the safety and well-being of all stakeholders. FIGURE 2.6 represents the concepts and principles of a 5S supply chain where all the interaction of core process of a

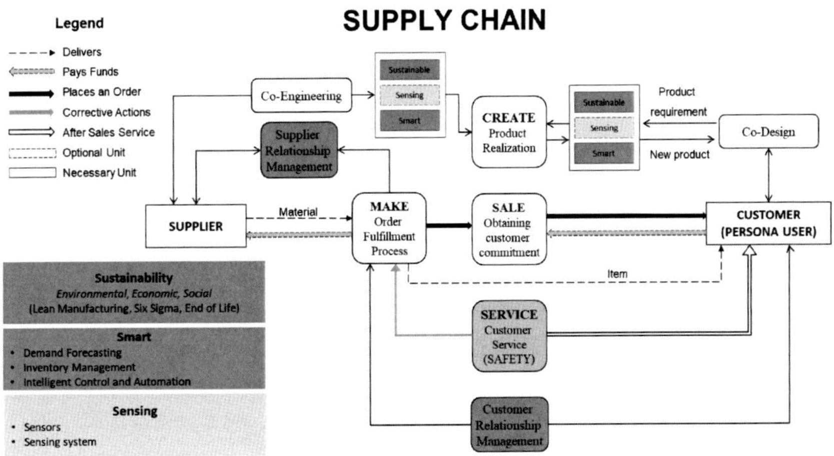

FIGURE 2.6
5S supply chain (social, sustainable, sensing, smart and safe).

company (Create, Sale, Make, Service) interacts with customers and suppliers to achieve a sustainable supply chain (zero footprint, zero emissions), sensing (use of sensors across the supply chain), smart (demand forecasting, inventory management, intelligent transport, supply chain management, control and automation), and ensures safe delivery of products and services.

The following table (TABLE 2.1) summarizes the principles of 5S for each of the systems being analyzed (material, product, process, manufacturing system and supply chain).

2.3 Life Cycle Engineering to Develop 5S Systems

Life Cycle Engineering (LCE) is a discipline that considers the entire life cycle of a system (i.e., material, product, process, manufacturing system or supply chain) from conception through disposal (End-Of-Life), to optimize its performance in terms of sustainability, efficiency, and cost-effectiveness. When applied to developing 5S Systems—incorporating Social, Sustainable, Sensing, Smart, and Safe principles—LCE provides a comprehensive framework that ensures these systems are designed and operated in a manner that is environmentally responsible, socially beneficial, technologically advanced, and inherently safe throughout their life cycle (FIGURE 2.7).

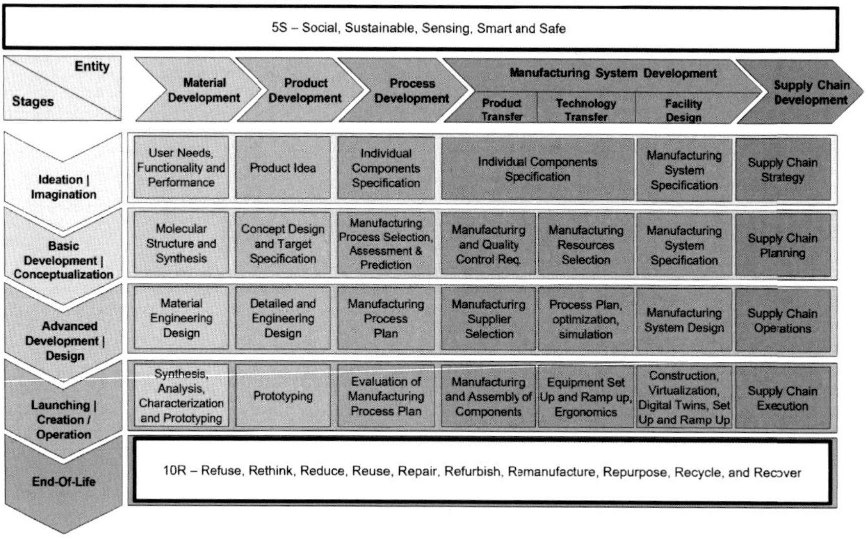

FIGURE 2.7
The framework of 5S Systems developed using life cycle engineering.

In the following sections, we will describe how the principles of 5S can be applied in each stage of the life cycle.

TABLE 2.1
5S Materials, Products, Processes, Manufacturing System and Supply Chain

Entity	Social	Sustainable	Sensing	Smart	Safe	Technologies
Material	People and Society requirements	Zero Carbon Footprint	Context variables: temperature, humidity, light, etc.	React to context and environment	Non-toxic Zero Hazard	Machine Learning Quantum Computing Mathematical modeling
Product	User's needs and Customer requirements	Zero Life cycle impact	Sensing of performance	Improvement of performance	Safety of use	IoT AI Smart Sensors
Process	Human centered	Resource Efficiency: Energy, Water, Emissions & Waste	Data Collection, Analysis and Use of sensors	Automation, Robotics, Predictive Maintenance and Optimization		Smart sensors Mathematical modeling
Manufacturing System	Human and Community centered	Environmental responsibility: Energy, Water, Emissions & Waste	Sensing capabilities, data analytics and real-time monitoring	Intelligent and Adaptive System	Healthy and safe environment for workers	AI, IoT, robotics, and automation
Supply Chain	Positive community impact throughout the supply chain	Environmental Stewardship and Resource Efficiency	Data-Driven Insights and Monitoring and Tracking	Technological Integration, Connectivity and Collaboration	Risk Management, Health, and Safety Standards	AI, machine learning, blockchain, and automation

2.3.1 Ideation—Imagination

Applying the 5S principles—Social, Sustainable, Sensing, Smart, and Safe—to the ideation—imagination phase of a project involves laying a comprehensive foundation that aligns the project's objectives with ethical, environmental, technological, and safety standards from the outset. These principles can guide the initial phase of project development:

- **Social:**
  - ***Stakeholder Engagement:*** Begin with a broad stakeholder engagement process to understand the needs, expectations, and concerns of all parties involved or affected by the project. This includes customers, employees, community members, and suppliers.
  - ***Social Impact Assessment:*** Conduct an assessment to understand the potential social impacts of the project, identifying opportunities to contribute positively to society and mitigate any negative impacts.
  - ***Inclusivity Goals:*** Set goals for inclusivity and accessibility, ensuring the project benefits a diverse range of users and does not exacerbate social inequalities.
- **Sustainable:**
  - ***Sustainability Objectives:*** Define clear sustainability objectives, focusing on minimizing environmental impact through efficient use of resources, reducing emissions, and promoting circular economy principles.
  - ***Life cycle Perspective:*** Adopt a life cycle perspective from the start, considering how materials, production, operation, and end-of-life stages can be managed to enhance environmental sustainability.
  - ***Regulatory Compliance and Beyond:*** Ensure the project not only complies with current environmental regulations but also anticipates future sustainability standards and public expectations.
- **Sensing:**
  - ***Data Strategy:*** Develop a data collection and management strategy, identifying what data will be valuable for monitoring performance, optimizing operations, and enhancing user experience.
  - ***Technology Integration:*** Consider how sensing technologies can be integrated into the project to collect data on usage patterns, environmental conditions, and other relevant metrics.
  - ***Feedback Mechanisms:*** Plan for feedback mechanisms that use collected data to inform ongoing project development and adaptation, ensuring the project remains responsive to both human and environmental needs.

- **Smart:**
  - ***Technology Roadmap:*** Outline a technology roadmap that incorporates smart solutions, such as AI, IoT, and machine learning, to improve efficiency, adaptability, and user engagement.
  - ***Interoperability:*** Ensure the smart components of the project are designed to be interoperable with existing and future technologies, facilitating seamless integration and scalability.
  - ***Privacy and Security:*** Address privacy and security considerations from the outset, especially in relation to the collection, storage, and use of data.
- **Safe:**
  - ***Safety Assessment:*** Conduct a comprehensive safety assessment, identifying potential risks to users, employees, and the environment, and planning measures to mitigate these risks.
  - ***Health and Safety Standards:*** Commit to adhering to or exceeding health and safety standards relevant to the project, establishing protocols that ensure ongoing vigilance and responsiveness to safety concerns.
  - ***Emergency Preparedness:*** Develop plans for emergency preparedness and response, considering how the project can maintain safety in the face of unexpected events.

2.3.2 Basic Development—Conceptualization

The application of 5S principles to basic development—conceptualization, requires you to integrate these core values into the very foundation of your system. A strategic approach to do that is the following:

- **Social:**
  - ***Community and User Analysis:*** Understand the needs, expectations, and cultural contexts of the potential users and affected communities. Incorporate social value propositions into the concept, such as improving quality of life or solving specific social challenges.
  - ***Stakeholder Involvement:*** Involve a wide array of stakeholders in the conceptualization process to ensure diverse perspectives are considered, fostering inclusivity and equity.
- **Sustainable:**
  - ***Eco-Design Principles:*** At the conceptual stage, integrate eco–design principles that minimize environmental impacts across the system life cycle. This could include using renewable materials, designing for longevity, and planning for end-of-life recycling or disposal.

- ***Sustainability Goals:*** Set explicit sustainability targets such as reducing energy consumption, lowering emissions, and minimizing waste.

- **Sensing:**
  - ***Data Collection Plans:*** Define what data will be valuable for optimizing performance and how it can be collected responsibly and securely. This involves identifying key points in the system's operation where sensors could provide insights for improvement or innovation.
  - ***Feedback Loops:*** Design feedback loops into the system's functionality, allowing for the collection of user input and environmental conditions that inform continuous product development.

- **Smart:**
  - ***Intelligent Features Identification:*** Determine which intelligent features, like automation, adaptability, and connectivity, can enhance the system's performance and user experience.
  - ***Technology Roadmap:*** Create a roadmap for integrating smart technologies, ensuring that the system can evolve with technological advances and remain compatible with future systems.

- **Safe:**
  - ***Risk Assessment:*** Conduct an early risk assessment to anticipate potential safety issues and integrate design elements that mitigate these risks, ensuring user and environmental safety.
  - ***Safety Standards Compliance:*** Plan for compliance with current and foreseeable safety standards, considering both physical and cybersecurity aspects, especially if data collection and connectivity are involved.

2.3.3 Advanced Development—Design

The design and advanced development stages using 5S principles means ensuring that these foundational values are embedded in the system from the early design phase through to the sophisticated stages of development. The integration of these principles effectively can be done as follows:

- **Social:**
  - Involve diverse user groups in the design process to ensure the system is accessible and meets a wide range of needs. Consider the societal impact of the system and design for empowerment, education, or problem-solving in communities.
  - Design for inclusivity and accessibility, ensuring the system is usable and beneficial for a broad range of users and communities.

  - Consider the social implications of the system, including job creation, community engagement, and potential impacts on local economies.
  - Test the system with a focus on social impact and adjust based on feedback to ensure it delivers on its promises to support and enhance users' lives and communities.

- **Sustainable:**
  - Refine the use of materials and processes for greater sustainability. For example, if the system uses energy, make sure it operates efficiently and consider incorporating renewable energy sources.
  - Choose eco-friendly materials and design for energy efficiency, durability, and end-of-life recyclability. Consider the system's entire life cycle to minimize its environmental footprint.
  - Apply principles of sustainable design, such as using renewable resources, minimizing energy consumption, and designing for recyclability or biodegradability.
  - Assess the environmental impacts using tools like Life Cycle Assessment (LCA) to identify and mitigate potential negative effects.
  - Design the system for easy disassembly and recycling, facilitating the recovery of materials and components.

- **Sensing:**
  - Utilize the data collected from the sensors to fine-tune the system's performance. Ensure that the sensing capabilities align with privacy regulations and ethical data use.
  - Design systems with built-in sensors that can provide valuable data on system performance and user interaction, which can inform future improvements or enhance the user experience.
  - Integrate sensing technologies to monitor the system's performance and environmental impact, enabling data-driven decisions throughout the life cycle.
  - Design for data collection and analysis to optimize operation and maintenance, enhance user experience, and monitor

- **Smart:**
  - Implement AI and machine learning algorithms where appropriate to enhance the system's functionalities. Make sure the system's smart features lead to tangible benefits for users.
  - Integrate smart features that anticipate user needs, adapt to user behaviors, and can receive updates for future improvements. Ensure that the system can interface seamlessly with other devices or systems.
  - Incorporate smart technologies to improve efficiency, adaptability, and user interaction, ensuring the system can evolve over time with technological advancements.

  - Ensure that smart components are designed for energy efficiency and longevity, with the ability to be updated or recycled.

- **Safe:**
  - Conduct comprehensive risk assessments as the system moves through development. Continuously iterate on the design to mitigate any new risks that arise, ensuring the highest levels of safety.
  - Ensure that the system's design includes all necessary safety features and complies with relevant safety standards. Plan for rigorous safety testing and incorporate user feedback to identify potential safety issues.
  - Design with safety in mind, considering both immediate and long-term health and safety impacts on users and the environment.
  - Include mechanisms for monitoring and addressing safety concerns throughout the system's life cycle, from construction and operation to decommissioning.

By diligently applying the 5S principles during the advanced development-design, the system will be aligned with modern demands for technology that is responsible, responsive, and resilient. This approach not only creates a better system but also fosters trust and loyalty among users and stakeholders, setting the stage for long-term success.

2.3.4 Launching—Creation and Operation

Using the 5S principles to create and operate a system requires a thorough application of these values at each stage.

- **Social:** Communicate the social value of the system. Engage with communities and stakeholders to highlight the benefits and ensure the launch addresses their interests and needs.
- **Sustainable:** Launch the system with transparency about its environmental credentials. Use sustainable packaging and marketing materials and consider the carbon footprint of the launch event itself.
- **Sensing:** Use data collected during the testing phase to optimize the launch strategy and prepare to collect user data (with proper privacy measures) post-launch for continuous improvement.
- **Smart:** During the launch, showcase the smart functionalities and demonstrate how the system uses technology to enhance user experience and performance.
- **Safe:** Reassure users of the safety features and protocols in place. Ensure safety information is clear and accessible to all users from the outset.

2.3.5 Creation

- **Social:** Create systems that address or do not exacerbate social issues. Ensure that the creation process involves fair labor practices and contributes positively to the communities involved.
- **Sustainable:** Use of materials and processes that minimize environmental impact. The creation phase should emphasize the efficient use of resources and the minimization of waste.
- **Sensing:** Build in sensors and data collection capabilities to monitor the system's performance and usage, ensuring that they align with the intended design and efficiency goals.
- **Smart:** Incorporate intelligent features that allow for automation, ease of use, adaptability, and interconnectivity with other systems and devices.
- **Safe:** Prioritize safety in design to prevent accidents and malfunctions. Establish rigorous testing protocols to ensure the product or system operates safely under all conditions.

2.3.6 Operation

- **Social:** Maintain ethical practices in all business operations. Foster a culture of social responsibility within the organization and in interactions with partners and customers.
- **Sustainable:** Operate with a commitment to sustainability, continuously seeking to reduce the environmental impact and encourage sustainable use among consumers.
- **Sensing:** Utilize the sensing capabilities to monitor the system's performance continuously. Gather and analyze data to make informed decisions about maintenance, updates, and user support.
- **Smart:** Keep the system up to date with the latest smart technologies. Use data and user feedback to improve functionality and integrate with emerging technologies and systems.
- **Safe:** Implement a comprehensive safety management system. Monitor the system's safety performance and respond promptly to any issues that arise.

By systematically applying the 5S principles throughout the creation, launching, and operations of a system, you can ensure it succeeds in the market and adheres to high standards of social responsibility, environmental sustainability, technological advancement, and safety. Also, you can achieve the following:

- **Sustainable Manufacturing:** Implement sustainable manufacturing practices, reducing waste, conserving energy, and minimizing the carbon footprint.

- **Social Responsibility:** Ensure ethical practices in the supply chain, fair labor practices, and positive community relations during operation.
- **Maintenance and Upgrading:** Plan for regular maintenance and updates to keep the system safe, efficient, and technologically up to date, extending its useful life.

2.3.7 End-of-Life Management

The end-of-life (EOL) management of systems using the 5S principles requires a strategic approach to ensure that the closing phase of the life cycle is handled with the same care and consideration as the creation and operational phases. FIGURE 2.8 shows how the 5S principles integrate with the 10R framework enhance this approach by introducing specific strategies to maximize resource efficiency and sustainability:

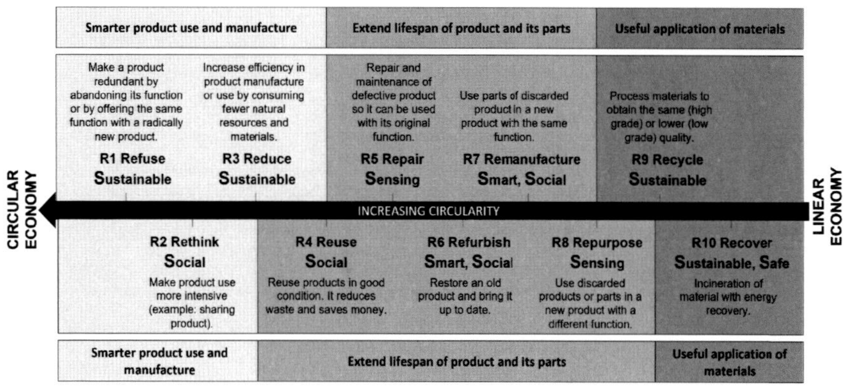

FIGURE 2.8
The 10R framework applied into the 5S principles.

- **Social:**
  - ***Community Programs (Rethink, Reuse):*** Develop community-focused programs for systems take-back or recycling, offering education on the importance of responsible EOL management. Encourage sharing models to make systems use more intensive.
  - ***Job Creation (Refurbish, Remanufacture):*** Create jobs not only in recycling but also in refurbishing and remanufacturing sectors, providing valuable services that contribute positively to the economy.
  - ***Equitable EOL Solutions:*** Ensure that EOL processes do not disproportionately affect underserved or marginalized communities, often the most impacted by waste and pollution.

- **Sustainable:**
  - ***Recycling and Upcycling (Recycle):*** Facilitate easy recycling or upcycling of systems, designing them from the start to be disassembled and repurposed, reducing the amount of waste that goes to landfills.
  - ***Materials Recovery (Recover):*** Recover valuable materials for reuse in new systems, reducing the demand for virgin materials and minimizing the environmental impact of resource extraction.
  - ***Biodegradability (Refuse):*** Use materials that are biodegradable under appropriate conditions for applicable systems, making some waste elements redundant by abandoning non-environmentally friendly functions.
  - ***Sustainable Disposal:*** For components that cannot be recycled, ensure environmentally safe disposal methods are used.
  - ***Life cycle Closure (Reduce):*** Evaluate the system's end-of-life impact and use the insights gained to inform the design of future systems, aiming for increased efficiency in product manufacture and use.
- **Sensing:**
  - ***Data Collection (Repair, Repurpose):*** Utilize sensors to inform users and manufacturers when a system is reaching its EOL, providing data that can help manage disposal or recycling efficiently. Data can also guide whether a system should be repaired or repurposed.
  - ***Condition Monitoring:*** Monitor the condition of systems to facilitate EOL decision-making, determining if they can be repaired, repurposed, or need to be recycled.
- **Smart:**
  - ***Smart Decommissioning (Refurbish, Remanufacture):*** Implement smart systems that guide users through the decommissioning process, perhaps through an app or online service that can recommend the best EOL options including refurbishment and remanufacture.
  - ***Resource Optimization:*** Use smart logistics to optimize the collection and redistribution of EOL systems, ensuring they are dealt with efficiently and effectively.
- **Safe:**
  - ***Safe Disposal (Recover):*** Ensure systems are disposed of in a manner that does not pose a risk to the environment or human health, adhering to all safety regulations and guidelines. This includes managing the incineration of materials with energy recovery where appropriate.
  - ***Hazardous Materials Management:*** Carefully manage and dispose of any hazardous materials contained in systems, preventing contamination or harm.

Thus, by integrating the 10R framework into the 5S principles in EOL management, organizations can enhance their approach to creating a closed-loop system life cycle. This not only contributes to a more sustainable economy but also fosters a healthier planet, reinforces the company's commitment to ethical practices and environmental responsibility, enhances its reputation, and often leads to cost savings and innovation opportunities.

Organizations can create a closed-loop system life cycle by considering the 5S principles in EOL management, contributing to a more sustainable economy and a healthier planet. This approach reinforces the company's commitment to ethical practices and environmental responsibility, enhances its reputation, and often leads to cost savings and innovation opportunities.

2.4 Conclusions

This chapter explored the Life Cycle Engineering (LCE) framework applied to create 5S systems with its principles—Social, Sustainable, Sensing, Smart, and Safe. The framework and 5S principles collectively shape systems (material, product, process, manufacturing system, and supply chain) that are optimized for performance and adhere to ethical, environmental, and safety standards throughout their life cycles.

The LCE framework, focusing on the entire life cycle of a system from ideation to disposal, the 5S principles complements the LCE framework by ensuring that each stage of a system's life adheres to socially beneficial, environmentally responsible principles, technologically advanced, and inherently safe principles. This integration enhances the sustainability and efficiency of systems, making them more adaptive and capable of meeting modern requirements and future challenges.

1. **Life cycle Perspective with 5S Integration:** By embedding the 5S principles into the life cycle stages dictated by LCE, systems achieve a higher performance:
    - Creating design systems that are efficient, considers environmental impact, promote social equity, and leverage technological advancements.
    - Ensuring operations and disposal processes adhere to safety and sustainability standards, thus extending the ecological and social benefits throughout the system's life.
2. **Enhanced Sustainability and Ethical Impact:** The 5S systems ensures that all operational activities consider their long-term impacts on the environment and society. This results in:
    - Systems that reduce waste and emissions, use energy more efficiently, and promote the 10Rs (Refuse, Rethink, Reduce,

Reuse, Repair, Refurbish, Remanufacture, Repurpose, Recycle and Recover) of materials.
- Practices that foster fair labor, support community development, and ensure equitable access to the benefits of technological advancements.

3. **Adaptation and Responsiveness:** Through the Sensing and Smart principles, systems developed under the LCE framework and 5S principles can dynamically respond to changing conditions and optimize performance over time. This adaptability is crucial for:
   - Continuous improvement based on real-time data and feedback.
   - The ability to predict maintenance needs and adapt to variations demand or resource availability variations to Safety and Security: The Safe principle in the 5S systems ensures that every stage of a system's life cycle incorporates robust safety and security measures, protecting users and the environment. This includes:
   - Designing systems that prevent accidents and mitigate risks.
   - Implementing emergency preparedness and robust risk management strategies across all operational phases.

2.4.1 Future Implications of LCE Framework and 5S Systems

The combined application of the LCE framework to create 5S systems sets a new paradigm for designing and operating systems that are prepared for future complexities. As organizations increasingly adopt these concepts, they can expect:

- **Improved Operational Efficiency:** Systems that are not only optimized for current conditions but are also adaptable to future technological and environmental changes.
- **Enhanced Reputation and Compliance:** Adhering to high ethical and environmental standards improves organizational reputation and aligns with global regulatory expectations.
- **Innovative and Sustainable Growth:** By fostering a culture that values continuous improvement and responsible practices, organizations can drive innovation and sustainable growth.

Hence, the use of the LCE framework using 5S principles represents a forward-thinking approach emphasizing resilience, adaptability, and responsibility. This approach enable organizations to face current and future challenges, ensuring that their systems benefit both people and the planet while remaining economically viable. This holistic approach is essential for achieving long-term operational and strategic excellence.

3

Materials: Discovery, Creation and Development

This chapter delves into the forefront of materials innovation, emphasizing sustainability as a core driver. It examines how sustainable materials development integrates advanced technologies and life cycle engineering (LCE) to create environmentally friendly materials. This strategic approach ensures sustainability from ideation to end-of-life management by focusing on minimizing waste and conserving resources.

Advanced digital technologies such as machine learning, materials informatics, and high-throughput screening are utilized to optimize material properties and accelerate discovery. During the operational phase, sustainable practices in production and recycling are implemented, significantly reducing energy use and carbon emissions.

A case study of Honeywell Aerospace illustrates the practical application of this framework, detailing the energy-efficient production of turbine blades using additive manufacturing. This example underscores the potential of sustainable practices to mitigate climate change and drive innovation.

The chapter highlights the future of sustainable manufacturing in materials development, emphasizing the integration of digital technologies, circular economy principles, and policy frameworks. This comprehensive approach addresses global environmental, economic, and social challenges, positioning industries to thrive sustainably in the evolving market landscape.

3.1 Introduction

The evolution of materials science has historically marked significant eras in human development, from the Stone Age to the Silicon Age. Humanity is on the brink of a new era where sustainability is a core driver of innovation. The sustainable development of materials is about creating materials with lower environmental footprints but also about integrating these materials into a circular economy where resources are reused and recycled, minimizing waste and conserving natural resources [126].

DOI: 10.1201/9781003501619-3

Over the past five decades, global material consumption has increased dramatically. Between 1970 and 1995, worldwide consumption of raw materials doubled. This surge has led to significant environmental impacts, such as deforestation, biodiversity loss, and increased greenhouse gas emissions [141].

- **Deforestation:** The rate of deforestation in tropical regions is approximately one acre per second.
- **Biodiversity Loss:** Habitat destruction has contributed to species extinction at rates much higher than natural background rates.
- **Overfishing:** 75% of marine fisheries are now overfished or fished to capacity.
- **Water Scarcity:** Freshwater withdrawals have doubled between 1960 and 2000, stressing major rivers like the Colorado and the Nile.
- **Greenhouse Gas Emissions:** Global emissions increased by 70% between 1970 and 2004, largely due to material extraction and processing activities.

Traditional environmental management focuses primarily on end-of-pipe solutions—controlling emissions and managing waste after it has been created. In contrast, Sustainable Materials Management (SMM) takes an LCE approach, addressing environmental impacts at every stage of the material life cycle. TABLE 3.1 depicts the key differences between traditional and sustainable material management.

Furthermore, in a world grappling with the brink of resource depletion and environmental degradation, sustainability emerges not just as a strategy but as a necessity.

Materials are the backbone of industrial processes and consumer products. In the circular economy, the life cycle of materials is extended through continuous reuse and recycling, transforming how production and consumption are conceptualized. Materials like steel, aluminum, and PET plastics are recycled and reincarnated into new forms, reducing the need for virgin materials and minimizing environmental impact.

For instance, aluminum is a metal used ubiquitously, from soda cans to automotive parts. Recycling aluminum can save up to 95% of the energy required for its primary production, significantly reducing carbon emissions. Similarly, recycling steel in packaging has reached a remarkable efficiency rate, saving millions of tonnes of carbon emissions annually. The substantial CO_2 emission reductions achieved by recycling other materials like PET plastics play a pivotal role in reducing the carbon footprint of packaging industries.

As industries and consumers increasingly prioritize sustainability, the demand for recyclable materials such as copper and iron has surged. This shift is not merely a trend but a transition towards more sustainable industrial practices, reflecting a global consciousness geared towards environmental preservation. The increasing demand drives technological and infrastructural advancements in recycling facilities, underscoring a significant move towards more robust, sustainable material management.

TABLE 3.1
Materials Management: Traditional vs Sustainable Approach. Adapted from [141]

Key Differences	Traditional Management	Sustainable Management
Life cycle Engineering	Usually focuses only on what to do with wastes once they are generated.	Focuses broadly on impacts and policies relating to all the life cycle stages from ideation to launching—including such upstream considerations as using less material, using less environmentally intensive materials, or making products more durable, as well as downstream solutions such as reuse and recycling.
Inputs and Outputs	Concerned mainly with outputs to the environment (air, water, land) and usually only those from waste and only where the waste is managed.	Concerned with inputs and outputs from/to the environment, including the use of materials, energy, and water, plus multiple environmental impacts. It is not geographically constrained.
Goal Orientation	Aims to minimize and manage waste.	Seeks the most productive and sustainable use of resources, minimizing overall environmental impacts and fostering systemic sustainability.
Responsible Parties	Focuses on waste generators.	Involves all stakeholders throughout the material life cycle, including industry, manufacturers, consumers, and policymakers.

For instance, biodegradable materials offer a promising solution to the waste problem, especially in sectors like packaging. Materials such as PLA, derived from renewable resources, reduce dependency on fossil fuels and offer the intriguing possibility of 'closing the loop'—returning to nature without leaving harmful residues. However, while biodegradable materials are favorable, they are not without challenges. The production and degradation conditions must be carefully managed to ensure true sustainability, making their integration into the circular economy complex yet potentially rewarding.

3.1.1 The Significance of Sustainable Material Development in the Modern World

Sustainable material development is of paramount significance in the modern world. It addresses pressing environmental, economic, and social challenges. Integrating sustainability into materials science and engineering is a crucial response to global imperatives such as climate change, resource depletion, and environmental degradation [126].

One of the foremost reasons for emphasizing sustainable material development is its potential to mitigate environmental impacts. Due to the extraction, processing, and disposal phases, traditional materials often entail significant ecological footprints. By incorporating sustainability metrics such as green chemistry and materials circularity, materials development can significantly reduce carbon footprints, minimize waste, and decrease pollution [126].

Material production is a significant source of greenhouse gas emissions, accounting for 23% of global emissions between 1995 and 2015. This includes emissions from producing machinery, vehicles, and other products. Life Cycle Assessment (LCA) is a method used to evaluate the environmental impact of a material throughout its entire life cycle, from raw material extraction to disposal. It helps identify stages where emissions can be reduced. Despite its effectiveness, LCA is complex and time-consuming, prompting the need for simpler evaluation methods. Therefore, a simpler approach to evaluate a material's impact on the environment is through two main indicators: embodied energy and carbon footprint [84]:

- **Embodied Energy:** It is the total energy required for the extraction, processing, production, and delivery of a material or product. It encompasses all the energy inputs up to material use but excludes operational and disposal energy costs. Minimizing embodied energy is essential for reducing a product's overall environmental impact. Although it does not cover the operation and disposal from the LCA, this indicator simplifies the evaluation of the material's impact. FIGURE 3.1 depicts the energy values typically required to produce different engineering materials [11].
- **Carbon Footprint:** The carbon footprint of a material refers to the total greenhouse gas emissions generated throughout its life cycle. This includes all CO_2 emissions from the material's extraction, processing, production, and transportation. Reducing the carbon footprint is vital to mitigating the impacts of climate change. In that sense, recycling has helped in reducing CO_2 emissions for several recyclable materials, for example [130]:
  - Recycling 1 kg of aluminum saves 8.14 kg of CO_2.
  - Recycling 1 kg of copper saves 2.66 kg of CO_2.
  - Recycling 1 kg of glass saves 0.31 kg of CO_2.
  - Recycling 1 kg of metals saves 5.86 kg of CO_2.
  - Recycling 1 kg of paper saves 0.46 kg of CO_2.

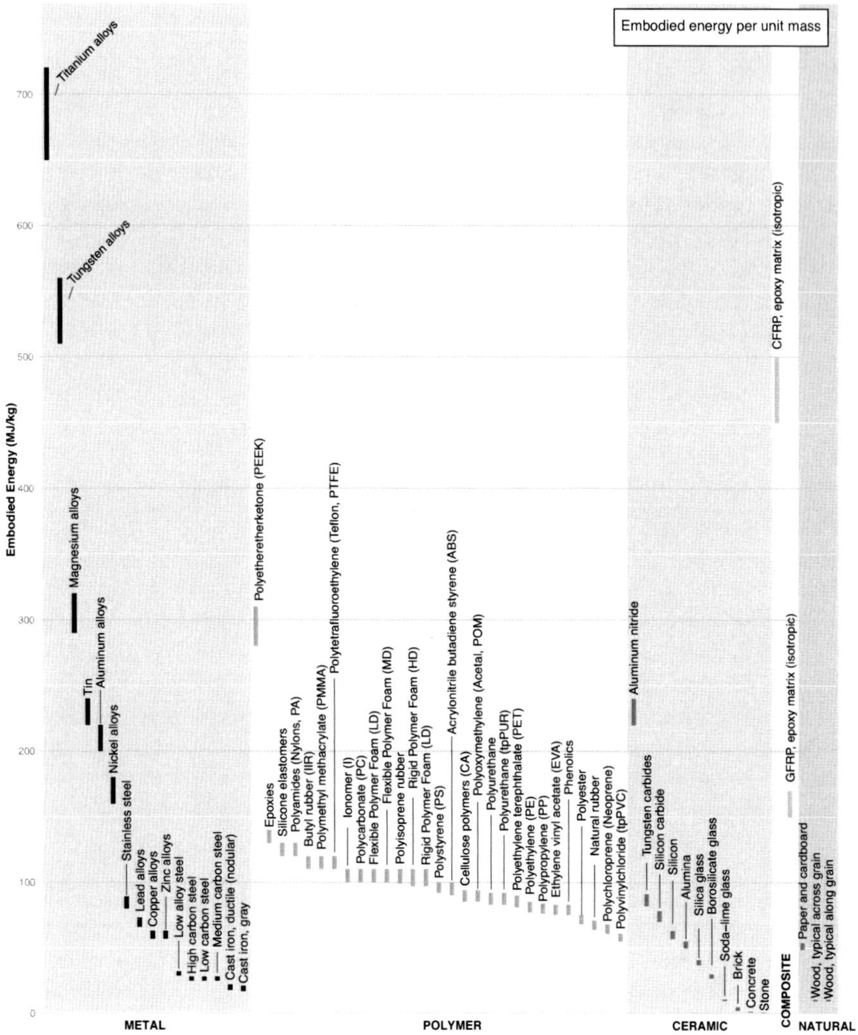

FIGURE 3.1
Embodied energies of different engineering materials based on Michael Ashby [11].

- Recycling 1 kg of plastic saves 1.02 kg of CO_2.
- Recycling 1 kg of scrap metals saves 3.57 kg of CO_2.
- Recycling 1 kg of steel saves 0.86 kg of CO_2.
- Recycling 1 kg of textiles saves 3.37 kg of CO_2.

Hence, recycling is important because it reduces materials' embodied energy and carbon footprint. Prioritizing materials with lower environmental burdens can achieve sustainable development.

In addition, understanding how materials move through the economy and the environment is critical to managing their impacts. The material life cycle begins with the extraction, processing, manufacturing, use, and finally disposal or recycling of natural resources. This process requires significant energy and water inputs at each stage, resulting in various environmental impacts, including air, water, and land emissions.

Therefore, to transition to sustainable materials management, comprehensive strategies must be adopted that address the root causes of material waste and inefficiency. The following recommendations outline a path toward this type of management:

- Encourage efforts to manage materials and products on a life-cycle basis. This involves:
  - Selecting materials and products for demonstration projects to showcase the benefits of integrated materials management strategies.
  - Expanding the focus of environmental programs to encompass life-cycle materials management fully.
  - Promoting greener products, product stewardship, and product-to-service transformations.
- Strengthen the capacity of government programs to support SMM by:
  - Establishing and improving databases to promote materials management.
  - Enhancing decision-making tools to support life-cycle management.
  - Integrating materials management strategies into regulatory development, permitting, and partnership programs.
- Foster a national conversation about life-cycle materials management to raise awareness and encourage action by:
  - Engaging multiple stakeholders in dialogue about materials management.
  - Exploring economic instruments to encourage better materials management.
  - Sharing knowledge and best practices widely.

The social implications of sustainable material development are equally profound. Sustainable practices in materials science can lead to healthier communities by reducing the harmful effects of toxic chemicals and pollutants. Engaging diverse stakeholders, including policymakers, public health experts, and local communities, ensures that materials development considers human health and environmental justice. Furthermore, sustainable material development can promote equitable access to advanced technologies, such as clean energy solutions, thereby improving quality of life and fostering inclusive growth [126].

Education and workforce development are foundational to successfully implementing sustainable material development. A multidisciplinary approach to education, integrating sustainability into the materials science and engineering curriculum, equips future professionals with the necessary skills and knowledge. Developing a workforce that is technically proficient, environmentally conscious, and capable of cross-disciplinary collaboration is critical for driving innovation in sustainable materials [126].

Effective policy and governance play a crucial role in advancing sustainable material development. Legislative frameworks such as Extended Producer Responsibility (EPR) and initiatives promoting transparency and accountability in the materials supply chain are essential for driving sustainable practices. Policies that encourage recycling, reuse, and the development of alternative materials can significantly reduce the environmental footprint of industries and promote a circular economy [126].

3.1.2 Overview of How Digital Technologies are Transforming Material Development

Material development traditionally spans decades from concept to commercialization. Historically, material development processes relied heavily on researchers' empirical knowledge and instincts, who designed experiments and interpreted results based on their experience. This traditional method was often slow, expensive, and subjective, with significant challenges in repeatability. Recent advances in digital technologies, especially machine learning, have transformed this landscape, promising shorter development cycles and more precise material properties prediction.

Digital technologies, notably machine learning, have significantly compressed the timeline for the discovery and development of materials. These algorithms reduce the research and development cycle time by analyzing vast datasets to identify potential materials and predict their properties swiftly and accurately. This rapid analysis capability enables industries to respond quickly to technological advancements and evolving market demands.

Enhanced predictive models powered by machine learning can now accurately predict material behaviors under various conditions. Leveraging historical data, these models can forecast performance metrics such as durability, stability, and mechanical properties. Optimizing the research and development process not only saves time but also reduces resource consumption, making it more efficient and less costly.

Another transformative aspect is the integration of digital technologies with extensive material databases. These databases, rich in detailed information about material properties and performance, support machine learning algorithms in developing insights that guide decision-making in material design and application. This robust framework ensures that the materials selected are optimized for their specific applications, enhancing the effectiveness of the development process.

The development of innovative research tools such as inverse design and high-throughput computational screening exemplifies the impact of digital technologies in materials science. Inverse design allows researchers to specify desired outcomes, and the system recommends chemical compositions and processing routes that might achieve those properties. Conversely, high-throughput computational screening automates testing thousands of material formulations under various conditions, significantly enhancing the efficiency and breadth of material testing.

Digital technologies facilitate a deeper understanding of Processing-Structure-Properties-Performance (PSPP) relationships. This comprehensive insight helps optimize materials for specific functions, creating materials with superior performance and tailored functionalities suitable for a wide range of applications.

Materials informatics (MI), driven by machine learning and big data, addresses complex challenges in material development. It tackles issues such as predicting material behavior under untested conditions, discovering new materials for high-tech applications, and enhancing material sustainability. These technologies solve existing problems and anticipate future material needs, thereby driving forward innovation in material science.

For instance, MI employs computational science to predict the performance of materials using historical experimental data, including material compositions, mix ratios, and process conditions. This shift from an intuitive to a data-driven approach streamlines the development process and extends its applicability across diverse sectors.

MI exemplifies a revolutionary shift in material development, employing information science principles and computational methods. This sophisticated approach utilizes large-scale datasets, which include experimental results and materials data, enhanced by cutting-edge technologies such as AI/ML and deep learning at the forefront. A comparison between traditional development and materials informatics is as follows:

- **Designing candidates for experiments**
  - ***Traditional Development:*** Historical experimental data + designing experiments based on experienced researchers.
  - ***Materials Informatics:*** Historical experimental data + AI usage such as simulations (descriptor transformation), text mining (extract feature quantities from patents and papers), and image analysis.
- **Simulation**
  - ***Traditional Development:*** Simulation on all materials.
  - ***Materials Informatics:*** Analysis based on computational science.

- **Experiments**
  - ***Traditional Development:*** Prototyping all experiment candidates.
  - Employ AI to prototype and evaluate only selected candidates. Hence, the experiments are reduced.

3.1.2.1 Materials Discovery

Historically, materials discovery has been labor-intensive and time-consuming, relying heavily on empirical methods and serendipity. However, the advent of digital technologies has revolutionized this field, transforming it into a more systematic, efficient, and data-driven endeavor. This new approach leverages computational power, ML, and vast datasets to accelerate identifying and developing new materials with desired properties [62].

Digital technologies in materials discovery encompass a range of tools and methods that enable researchers to predict and design new materials before they are synthesized in the laboratory. These technologies include:

- **High-Throughput Computational Screening:** Automated simulations that evaluate the properties of thousands of candidate materials quickly.
- **Machine Learning and Artificial Intelligence:** Algorithms that learn from existing data to predict the properties of new materials, identify patterns, and suggest novel compounds.
- **Databases and Informatics Platforms:** Comprehensive repositories of materials data that facilitate the sharing and analysis of information across the scientific community.
- **Generative Design Models:** Advanced models that propose new materials compositions and structures based on specified criteria.

These technologies significantly reduce the time and cost associated with traditional experimental methods, enabling faster innovation and discovery. Furthermore, MI is a key component of digital materials discovery, integrating materials science with information technology. It utilizes large-scale datasets, advanced algorithms, and computational models to predict materials properties, generate new structures, and design novel materials. Here are the primary elements of MI [62, 78]:

- **Data Availability and Access:** Comprehensive datasets from experimental and computational sources are the foundation of materials informatics. These datasets provide the necessary information for training ML models and validating predictions.
- **Machine Learning Models:** ML models are used to predict various properties of materials, such as mechanical strength, electrical conductivity, and thermal stability. These models can be trained using either compositional data (which describes the elements present in a material) or structural data (which describes the arrangement of atoms within a material).

- **Generative Design:** Generative models, such as Variational Autoencoders (VAEs) and Generative Adversarial Networks (GANs), can create new material compositions and structures. These models explore vast chemical and structural spaces to propose potential new materials that meet specific criteria.
- **Web-Based Platforms:** Platforms like MaterialsAtlas.org provide user-friendly interfaces for accessing tools for materials discovery. These platforms offer functionalities such as composition and structure validation, property prediction, and screening of hypothetical materials, making advanced tools accessible to researchers without deep computational backgrounds.

In that sense, MI involves critical components and tools that facilitate the discovery process [62, 78]:

- **High-Throughput Screening Tools:** These tools enable the rapid evaluation of a large number of materials for specific properties. They are essential for identifying promising candidates from vast chemical spaces.
- **Predictive Models:** Predictive models use existing data to forecast the properties of new materials. These models can be based on simple linear regression techniques or more complex deep learning architectures like Graph Neural Networks (GNNs).
- **Generative Models:** Generative models create new material compositions and structures that are likely to exhibit desired properties. These models are trained on existing data and can propose innovative materials that might not have been considered otherwise.
- **Web Apps and Databases:** Web apps and databases provide easy access to tools and data for materials discovery. Examples include Materials Project, AFLOWLIB, and JARVIS, which offer extensive datasets and analysis tools for materials research.

Some of the available web apps for materials discovery are depicted in TABLE 3.2.

Materials discovery through digital technologies represents a paradigm shift in developing new materials. By leveraging it, researchers can explore vast chemical spaces efficiently, predict material properties accurately, and design novel materials with unprecedented speed. This approach accelerates the discovery process and opens up new possibilities for innovation in materials science.

The chapter delves into Life Cycle Engineering (LCE), which optimizes the entire life cycle of materials from raw material extraction to end-of-life disposal. LCE promotes bio-based and recycled materials, waste and energy reduction, and designs that facilitate recycling and reuse. The benefits of LCE in enhancing environmental sustainability and business efficiency are discussed, along with the challenges and limitations of implementing LCE principles in material development.

TABLE 3.2
Some Web Apps for Materials Discovery [62]

App Name	URL	App Functions
MaterialsAtlas	www.materialsatlas.org	Composition/structure validation, property prediction, screening of materials, ML, composition enumeration, and more
Materials Project	materialsproject.org	Crystal toolkit, structure predictor, phase diagram, Pourbaix Diagram, reaction calculator, interface reaction, nanoporous materials analysis, and synthesis description search
Aflowlib	aflowlib.org	Elastic, thermal, prototype, chull, aflow-ML for superconductor Tc, free energy and entropy, metal/insulator classification, band gap energy, bulk/shear moduli, Debye temperature, and heat capacities
OQMD	oqmd.org/analysis	Phase diagram, structure visualizer, and ground state analysis
JARVIS	jarvis.nist.gov	Web ML tools for diverse property predictions (regression/classifications)
Crystal.AI	crystals.ai	Prediction models of formation energy, bandgap, elastic constants, perovskite/garnet stability, and coordination from X-ray absorption spectroscopy
Matgenie	matgenie.materialsvirtu-allab.org	Structure file format conversion, symmetry analysis, structure similarity comparison, XRD calculation, and surface generation
Materials Cloud	materialscloud.org/ work/tools	QE input generator, chemical shift, molecular polarizability, phonon visualizer, synthesis condition finder, predicting oxidation states, atomic environment finder, electron transport, and simulation in cloud (AiiDA)
Bilbao Crystallographic Server	www.cryst.ehu.es	Show Wyckoff positions, symmetry, and structure utility
SUNCAT	catalysis-hub.org	Database and tools for interface science and catalysis design

This chapter also explores the application of the 5S framework—Social, Sustainable, Sensing, Smart, and Safe—in material development, ensuring materials are ethically produced, environmentally responsible, technologically advanced, and safe. Each principle is detailed, emphasizing ethical sourcing, eco-friendly materials, integrated sensing capabilities, adaptive intelligence, and rigorous safety standards. This holistic approach guarantees that materials meet modern sustainability and safety requirements while fostering innovation and addressing social needs.

Additionally, the chapter covers advanced technologies that streamline material research and production, including computational materials science, high-throughput screening, and additive manufacturing. A case study on the sustainability impacts of 3D printing technologies illustrates the practical application of LCE. The chapter concludes with a discussion on future trends in sustainable manufacturing, such as digital twins, smart manufacturing, circular economy principles, and collaborative research, highlighting the path toward a more sustainable and innovative future in materials development.

3.2 Materials as 5S Systems

Developing materials using the 5S framework—Social, Sustainable, Sensing, Smart, and Safe—integrates a holistic approach that ensures materials are efficient and functional, ethically produced, environmentally responsible, technologically advanced, and inherently safe. Thus, FIGURE 3.2 shows how each of the 5S principles can be applied to material development:

- **Social:**
  - ***Ethical Sourcing:*** Ensure that all materials are sourced from suppliers that promote fair labor practices and contribute positively to their communities. This includes avoiding materials sourced from conflict zones or regions known for human rights abuses.
  - ***Community Engagement:*** Engage with communities affected by the material production process. This could include supporting local initiatives, improving infrastructure, or ensuring that the material development process does not adversely affect local populations.
  - ***Inclusive Development:*** Develop materials that cater to a diverse user base, considering various socio-economic backgrounds and aiming to improve quality of life through increased accessibility and functionality.

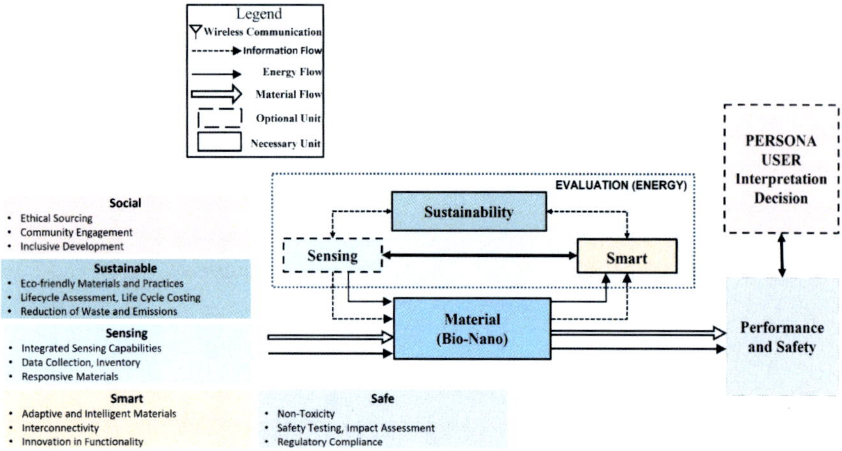

FIGURE 3.2
5S materials (social, sustainable, sensing, smart and safe).

- Sustainable
  - ***Eco-Friendly Materials:*** Choose materials that are renewable, recyclable, or have a minimal environmental footprint. Prioritize materials that can be produced with less water, energy, and raw material consumption.
  - ***Life cycle Assessment:*** Conduct comprehensive life cycle assessments (LCA) to evaluate the environmental impact of materials throughout their entire life cycle—from extraction and processing to usage and disposal.
  - ***Reduction of Waste and Emissions:*** Implement processes that minimize waste and emissions during material production. Innovate ways to reuse by-products or integrate closed-loop manufacturing processes.
- Sensing:
  - ***Integrated Sensing Capabilities:*** Develop materials incorporating sensing technologies, enabling them to monitor their structural health and environmental conditions or interact with other smart systems.
  - ***Data Collection:*** Utilize materials that can collect data for real-time monitoring and feedback, essential for adaptive uses such as in smart buildings or wearable technology.
  - ***Responsive Materials:*** Create materials that respond to environmental stimuli (like temperature or pressure changes) and provide useful feedback or adjustments, enhancing performance and safety.

- **Smart:**
  - ***Adaptive and Intelligent Materials:*** Develop materials that feature intelligent behaviors, such as self-healing properties, or that can change their properties based on external conditions.
  - ***Interconnectivity:*** Ensure that smart materials can easily integrate with other smart systems, enhancing the overall functionality of the products they are used in.
  - ***Innovation in Functionality:*** Push the boundaries of traditional material functionalities by integrating nanotechnology, biotechnology, or new chemical processes that allow for unprecedented features like enhanced conductivity, strength, or flexibility.
- **Safe:**
  - ***Non-Toxicity:*** Ensure all materials are safe for human use and environmental exposure. Avoid harmful chemicals and pollutants in the development and manufacturing stages.
  - ***Safety Testing:*** Rigorously test materials under various conditions to ensure they do not pose any risk during normal or extreme operational conditions.
  - ***Regulatory Compliance:*** Adhere to international safety standards and regulations. Proactively update safety measures and testing protocols as new standards and technologies emerge.

Incorporating these 5S principles into material development ensures that the materials meet industry standards and advance new sustainability, safety, and social responsibility standards. This comprehensive approach fosters innovation while addressing environmental concerns and social needs, positioning your materials to be competitive in markets that increasingly value responsible development.

3.3 Life Cycle Engineering Description

Life Cycle Engineering (LCE) is an integrative approach that seeks to optimize the entire life cycle of materials, from raw material extraction to end-of-life disposal or recycling. By emphasizing sustainability and efficiency, LCE becomes instrumental in materials development, ensuring decisions made at the design phase consider long-term impacts on both the environment and economic viability. The field of LCE in materials development is rapidly evolving, driven by the need for sustainable solutions across various industries. From advanced materials research and practical industry applications to educational initiatives and innovative technologies, the integration of LCE principles is paving the way for a more sustainable future.

In the design phase, LCE involves strategically selecting materials based on environmental impact assessments, availability, and their potential for recycling. This often leads to the preference for bio-based or recycled materials, significantly reducing the environmental burdens associated with using virgin raw materials. During the manufacturing process, LCE promotes the adoption of technologies that minimize waste and energy use, enhancing material efficiency and reducing the overall carbon footprint of production activities.

One critical aspect of LCE in materials development is planning for products' end-of-life. This approach advocates for designs that facilitate easy disassembly and reuse or recycling of materials rather than disposal. By integrating strategies like modular design and compatibility with existing recycling technologies, LCE helps extend the life cycle of materials and mitigates their environmental impact.

3.3.1 Benefits and Limitations of Applying Life Cycle Engineering Principles to Materials Development

Applying LCE principles in materials development enhances environmental sustainability and optimizes business efficiency. LCE reduces the ecological footprints of materials throughout their life cycle—from extraction to disposal—by promoting better waste management and sustainable resource use. It improves cost efficiency and economic viability through optimized resource utilization and enhanced recycling processes, thereby reducing the total life cycle costs of materials.

Incorporating LCE principles leads to the creation of products and services designed to minimize environmental impacts and optimize economic, material, and energy resources during manufacturing. Additionally, it fosters the development of intelligent products that improve users' quality of life while promoting responsible disposal practices that favor the circular economy. This proactive approach ensures compliance with stringent environmental regulations, enhancing market competitiveness and fostering customer loyalty. LCE also drives innovation by integrating environmental considerations early in the material development process, which can lead to novel materials and processes that align with future sustainability trends and consumer demands.

Despite the significant advantages, the implementation of LCE presents several challenges. The initial investment for deploying LCE can be high, requiring significant spending on advanced technologies, research, and development, and personnel training, which can be particularly burdensome for small and medium-sized enterprises. The application of LCE principles also has limitations; developing optimized, innovative products and services may inadvertently lead to environmental impacts at certain stages of the life cycle. For example, while optimizing one stage of the cycle, another might be negatively impacted by the use of materials that are not environmentally friendly, or the necessary use of toxic agents in developing eco-friendly products.

Additionally, comprehensive life cycle assessments, crucial for effective LCE, involve complex and time-consuming data collection and analysis. The need for detailed life cycle data, which is often challenging to gather and keep updated, can hinder the ability to make informed environmental decisions. For instance, converting food resources like corn into industrial materials such as bioethanol, while beneficial in some respects, also leads to the generation of carbon dioxide emissions—a significant greenhouse gas. Moreover, LCE's dependency on the latest technologies can make it vulnerable to rapid obsolescence, complicating long-term implementation.

3.3.2 How Life Cycle Engineering Contributes to Sustainable Manufacturing in Materials Development

LCE is instrumental in driving sustainable manufacturing practices within the materials development sector. By providing a systematic approach to monitor the environmental profile of materials across their entire life cycle, LCE helps manufacturers optimize the use of resources and minimize waste. This approach ensures that all manufacturing stages—from raw material extraction, production and use, to end-of-life management—are conducted to reduce environmental impact and enhance efficiency.

Central to LCE is its ability to support the circular economy principles. It promotes designing materials and products that can be easily recycled or reused, thus extending their life cycle and reducing the reliance on virgin raw materials. This conserves resources and decreases the environmental degradation associated with material extraction and waste disposal. The modular design and compatibility for recycling that LCE encourages are vital for maintaining materials within the economic cycle for extended periods, thereby minimizing the overall ecological footprint of manufacturing processes.

Moreover, LCE is a critical tool in sustainable manufacturing as it generates valuable data that inform decision-making throughout the materials development and discovery process. This information is essential for designing processes that are not only environmentally friendly but also economically and socially viable. Sustainable manufacturing requires optimizing these processes to reduce pollution and industrial waste, especially as new technological developments are scaled for large-volume production.

The application of LCE in materials development also drives innovation by pushing for the discovery and use of new, sustainable materials that comply with stringent environmental standards. As product innovations continue to emerge from technological advancements, the data provided by LCE ensures that these new products are developed with minimal environmental burden. This forward-thinking approach helps manufacturers comply with international environmental regulations and meet the growing consumer demand for sustainable products.

Life Cycle Engineering enhances sustainable manufacturing by improving resource efficiency, supporting the circular economy, and providing critical data for environmentally friendly product design. By integrating LCE into materials development, companies can achieve significant environmental, economic, and social benefits, positioning themselves as leaders in the increasingly important field of sustainable manufacturing.

3.4 Life Cycle Diagram Flow

The diagram from FIGURE 3.3 illustrates the comprehensive life cycle of material development, emphasizing the structured progression through distinct stages: Ideation/Imagination, Basic Development/Conceptualization, Advanced Development/Design, and Launching/Creation/Operation. The process begins with identifying user needs and performance criteria, progresses through designing and synthesizing materials to meet these requirements, and culminates in materials' synthesis, analysis, characterization, and prototyping. Each stage builds upon the previous one, ensuring a seamless transition from conceptualization to practical application, strongly emphasizing aligning the materials with functional and sustainability goals.

3.4.1 Entity Description: Material

Material selection is based on cost, quality, and performance. Product design must determine which types of raw materials will be used, such as whether recycled materials can be used to reduce waste and lessen the environmental impact. The material to be used is determined based on the use of the final product. For example, in the production of plastic packaging, national and international authorities are increasingly establishing regulations on environmental laws regarding the sustainability of these products. In the initial stages of the LCE, information must be collected on potential alternatives for a better design of these plastic packaging, which could be composed of petroleum-derived polymers, biodegradable and/or compostable materials, or a combination of these, in a way that boosts the circular economy. It is also possible to use recycled materials in some cases, but this depends on the final use of the packaging. For example, in the food packaging sector, some companies prefer their packaging not to be made of recycled materials. Moreover, food packaging must comply with approvals

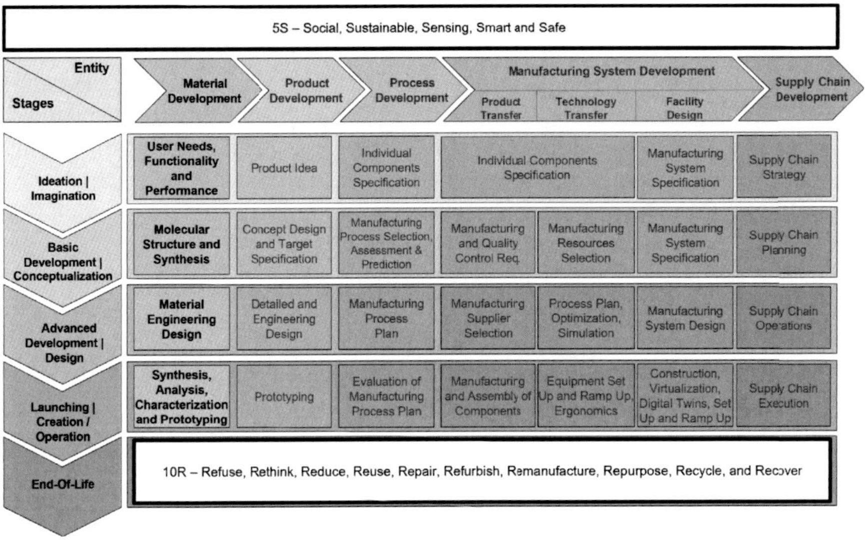

FIGURE 3.3
Life cycle diagram flow: material development and its stages.

from other administrations. Therefore, the LCE must project costs as they may be higher to ensure the required quality and performance.

Material Development encompasses the systematic creation and enhancement of materials, ensuring they meet specific user needs and performance criteria. This entity focuses on the detailed design and synthesis of molecular structures, translating these designs into tangible materials with precise properties. Throughout the development process, considerations of scalability, cost-effectiveness, and environmental impact are paramount. This meticulous approach ensures that the materials developed are suitable for their intended applications and align with broader sustainability and ethical standards.

In the Basic Development/Conceptualization stage, the emphasis shifts to designing and engineering the molecular structure and synthesis of materials to achieve desired properties and functions. This phase establishes the product's concept design and target specifications, ensuring it aligns with the material characteristics. Manufacturing process selection, assessment, and prediction are carried out to determine the best methods for producing the components. Concurrently, manufacturing resources are selected, and system specifications are developed to support the production process. The supply chain planning continues, integrating the new processes and resource requirements to ensure a cohesive and efficient workflow.

The Advanced Development/Design stage involves refining the materials' form, structure, and dimensions. Material engineering design is carried out to create precise drawings and specifications for manufacturing. Detailed engineering design ensures the materials align with the production processes, facilitating a smooth transition into manufacturing. A comprehensive manufacturing process plan is developed, detailing the steps and resources needed for production. Supplier selection, process plan optimization, and simulation are also addressed to ensure efficiency and quality. The supply chain operations are planned, ensuring a smooth flow of materials and components through the system.

The Launching/Creation/Operation stage focuses on synthesizing, analyzing, characterizing, and prototyping materials. This stage involves creating or improving new materials to achieve the desired properties. Analyzing the materials helps understand their composition and structure, while characterization evaluates their physical, chemical, and mechanical properties to ensure they meet design specifications. Prototyping creates sample versions of the materials for real-world testing, providing valuable insights before full-scale production. Concurrently, the manufacturing process plan is evaluated for efficiency and effectiveness, and components are manufactured and assembled, addressing equipment setup and ergonomics.

Finally, the End-of-Life stage focuses on the sustainable disposal and recycling of materials, adhering to the 10R principles (Refuse, Rethink, Reduce, Reuse, Repair, Refurbish, Remanufacture, Repurpose, Recycle, and Recover). This stage ensures that materials and products are managed responsibly throughout their life cycle. By integrating these principles, the development process aligns with the 5S principles (Social, Sustainable, Sensing, Smart, and Safe), ensuring materials, products, processes, manufacturing systems, and supply chains are efficient, functional, ethically produced, environmentally responsible, technologically advanced, and inherently safe. This comprehensive approach fosters innovation while addressing environmental concerns and social needs, positioning materials and products to be competitive in markets that increasingly value responsible development.

3.4.2 Stages

The development of materials can be broken down into several critical stages, each focusing on different aspects of the process from ideation to the end of life. These stages ensure that materials are designed to meet functional requirements and developed with sustainability and efficiency in mind.

3.4.2.1 Idea—Imagination

- The initial stage involves a thorough analysis of user applications, environments, and performance needs. Material choices are guided by functionality requirements, such as structural strength or conductivity, and performance criteria, including durability, environmental resistance, longevity, and adherence to standards. This foundational understanding sets the stage for the subsequent design and development processes, ensuring that the materials developed are tailored to meet specific user and application needs.

 - ***Activity Inputs:*** User application requirements, environmental conditions and constraints, performance criteria and standards, market research and user feedback, sustainable materials identification that meet user needs with a lower environmental footprint, consideration of the potential for recyclability and reusability in the initial material selection.
 - ***Activity Outputs:*** Detailed user needs and application specifications, preliminary material selection based on functionality and performance criteria, initial design briefs, and performance goals.

3.4.2.2 Basic Development—Conceptualization

- The goal is to achieve desired properties and functions through meticulous molecular structuring and synthesis. The process emphasizes creating materials with specific characteristics that align with product requirements. Key considerations include scalability, cost-effectiveness, and minimizing environmental impact, ensuring the materials can be produced efficiently and sustainably. Thus, this stage must consider employing green chemistry principles to minimize hazardous substances and waste and optimize synthesis methods for energy efficiency and reduced environmental impact.
  - ***Activity Inputs:*** Material properties and performance goals from the ideation stage, molecular design techniques and tools, synthesis methods and protocols, environmental impact assessments.
  - ***Activity Outputs:*** Molecular models and blueprints, synthesized material samples, evaluation reports on scalability, cost, and environmental impact, refined material specifications.

3.4.2.3 Advanced Development—Design

- This stage involves establishing the materials' form, structure, and dimensions. Detailed drawings and specifications for manufacturing are created to ensure that the materials match production processes. This phase is crucial for translating molecular designs into practical, manufacturable materials, ensuring precision in design and compatibility with production systems. It sets the stage for the practical application and testing of the materials in real-world conditions. Therefore, incorporating sustainability requires design for manufacturability with minimal waste, maximal resource efficiency, and ensuring compatibility with sustainable manufacturing processes and technologies.

  - ***Activity Inputs:*** Molecular design and synthesized samples from the conceptualization stage, engineering design tools and software, manufacturing process capabilities, production system requirements.
  - ***Activity Outputs:*** Detailed engineering drawings and specifications, material prototypes and test plans, manufacturing process alignment documents, initial quality and performance testing results.

3.4.2.4 Launching—Creation/Operation

- This stage focuses on synthesizing new materials or improving existing ones to achieve the desired properties. Analyzing these materials helps understand their composition and structure, while characterization evaluates their physical, chemical, and mechanical properties to ensure they meet design specifications. Prototyping involves creating sample or trial versions of the materials for testing in real-world conditions before full-scale production. This stage ensures that materials are thoroughly vetted and refined, ready for practical application. Furthermore, a way to incorporate sustainability is by ensuring the synthesis process uses renewable energy sources where possible and by implementing testing for environmental impact and sustainability metrics.
  - ***Activity Inputs:*** Detailed designs and specifications from the engineering design stage, synthesis protocols and materials, analytical tools and characterization equipment, prototyping facilities and testing environments.
  - ***Activity Outputs:*** Synthesized material batches, comprehensive analysis and characterization reports, prototypes for testing and validation, finalized material specifications, and readiness for production.

3.4.2.5 End of Life: 5S—Social, Sustainable, Sensing, Smart, Safe

Incorporating the 5S principles ensures that materials are functional and high-performing, ethically produced, environmentally responsible, technologically advanced, and safe. Each principle has specific activity inputs and outputs that contribute to sustainable materials management.

- **Social:** Social considerations at the end of a material's life cycle involve refusing materials from unethical sources and rethinking ways to source materials ethically and sustainably. Engage communities in material repair and refurbishment programs, collaborating with local communities to refurbish and repurpose materials, ensuring social responsibility throughout the material life cycle.

- ***Activity Inputs:*** Ethical sourcing guidelines: Ensure materials are sourced responsibly, considering labor practices and community impact.
- ***Activity Outputs:*** Ethically sourced materials: Materials procured following fair labor practices and ethical guidelines. Community engagement reports: Documentation of community impact assessments and engagement initiatives.

- **Sustainable:** Sustainable end-of-life management includes implementing strategies to reduce material usage and waste. Design materials that can be easily reused in different applications and develop efficient recycling processes to minimize environmental impact. Additionally, implement systems to recover valuable materials at the end of their life cycle, ensuring sustainability.
  - ***Activity Inputs:*** Environmental impact assessments and life cycle analysis: Evaluate the environmental impact from production to disposal.
  - ***Activity Outputs:*** Life cycle assessment reports: Detailed analysis of the environmental impact of materials throughout their life cycle. Waste reduction strategies: Plans to minimize waste during production and use.
- **Sensing:** Utilize sensing data to improve material reuse and recycling processes, leveraging real-time information to enhance the efficiency of these processes. Use sensor data to refine the remanufacturing process, ensuring materials can be effectively reused and repurposed.
  - ***Activity Inputs:*** Advanced sensing and monitoring technologies: Integrate sensors to monitor material conditions in real-time.
  - ***Activity Outputs:*** Integrated sensing capabilities: Materials equipped with sensors for real-time data collection and monitoring. Adaptive response systems: Technologies enabling materials to respond to environmental changes.
- **Smart:** Smart end-of-life management involves designing materials that can be repurposed for different functions after their initial use, ensuring they remain valuable throughout their life cycle. Develop smart materials that are easier to disassemble and remanufacture, enhancing their sustainability and functionality.
  - ***Activity Inputs:*** Smart material design protocols: Guidelines for designing materials with intelligent and adaptive features.
  - ***Activity Outputs:*** Smart, adaptive material prototypes: Innovative materials that can adapt to their environment and improve performance. Enhanced functionality: Materials with integrated smart technologies for better performance and usability.

- **Safe:** Safe end-of-life practices ensure the safe handling and refurbishment of materials, implementing safe recycling processes to prevent environmental contamination. Safely recover valuable components from materials at the end of their life cycle, ensuring they do not pose any risk to human health or the environment.
  - ***Activity Inputs:*** Safety standards and testing protocols: Ensure materials meet rigorous safety requirements.
  - ***Activity Outputs:*** Safety compliance certifications: Documentation proving materials meet all safety standards. Non-toxic materials: Materials designed and tested to be safe for human use and environmental exposure.

Integrating these 5S principles and end-of-life considerations into each stage of material development ensures that the materials are not only functional and high-performing but also ethically produced, environmentally responsible, and safe. This comprehensive approach aligns with modern sustainability and ethical standards, paving the way for a more sustainable future.

3.4.3 Techniques and Methods Required During the Development of Materials

The techniques and methods required for materials discovery, creation, and development in the context of sustainable practices and digital technologies include a combination of experimental, computational, and data-driven approaches. Here is an overview of the essential methods:

- **High-Throughput Computational Screening**
  - ***Purpose:*** Quickly evaluates thousands of materials to identify those with desirable properties.
  - ***Methodology:*** Utilizes computational simulations to predict material properties without the need for physical experiments.
- **Machine Learning and Artificial Intelligence**
  - ***Purpose:*** Enhances the prediction and optimization of material properties.
  - ***Methodology:*** AI algorithms learn from historical data to predict new materials' properties, optimize material compositions, and discover potential applications.
- **Materials Informatics**
  - ***Purpose:*** Integrates materials science with information technology to enhance the discovery and development of materials.

  - ***Methodology:*** Employs large datasets and advanced algorithms to predict properties, optimize processes, and suggest novel materials based on desired characteristics.

- **Generative Design Models**
  - ***Purpose:*** Automates the design process to propose novel material structures and compositions.
  - ***Methodology:*** Uses algorithms like Variational Autoencoders (VAEs) and Generative Adversarial Networks (GANs) to explore vast design spaces and create optimized material solutions.

- **Databases and Informatics Platforms**
  - ***Purpose:*** Facilitates access to comprehensive materials data and computational tools.
  - ***Methodology:*** Platforms like the Materials Project provide tools for structure prediction, property validation, and materials screening.

- **Predictive Models and Simulation**
  - ***Purpose:*** Predicts material behavior under various conditions and optimizes processes.
  - ***Methodology:*** Uses statistical and machine learning models to forecast material properties and performance, reducing the need for extensive physical testing.

- **Web Apps and Collaborative Platforms**
  - ***Purpose:*** Provides tools for remote collaboration, design, and testing.
  - Platforms like AFLOWLIB and JARVIS offer user-friendly interfaces for material prediction and design, enabling broad access to advanced computational tools.

The implementation strategy for sustainable materials development integrates several key elements to navigate the complexities of this field while leveraging the latest advancements in digital technologies. Firstly, it is crucial to integrate digital tools such as artificial intelligence and machine learning from the initial stages of material development. These technologies streamline the discovery process and enhance the accuracy of predictions, making it possible to identify and optimize new materials rapidly.

Secondly, sustainable and ethical practices are foundational to this approach. Employing green chemistry principles and conducting thorough life cycle assessments ensures that the materials developed have minimal environmental impacts and are safe for end use. This commitment to sustainability extends throughout the material's life cycle, from initial development to final disposal.

Collaboration and data sharing are also vital components of the strategy. Utilizing web-based platforms enables the sharing of results, methodologies, and data across the global scientific community. This openness accelerates innovation, fosters transparency, and allows for collective progress in material science.

Leveraging high-throughput experimental setups and virtual prototyping technologies further enhances advanced testing and prototyping. These tools allow for the efficient testing of materials under various conditions, significantly saving time and resources while providing robust data for further refinement.

Lastly, an iterative approach to feedback and refinement is essential. Insights gained from each development phase are used to continuously refine materials and processes, ensuring they meet the stringent specifications and performance criteria required for modern applications.

By adopting these strategies, innovative and environmentally friendly materials can be effectively developed, meeting today's needs without compromising future generations and embodying sustainable development principles in materials science.

3.4.4 Technologies to Employ for the Accelerated Development of Materials

A suite of advanced technologies that optimize and expedite every phase of material research and production increasingly facilitates the accelerated development of materials. TABLE 3.3 shows the key technologies employed to streamline the development process, enhance material properties, and ensure sustainability [62, 78, 141].

TABLE 3.3
Main Digital Technologies to Streamline the Development Process

Digital Technology	Description	Application
Computational Materials Science	Computational materials science involves using computer simulations to study and predict the properties of materials before they are synthesized. This technology uses principles from physics and chemistry to model interactions at the atomic and molecular levels, providing insights into material behaviors under various conditions.	Particularly useful in predicting material stability, reactivity, electronic properties, and mechanical strength, allowing researchers to narrow down potential material compositions for specific applications.

Digital Technology	Description	Application
High-Throughput Screening (HTS)	HTS technology automates the testing of numerous material samples under varied conditions to determine their properties and performance metrics quickly. This approach uses robotics, data processing software, and sensitive instrumentation to conduct experiments simultaneously.	Used to screen thousands of materials for applications such as photovoltaics, batteries, and catalysts, significantly cutting down the time from conceptualization to market deployment.
Machine Learning and Data Analytics	Machine learning algorithms interpret complex data from material testing and simulations to predict outcomes and optimize material formulations. These models improve over time as they process more data, continuously enhancing their predictive accuracy.	Used for predicting the properties of new alloys, polymers, and composites, facilitating the design of materials with tailored properties for specific applications.
Additive Manufacturing (3D Printing)	Additive manufacturing, or 3D printing, builds materials layer by layer, allowing for the precise control of material structure and composition. This technology reduces waste and enables the creation of complex geometries not possible with traditional manufacturing methods.	Widely used for prototyping, manufacturing bespoke parts in aerospace and biomedical industries, and developing new materials like advanced ceramics and metal alloys.
Advanced Characterization Techniques	Advanced characterization techniques, such as electron microscopy, X-ray diffraction, and spectroscopy, provide detailed insights into the microstructure and composition of materials. These techniques are essential for understanding the relationship between material structure and properties.	Critical for characterizing nanomaterials, semiconductors, and biomaterials, ensuring that materials perform as expected in real-world applications.

Digital Technology	Description	Application
Synthetic Biology and Bioengineering	Synthetic biology and bioengineering apply principles from biology to develop materials that mimic or surpass the functionalities found in nature. These technologies focus on creating sustainable materials by utilizing biological processes.	Used in developing biodegradable plastics, bio-based polymers, and environmentally friendly dyes and pigments, contributing to sustainability and opening new avenues in material functionality.
Internet of Things (IoT) and Sensors	IoT technology integrates sensors and network connectivity into materials and devices, allowing for real-time data collection and monitoring. This technology enables the continuous assessment of material performance under operational conditions.	Instrumental in developing smart materials and structures that can adapt to environmental changes, monitor structural health, and provide data for predictive maintenance in sectors such as construction and automotive.
Quantum Computing	Although still in its early stages, quantum computing promises to revolutionize material science by solving complex molecular and material simulations much faster than conventional computers.	Potential applications include discovering new pharmaceuticals, catalysts, and energy-predictive maintenance data in the construction and automotive sectors beyond classical computers' reach.
Blockchain Technology	Blockchain technology provides a secure, immutable, and decentralized ledger for recording transactions and tracking assets in a business network. In the context of materials development, blockchain offers unparalleled traceability and transparency.	Can be used to document the provenance and life cycle of materials, from raw material sourcing through manufacturing to end-of-life disposal, ensuring authenticity and compliance with environmental and regulatory standards.
Optimization Techniques	Optimization techniques involve mathematical methods and algorithms designed to find the best solution from a set of available alternatives under given constraints. These techniques are crucial for enhancing the performance and efficiency of material systems.	Used in materials selection, process design, and the configuration of material properties to meet specific requirements, optimizing critical parameters such as weight, strength, and thermal conductivity.

Digital Technology	Description	Application
Big Data Analytics	Big Data Analytics involves examining large and varied data sets to uncover hidden patterns, correlations, and insights. This technology processes vast amounts of data from experiments, simulations, and real-world applications to accelerate material discovery and development.	Crucial for predictive modeling and decision-making in materials science, enabling researchers to identify optimal material compositions and processing conditions quickly.

Using these technologies, the field of materials science is positioned to achieve significant advancements in developing new materials. Each technology offers unique capabilities that, when combined, can significantly reduce development time, optimize material properties, enhance sustainability, and create innovative products that meet the demands of various industries.

3.4.5 The Ten Core Technologies Applied to the Entity

Integrating the ten core technologies into this entity is applied strategically, as depicted in FIGURE 3.4.

1. **Ideation—Imagination:** During the initial stage, user needs, functionality, and performance are defined.
   (a) AI/ML predicts performance needs by analyzing user data and historical trends.
   (b) Big Data Analytics identifies patterns in user preferences and environmental constraints to shape product ideation.
   (c) Simulation Tools virtually assess the feasibility of material concepts, saving time and resources.
   (d) AR/VR offers immersive visualizations of early concepts, enabling stakeholders to evaluate usability and potential design challenges.
2. **Basic Development—Conceptualization:** In material synthesis and property engineering.
   (a) Simulation Tools aid in designing molecular structures with specific functional goals.
   (b) Cloud Computing provides the infrastructure for real-time collaboration and scalable material simulations.

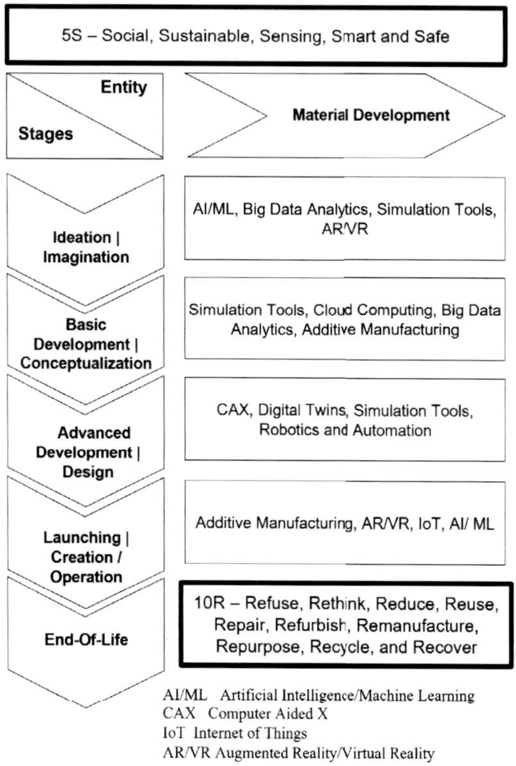

FIGURE 3.4
Core technologies applied to this entity.

(c) Big Data Analytics processes data from simulations to refine material properties.
(d) Additive Manufacturing enables rapid prototyping of molecular designs for physical evaluation.

3. **Advanced Development—Design:** In establishing the structure, form, and manufacturing specifications.

(a) Computer-Aided Tools (CAX) create precise engineering designs that integrate performance and production needs.
(b) Digital Twins virtually replicate materials and processes to optimize designs and predict production challenges.
(c) Simulation Tools validate designs under simulated environmental and operational conditions.
(d) Robotics and Automation streamline the precision assembly and testing of material prototypes.

4. **Launching—Creation/Operation:** In the deployment and operationalization of the material system.
   (a) Additive Manufacturing enables the scaled production of materials while maintaining precision.
   (b) AR/VR supports immersive training to handle and integrate new materials.
   (c) IoT facilitates real-time monitoring of material performance under operational conditions.
   (d) AI/ML dynamically optimizes operational processes based on collected data.

3.5 Case Study Applying the Life Cycle Engineering to Materials Development

The first stage in the LCE process for any 3D printed product focuses on the raw material source, extraction processes, and supplier operations. The raw material must fulfill technical specifications with built-in flexibility to reduce potential problems during production. Designers must consider the following aspects during the product production process depending on material specifications and available manufacturing technologies:

1. **Material properties requirements:** The material and technical specifications must reflect the properties needed to produce the part and to fulfill design standards.
2. **CAS register number and purity:** The CAS register number is a unique identification number of the Chemical Abstract Service (CAS) assigned to all chemical compounds, materials, alloys, minerals, and substances identified in the scientific literature. The CAS number is not indicative of the chemical or material's functional or performance characteristics, but serves as a unique identifier for classification and regulatory purposes.
3. **Implicit and explicit material properties:** The implicit material properties are those of the end product. In contrast, the explicit properties are listed on the datasheet and the analysis certificate. The implicit and explicit property values must be revised carefully to identify possible differences. The same raw material properties provided by different suppliers could behave differently during production due to small raw material production during batch production, i.e., two materials could have the same tensile strength under the conditions specified by the standard used for testing but different behavior when subjected to temperature changes, which means that only one of the materials is usable.

4. **Entry controls:** Analyses for raw materials input controls are important to verify whether the material meets the required specifications. This allows for avoiding potential problems during production. Although entry controls demand sophisticated equipment and qualified personnel, it would be expensive to stop the production process when material specification discrepancies during experimental tests of the end product are detected. Therefore, it is highly desirable to implement inspection controls that will reveal nonconformities between raw material batches.
5. **Testing raw materials:** Conducting tests on a raw material is relevant to gain insight into its properties and limitations. At the start of additive manufacturing technologies to produce metallic components, engineers would look at the metallic powders to verify their morphology and size distribution and whether the metallic powders manufactured from chemical processes were produced with a uniform round-shape-like form and normal size distribution. The belief is that these powder characteristics could influence the mechanical properties and, in the end, avoid the presence of porosity in the additive-manufactured part. Therefore, engineers must verify the metallic powder batch characteristics, knowing that powder size distributions and morphology are important in additive manufacturing technologies.

The study illustrates the utilization of LCE in evaluating the sustainability impacts of 3D printing technologies by considering the production process of eight turbine blades in Honeywell Aerospace's plant located in Mexico. This process was based on ISO 14040:2006 standards to evaluate the environmental impacts of additive manufacturing (AM) compared to CNC manufacturing technologies. The main focus of the study was on energy consumption during the turbine blade production processes. FIGURE 3.5 shows the general geometry of the built turbine blades by casting and additive manufacturing processes. However, the study does not include factors such as human consumption and transportation. The life cycle assessment of CNC process considers the casting of Inconel 718.

TABLE 3.4 shows the energy consumed directly or indirectly per hour during the fabrication of eight blades considering the following manufacturing processes: grinding, first nitric acid line, fluorescent penetrant inspection (FPI), shot peening, part marking, second nitric acid line, and packaging processes. These results provide information about the energy consumption of the major fabrication processes, such as casting, grinding, nitric clean lines, and part finish operations, involved in achieving the desired turbine blades.

To estimate the energy consumption of the sintering laser melting additive manufacturing process (SLM), one must remember that this additive manufacturing technology employs metallic powder layers (25–50 mm) spread

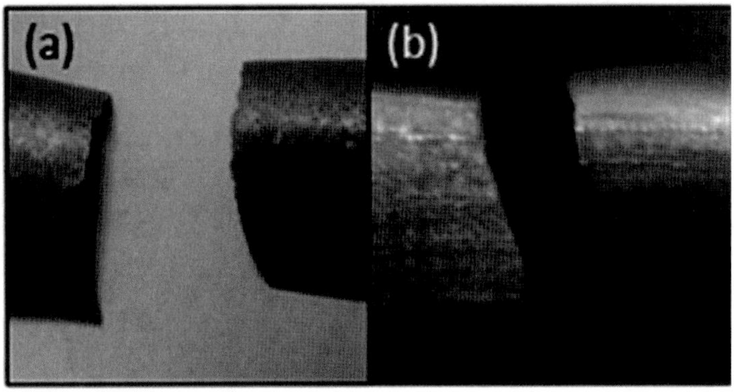

FIGURE 3.5
Schematic of the turbine blades built by: (a) casting, (b) SLM.

out on a building platform. The production of metallic powder used by the AM process demands considerable energy. Besides, SLM 125 HL equipment, shown in FIGURE 3.6, requires an external water/water or water/air cooling system. Furthermore, the SLM 125 HL machines have a process chamber that demands at least 2L/min of inert gases (Argon or Nitrogen), a power supply of 400 V, 50/60 Hz, 3 phases, and 32 A.

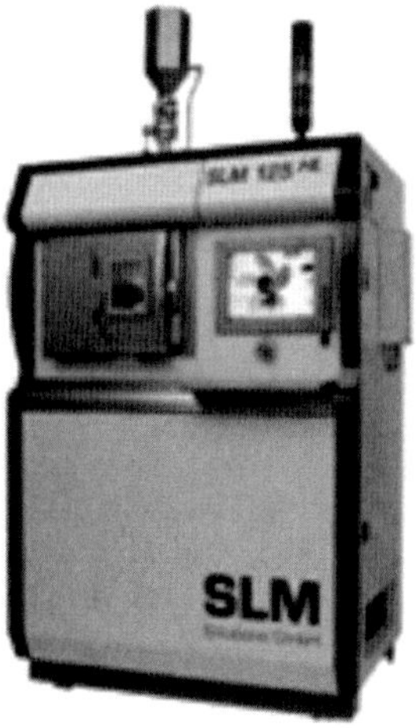

FIGURE 3.6
Illustration of the additive manufacturing SLM 125 HL equipment used to measure energy consumption during the building process of eight turbine blades.

The energy consumption for the internal air dryer (230 V) system and the energy needed to achieve the working pressure of 3.5 bars required by the cooling unit, among others, were measured during the fabrication of the eight turbine blades. TABLE 3.5 shows the general output energy consu-

TABLE 3.4
Fabrication Processes Energy Consumption of Eight Blades Produced with Inconel 718 Super Alloy

	Energy Consumption (kWh)
Investment Casting per 1 kg	
Wax Patterns Casting	4.94
Slurry and Grain	616.46
Oven	673.76
Grinding/Cleaning	4.58
Casting	79.63
Metal in Bar Stock Form	77.68
Grinding and Hand Finish	
Grind Root Form	19.88
Grind Tip/Slot	19.88
Grind Z Form	19.88
Highly Accelerated Stress Screening (HASS)	0.6
Coordinate Measuring Machine (CMM)	5.27
Hand Finish	0.98
Nitric Clean Lines	
Furnace	1.69
Pumping System	1.87
Scrubber	5.92
Feeding Boards	7.49
Tank (per 10)	1.26
FPI Process	
Penetrating Impregnation	0.2
Prewash	0.05
Emulsifying Tank	0.19
Post-wash	1.08
Resistance Drying Furnace	12.91
Revealed cabin	0.2
Inspection Station	0.2
Final Operation	
Shot Peening Process	4.44
Part Marking	0.23
Finish Virtual Inspection	0
Nitric Clean Second Line	7.29
Packaging	0
Total	**1568.56**

mption values for the SLM additive manufacturing process, which encompasses a variety of processes to build a part measured during the production of 1 kg of In100.

Additional energy consumption measurements are needed to analyze the result's sensitivity, which includes adjusting the additive manufacturing process parameters such as laser power, exposure time, point distance, and

TABLE 3.5
SLM Output Energy Consumption Values Measured During the Building of Eight Turbine Blades with 1 kg of Metallic In100 Powders

Life Cycle Inventory	Energy kWh
Powder Production	15.44
Metal in Bar Stock Form	77.68
AM Blades SLM Process	666.66*
Selective Laser Melting Operations	
Feeding Boards	28.141
Cooling Unit	1.552
Powder Sieving Station	0.03
Inert Gas Generator	0.515
Total	**759.78**

*Information provided by Honeywell Aerospace, Chihuahua.

building speed. Furthermore, some post-processing technologies, such as hot isostatic pressing, must be applied to get the manufactured component with the desired physical properties and geometry specifications.

Moreover, adopting LCE contributes to environmental conservation and affords manufacturers long-term advantages, including cost reduction and competitive differentiation. Embracing sustainable practices and refining material production processes through LCE represent critical steps toward cultivating a greener, more sustainable future [66, 67].

3.5.1 Walkthrough of the Entire Life Cycle Process: From Idea to Launching

This case study exemplifies the utilization of Life Cycle Engineering (LCE) in evaluating the sustainability impacts of 3D printing technologies, specifically in the production process of eight turbine blades at Honeywell Aerospace's plant in Mexico. The LCE approach ensures that every stage of the material's life cycle is optimized for sustainability, from the initial idea to the final launch.

- **Idea—Imagination:** The initial stage involves understanding the specific needs and constraints of the end-users, environmental factors, performance criteria, and market needs. The goal is to identify sustainable materials that meet functional requirements while minimizing environmental impact.
  - In this stage, the inputs include identifying user application requirements, understanding environmental conditions and constraints, establishing necessary performance criteria and standards, gathering market research and user feedback, and selecting sustainable materials that meet these criteria. The outputs from this stage are compre-

hensive user needs and application specifications, preliminary material selections that align with functional and performance criteria, and initial design briefs and performance goals that guide the next stages of development.

- **Basic Development—Conceptualization:** It designs and synthesizes materials to achieve desired properties and functions. This involves using advanced computational and experimental tools to create initial material samples and assess their feasibility in terms of scalability, cost, and environmental impact.
  - The inputs for this stage include material properties and performance goals derived from the idea stage, molecular design techniques and tools, synthesis methods and protocols, and environmental impact assessments. The outputs are detailed molecular models and blueprints, synthesized material samples for testing and evaluation, and evaluation reports that analyze scalability, cost, and environmental impact, providing a solid foundation for further development.
- **Advanced Development—Design:** This stage involves translating molecular designs into practical, manufacturable materials. Detailed engineering drawings and specifications are created to ensure that the materials align with production processes. This phase also includes developing prototypes and conducting initial quality and performance testing.
  - Inputs for this stage are molecular designs and synthesized samples from the conceptualization stage, engineering design tools and software, an assessment of manufacturing process capabilities, and production system requirements. The outputs include detailed engineering drawings and specifications, material prototypes and test plans, manufacturing process alignment documents, and initial quality and performance testing results, ensuring that the materials are ready for full-scale production.
- **Launching—Creation/Operation:** The focus here is on synthesizing, analyzing, characterizing, and prototyping materials. This stage ensures that materials are thoroughly vetted and refined, ready for practical application and full-scale production.
  - The inputs for this stage encompass detailed designs and specifications from the engineering design stage, synthesis protocols and materials for final production, analytical tools and characterization equipment, and prototyping facilities and testing environments. The resulting outputs are synthesized material batches ready for production, comprehensive analysis and characterization reports detailing the material's properties, prototypes for final testing and validation, and

finalized material specifications, indicating readiness for full-scale production.

Thus, throughout the LCE framework, regular monitoring and reporting mechanisms are established to maintain transparency with stakeholders and facilitate ongoing optimization of processes based on real-time data and feedback. Furthermore, the company seeks and employs financial strategies and instruments that support its long-term decarbonization goals, aligning financial investments with sustainable development objectives.

Additionally, during the end-of-life the 5S principles are incorporated to ensure that materials are managed responsibly throughout their life cycle, from initial production to end-of-life disposal or recycling.

In this stage, inputs include ethical sourcing guidelines and community impact assessments for the **Social** principle; environmental impact assessments and life cycle analysis for the **Sustainable** principle; advanced sensing and monitoring technologies for the **Sensing** principle; smart material design protocols for the **Smart** principle; and safety standards and testing protocols for the **Safe** principle. The outputs are ethically sourced materials and community engagement reports (Social), life cycle assessment reports and waste reduction strategies (Sustainable), integrated sensing capabilities and adaptive response systems (Sensing), smart, adaptive material prototypes and enhanced functionality (Smart), and safety compliance certifications and non-toxic materials (Safe).

3.6 The Future of Sustainable Manufacturing for Materials Development

The future of sustainable manufacturing for the Materials Development entity is poised to revolutionize how materials are designed, produced, and utilized, strongly emphasizing minimizing environmental impact and promoting circular economy principles. This evolution is driven by technological advances, increased environmental awareness, and stringent regulatory frameworks. Here are some key trends and advancements that will shape the future of sustainable manufacturing in materials development:

- **Advanced Digital Technologies:** Digital Twins and Smart Manufacturing: Digital twins, virtual replicas of physical systems, will be crucial in optimizing manufacturing processes. By simulating and analyzing production in real-time, manufacturers can predict and mitigate issues before they occur, reducing waste and improving efficiency. Smart manufacturing, powered by the IoT, AI, and machine learning, will enable real-time monitoring and adaptive control of production processes, further enhancing sustainability.

  - ***Materials Informatics:*** The integration of big data, AI, and machine learning into materials science, known as materials informatics, will accelerate the discovery and development of sustainable materials. By analyzing vast datasets, researchers can identify trends and predict the properties of new materials, significantly reducing the time and resources required for experimentation.

- **Circular Economy Principles:** The shift towards a circular economy will transform how materials are viewed and utilized. Emphasizing reuse, recycling, and remanufacturing, circular economy principles will ensure that materials are used for as long as possible, reducing the need for virgin resources and minimizing waste.
  - LCE will become integral to materials development, focusing on optimizing the entire life cycle of materials from raw material extraction to end-of-life disposal. By considering environmental impacts at every stage, LCE ensures that sustainability is embedded in the design, production, and disposal processes.

- **Sustainable Material Innovations:** Bio-based and Renewable Materials: The development of bio-based materials derived from renewable resources will gain momentum. These materials, which can replace conventional plastics and other non-renewable materials, offer reduced environmental footprints and biodegradability, contributing to more sustainable manufacturing practices.
  - ***High-Performance Recyclable Materials:*** Innovations in materials science will create high-performance materials that retain their properties after multiple recycling cycles. These materials will enable producing durable, high-quality products while supporting recycling and circular economy goals.

- **Energy Efficiency and Decarbonization:** Energy-Efficient Manufacturing Processes: Advances in manufacturing technologies will prioritize energy efficiency, reducing the carbon footprint of production processes. Techniques such as additive manufacturing and advanced machining will optimize material use and minimize energy consumption.
  - ***Renewable Energy Integration:*** Integrating renewable energy sources, such as solar and wind, into manufacturing operations will further reduce greenhouse gas emissions. Manufacturers will increasingly adopt clean energy solutions to power their facilities, aligning with global decarbonization efforts.

- **Policy and Regulatory Support:** Stricter Environmental Regulations: Governments and international bodies will continue to implement and enforce stricter environmental regulations, pushing manufacturers to adopt sustainable practices. Compliance with these regulations will drive innovation and investment in sustainable materials and processes.

  - ***Incentives for Sustainable Practices:*** Financial incentives, grants, and subsidies for sustainable manufacturing practices will encourage companies to invest in eco-friendly technologies and processes. These incentives will make it economically viable for manufacturers to adopt sustainable practices.

- **Collaboration and Knowledge Sharing:** Collaborative Research and Development: Collaboration between academia, industry, and government will be essential for advancing sustainable materials development. Joint research initiatives and public-private partnerships will facilitate the sharing of knowledge, resources, and expertise, accelerating innovation in sustainable manufacturing.
  - ***Global Standards and Best Practices:*** Establishing global standards and best practices for sustainable manufacturing will ensure consistency and transparency across the industry. Standardized metrics and reporting frameworks will enable manufacturers to benchmark their sustainability performance and identify areas for improvement.

The future of sustainable manufacturing in materials development is bright, driven by technological advancements, regulatory frameworks, and a global commitment to environmental stewardship. By embracing digital technologies, circular economy principles, innovative materials, energy efficiency, and collaborative efforts, the materials development entity can significantly reduce its environmental impact while maintaining competitiveness and fostering long-term sustainability. This holistic approach will benefit the environment, create economic opportunities, and enhance the quality of life for future generations.

3.7 Conclusion

Integrating sustainability at every stage of material development is essential to address environmental and social challenges. With a dramatic increase in global material consumption over the past five decades, significant environmental impacts such as deforestation, biodiversity loss, and greenhouse gas emissions have underscored the urgent need for Sustainable Materials Management. In that sense, LCE is instrumental in driving sustainable manufacturing and supporting the circular economy by optimizing resource use and minimizing waste. LCE provides valuable data for decision-making, ensuring that new materials and products are developed with minimal environmental impact and meet stringent sustainability standards. LCE focuses on reducing environmental impacts throughout the entire life cycle of materials, from ideation to end-of-life, promoting circular economy principles and sustainable practices.

Technological advancements, particularly in digital technologies like machine learning and materials informatics, have revolutionized the field, enabling faster and more accurate prediction of material properties. These innovations, coupled with a shift towards bio-based and recyclable materials, are critical in reducing materials' carbon footprint and embodied energy. Recycling is pivotal in mitigating greenhouse gas emissions and conserving resources, underscoring the importance of integrating sustainability into material production and use.

Moreover, adopting the 5S principles—Social, Sustainable, Sensing, Smart, and Safe—ensures that material development is efficient and functional, ethically produced, environmentally responsible, and safe. By embedding these principles into the life cycle of materials, industries can achieve significant environmental, economic, and social benefits.

As the world grapples with resource depletion and environmental degradation, integrating advanced digital technologies, sustainable practices, and circular economy principles is essential for the future of materials development. By embracing these strategies, industries can meet the growing demand for sustainable products and contribute to a greener, more sustainable future, ensuring long-term viability and competitiveness in a market increasingly focused on environmental responsibility.

4

Product: Ideation, Conceptualization, Engineering and Prototyping

This chapter discusses sustainable product development, emphasizing the integration of economic growth, environmental protection, and social responsibility. It is crucial for long-term business strategies to focus on creating products that are environmentally friendly, economically viable, and socially beneficial. By incorporating the 10R framework—reduce, reuse, recycle, and more—throughout the product life cycle, it aligns with Sustainable Development Goals (SDGs).

Developing new products is exciting but challenging, involving ideation, research, and prototyping, with each launch being unique. This chapter presents product development stages from a Life Cycle Engineering (LCE) framework perspective, using the 5S approach—Social, Sustainable, Sensing, Smart, and Safe. The stages are Ideation/Imagination (driving the product idea), Basic Development (concept design and target specification), Advanced Development (detailed product development), and Launching/Creation-Operation (prototyping).

A case study on Mio-Tracker, a single-channel electromyograph, illustrates the practical application of these principles, showing how integrating LCE and 5S frameworks can enhance product sustainability, usability, and marketability.

The chapter highlights the future of sustainable manufacturing in product development, emphasizing technological advancements, agile methodologies, and a customer-centric approach. Integrating advanced technologies like AI, IoT, and digital twins will streamline development processes, ensuring products are efficient, marketable, environmentally sound, and socially beneficial.

4.1 Introduction

Sustainable product development balances economic development, environmental protection, and social responsibility. As such, it is a crucial element of any business strategy that seeks to ensure the company's long-term

DOI: 10.1201/9781003501619-4

success. Product development means creating products that are both good for the environment and economically and socially beneficial. This concept is further explored in [33], which indicates that manufacturing companies can integrate Sustainable Development Goals (SDGs) into their operations by focusing on adopting the 10R throughout the product life cycle. Furthermore, sustainable product development determines around 80% of a product's total environmental impact during its life cycle, emphasizing the need for significant modifications across various product dimensions—including materials, architecture, services, business models, and ecosystems. Therefore, the role of decision-making is to support frameworks like life cycle engineering (LCE) and multi-criteria decision analysis in evaluating product variants against sustainability goals.

Managing life cycle information—from material selection to end-of-life stages—is crucial for aligning SPD processes with CE principles, enhancing product sustainability performance, and ensuring that comprehensive sustainability considerations inform the phases of product ideation, conceptualization, engineering, and prototyping.

LCE principles are integral to sustainable product development, emphasizing a comprehensive and systematic approach that considers a product's entire life cycle, from raw material extraction to end-of-life disposal. According to [23] and [21], sustainable development balances social demands around productivity and using natural resources. In this way, considering LCE as a tool that helps to assess the environmental impact of product design gives a step forward to look for what types of sustainable methods can be applied in which design phases are advantageous. This arrangement is very useful to organize the process better and separate individual deliverables into smaller tasks. Life Cycle Engineering enhances the approach towards sustainable product development:

- **Proactive Design and Development:**
  - ***Integration of Sustainability:*** LCE directly integrates sustainability considerations into product development. It involves designing products explicitly focusing on minimizing environmental impact across their entire life cycle, from material selection and manufacturing processes to usage and end-of-life disposal.
  - ***Innovative Solutions:*** By considering the life cycle perspective during the design phase, LCE encourages the development of innovative solutions that reduce waste, enhance recyclability, and improve energy efficiency.
  - ***Design for Sustainability:*** Optimize product designs to reduce energy consumption, waste generation, and emissions during manufacturing and use. Incorporate eco-design principles to minimize the environmental impact of product design. This includes designing for recyclability and optimizing energy efficiency during the manu-

facturing process. Design products with recyclability in mind. Consider the ease of disassembly and separation of materials for efficient recycling processes. Implement take-back programs to collect and recycle products at the end of their life cycle. This helps prevent products from ending up in landfills.

- **Holistic Approach:**
  - ***Comprehensive Strategies:*** LCE involves strategies such as Design for Environment (DfE), Design for Disassembly (DfD), and Design for Recycling (DfR). These strategies aim to optimize the product life cycle by reducing environmental impacts at source, promoting material recovery, and facilitating product reuse.
  - ***System Optimization:*** Beyond individual products, LCE examines system-level optimizations, considering how products interact within larger systems and finding ways to optimize these interactions for sustainability.
  - About materials selection, the consideration of renewable resources, recycled materials, and materials with minimal toxicity led us to choose biodegradability whenever possible for materials in product design. It is important to take a look at the possibilities of materials with lower environmental footprint levels, considering the entire life cycle of materials, including extraction, processing, use, and disposal, and the exploration of alternative and sustainable materials with lower environmental impacts.
- **Engineering Tools and Methodologies:**
  - ***Advanced Simulation and Modeling:*** LCE utilizes advanced engineering tools and simulations to predict the life cycle impacts of design choices. This can include everything from material strength and durability simulations to the modeling of energy consumption and emission outputs during use.
  - Implement energy-efficient processes during manufacturing to reduce energy consumption and greenhouse gas emissions. This involves optimizing material usage and reducing production scrap. Also, design products that are energy-efficient during their use phase. This can include optimizing electronic devices for energy conservation or developing products with a longer lifespan. Provide consumers with information on how to use products more sustainably, including proper maintenance and efficient usage practices.
  - ***Feedback Loops:*** Engineering solutions in LCE often involve creating feedback loops that continuously gather data from the product use phase, which can then be used to make iterative improvements in future design cycles.

- **Stakeholder Engagement:**
  - ***Cross-Disciplinary Collaboration:*** Effective LCE requires collaboration across various disciplines—engineering, environmental science, supply chain management, and marketing—to ensure that all aspects of product sustainability are considered.
  - ***Engagement with Supply Chain:*** LCE promotes deeper engagement with suppliers and other stakeholders along the product chain to implement sustainable practices more broadly, impacting the product's environmental footprint significantly.
  - Integrate Supply Chain Management (SCM) practices into the product development process under the Life Cycle Engineering Assessment framework. With this, organizations can effectively manage environmental and social impacts, optimize resource utilization, and create products that meet sustainability objectives while delivering value to customers and stakeholders. SCM encompasses the integration of various activities involved in sourcing, procurement, production, logistics, and distribution to ensure that products are delivered to customers efficiently and cost-effectively. Work with suppliers committed to ethical and sustainable practices. Consider factors such as fair labor practices, responsible resource extraction, and reduced transportation impacts. Opt for local suppliers to minimize transportation-related environmental impacts and support local economies.

Life Cycle Assessment (LCA) which evaluates the environmental impacts associated with each stage of the product's life cycle. This includes raw material extraction, manufacturing, transportation, product use, and end-of-life disposal. LCA helps identify hotspots where environmental impacts are most significant, guiding decisions to minimize overall ecological footprint. However, while LCA is primarily an assessment tool used to identify the environmental impacts of a product, LCE is an engineering discipline focused on implementing practical solutions to minimize those impacts throughout the product life cycle. LCE builds on the data and insights from LCA studies to inform sustainable engineering decisions, making it a more action-oriented approach that directly influences product design and manufacturing processes.

Besides, sustainable product development integrates environmental and social considerations into every product life cycle (PLC) stage. This approach addresses consumer demand for eco-friendly products and ensures long-term business viability. By recognizing the stages of a product's life cycle and implementing sustainable practices, companies can minimize environmental impact and promote social responsibility while maximizing the product's success and longevity in the market [4, 77]. Following the stages of the PLC are described:

1. **Introduction Stage:** The introduction stage is characterized by the launch of a new product. In sustainable product development, this phase ensures that the product design, materials, and manufacturing processes are environmentally friendly. Sales grow slowly, and profits are minimal due to high initial costs, but the focus is on creating awareness of the product's sustainable attributes. Businesses invest in advertising and promotions to highlight the eco-friendly aspects and stimulate trial among environmentally conscious consumers.
   - During the introduction stage, businesses should invest in advertising and promotions to build awareness of the product's sustainable features. Offering incentives, such as discounts or eco-friendly packaging, can encourage customers to try the new product. These efforts help establish a market presence and set the stage for future growth while promoting sustainability.
2. **Growth Stage:** In the growth stage, the product experiences rapid market acceptance and increasing profits. Sales increase significantly as more customers become aware of and purchase the product. Sustainable strategies during this phase include improving product features to enhance sustainability, such as using more recycled materials or reducing energy consumption. Companies expand distribution channels to reach a broader audience while committing to sustainable logistics and supply chain practices.
   - In this stage, improving product features and quality to enhance sustainability is crucial. Expanding distribution channels is also vital to reach more customers, ensuring that these channels adhere to sustainable practices. These strategies help capitalize on the increasing demand and drive further market penetration while maintaining an eco-friendly approach.
3. **Maturity Stage:** The maturity stage is when sales growth slows down as the product reaches peak market penetration. Profits stabilize or begin to decline due to increased competition. Efforts are directed towards defending market share by promoting the product's long-term environmental benefits. Extending the product's life cycle through sustainable innovations, such as offering repair services or take-back programs, becomes essential. Marketing efforts emphasize the product's durability and reduced environmental footprint.
   - To manage this stage, companies should innovate by adding new sustainable features or finding new uses for the product. Engaging in competitive pricing and promotions helps

maintain market share amidst stiff competition. Emphasizing the product's environmental benefits and longevity can attract environmentally conscious consumers and extend the product's life cycle.

4. **Decline Stage:** In the decline stage, sales and profits decline as the product loses consumer appeal or becomes obsolete. Sustainable strategies include exploring options for product recycling or repurposing to reduce waste. Companies may discontinue the product, ensuring proper disposal or recycling of remaining inventory. Alternatively, they may find ways to rejuvenate it by updating its design to incorporate new sustainable technologies or materials.
    - During this stage, businesses should focus on reducing costs and finding sustainable ways to dispose of or repurpose remaining inventory. Considering product modifications incorporating the latest sustainable technologies or finding niche markets can help extend the product's life. These strategies ensure that the product generates revenue while minimizing environmental impact.

Hence, sustainable product development is pivotal for businesses striving to achieve long-term success while addressing environmental, economic, and social responsibilities. Integrating sustainability throughout the product life cycle—guided by LCE principles—ensures that products minimize environmental impact from conception to disposal. By adopting strategies such as DfE, DfD, and DfR, companies can create products that are not only economically viable but also environmentally friendly and socially responsible. Engaging stakeholders across disciplines and supply chains further enhances sustainable practices, promoting a holistic approach to product development. Ultimately, sustainable product development fosters eco-friendly products that meet consumer demand, enhance business viability, and contribute to a more sustainable future.

4.1.1 The Significance of Sustainable Manufacturing in the Modern World

Modern customers are increasingly aware of and concerned about environmental issues. Sustainable manufacturing addresses these concerns by minimizing negative environmental impacts, conserving energy and natural resources, and ensuring the health and safety of employees and communities. By integrating sustainability into product development, companies can meet the growing demand for eco-friendly products, enhancing customer satisfaction and loyalty.

The concept of offering management emphasizes overseeing the entire life cycle of a product, from initial design to end-of-life disposal. Sustainable manufacturing aligns with this by promoting sustainable materials, energy-efficient production processes, and designs that facilitate recycling or reuse. This holistic approach ensures that every stage of the product life cycle contributes to sustainability goals, reducing the overall environmental footprint. Sustainable manufacturing drives innovation by encouraging companies to develop new materials, processes, and technologies that reduce environmental impact. This aligns with the agile and adaptable approach recommended for offering managers, who must quickly respond to market changes and customer needs. By prioritizing sustainability, companies can differentiate themselves in the market, offering innovative solutions that appeal to environmentally conscious consumers [18].

A seamless and positive customer experience is central to successful product management. Sustainable manufacturing enhances this experience by ensuring that high-quality products are functional and ethically produced. This builds brand trust and loyalty, as customers feel good about supporting companies prioritizing sustainability. Implementing robust feedback loops is crucial for managers to continuously gather and analyze customer sentiments. Sustainable manufacturing benefits from similar feedback mechanisms, where companies can track their products' environmental impact, gather customers' insights about their preferences for sustainable features, and make necessary adjustments. This continuous improvement cycle helps maintain alignment with customer values and regulatory requirements.

For various reasons, sustainable manufacturing holds immense significance in the modern world, reflecting the growing awareness of environmental, social, technological, safety, and economic issues. Starting with environmental conservation, sustainable product development incorporates the main principles and the application of life cycle engineering principles. Emphasizing the efficient use of resources, reducing waste, and minimizing environmental impact have become crucial in the face of resource depletion and environmental degradation. These sustainable practices aim to reduce greenhouse gas emissions, air and water pollution, and other environmental pollutants associated with manufacturing processes. Suppose the process includes sustainable manufacturing during product development. In that case, it is feasible to promote the adoption of renewable energy sources to power production processes, contributing to reducing carbon footprints and combating climate change. The use of low-carbon and energy-efficient technologies helps mitigate the impact of manufacturing on climate change. Besides this, the process must include considerations for fair labor practices, worker safety, and community well-being. Consumers and investors often view companies engaging in sustainable practices more favorably. By promoting supply chain transparency, sustainable manufacturing ensures that products are sourced and produced ethically and in compliance with human rights standards. The implementation of LCE represents cost savings and efficiency. Sustainable manufacturing practices often lead to increased energy and resource efficiency, resulting in cost savings for companies in the long run, with waste minimization through practices such as recycling and reusing materials. These practices can lead to cost reductions associated with waste disposal and raw material acquisition.

Another benefit is compliance with environmental regulations. Governments and international bodies are increasingly implementing strict environmental regulations. Sustainable manufacturing helps companies stay compliant and avoid penalties, enhancing market competitiveness. Modern consumers are increasingly concerned about the environmental and social impact of their products. Companies embracing sustainable manufacturing gain a competitive edge by meeting consumer preferences for eco-friendly products.

Consequently, some markets and industries now require adherence to sustainability standards for participation. Sustainable manufacturing practices enable companies to access these markets and meet regulatory requirements. For the long-term resilience of the product development process, sustainability efforts, including responsible sourcing and efficient supply chain management, contribute to the resilience of companies in the face of global disruptions and uncertainties. Additionally, these companies can adapt to changing norms. As sustainability becomes a norm in business practices, companies adopting sustainable manufacturing are better positioned to adapt to changing market dynamics and consumer expectations. Sustainable manufacturing addresses global challenges, including environmental degradation, climate change, and social inequalities. It helps protect the planet and contributes to businesses' long-term viability and success in an increasingly conscious and interconnected world.

By incorporating sustainability into its product development and manufacturing processes, Google can ensure its offerings align with its overall mission to minimize environmental impact. This includes using renewable energy in data centers and designing products with recycled materials. Google's commitment to sustainability extends beyond its product offerings to include significant investments in renewable energy projects, aiming to operate on carbon-free energy by 2030. This comprehensive approach reduces the environmental footprint of Google's operations and sets a precedent for other tech companies to follow.

For instance, Adobe transitions its product offerings to meet market demands; it can also focus on sustainable practices, such as reducing the carbon footprint of its software services and promoting digital solutions that decrease the need for physical materials. Adobe has integrated sustainability into its corporate strategy, setting ambitious goals to reduce greenhouse gas emissions and waste. The company's sustainability initiatives include sourcing renewable energy for its global operations and creating more energy-efficient software solutions. Adobe's holistic approach to sustainability underscores its commitment to reducing environmental impact while maintaining high standards of product innovation and customer satisfaction.

Unilever provides another compelling case study. The company has made sustainability a core part of its business model, aiming to decouple its growth from environmental impact while increasing positive social impact. Through its Sustainable Living Plan, Unilever has committed to sourcing 100% of its agricultural raw materials sustainably and halving the environmental footprint of its products. Unilever's focus on sustainability has driven innovation in product development, resulting in eco-friendly products that resonate with consumers' growing demand for sustainable choices.

Sustainable manufacturing is a crucial component of modern product development, resonating with the principles of customer-centricity, life cycle management, innovation, customer experience, and continuous improvement outlined in [18]. By embracing sustainability, companies can not only meet the evolving demands of consumers but also contribute to a more sustainable future, ensuring long-term success and market leadership. Sustainable manufacturing is crucial for addressing global challenges, including environmental degradation, climate change, and social inequalities. It helps protect the planet and contributes to businesses' long-term viability and success in an increasingly conscious and interconnected world.

4.1.2 Overview of How Digital Technologies are Transforming Product Design Process

Digital technologies are fundamentally transforming manufacturing processes, particularly in product development. This digital transformation, often referred to as Industry 5.0, involves the integration of advanced technologies to enhance efficiency, flexibility, and innovation throughout the product development life cycle. To succeed in today's market, companies must adopt a consumer-centric approach. This involves understanding and meeting consumer needs and aligning with their values. Companies increasingly leverage digital technologies to create and market products that foster deeper, longer-term consumer relationships. This includes developing smart, connected products that offer personalized and hyper-customized experiences [1].

For instance, P&G's Opte' beauty device and Philip Morris' iQOS tobacco heating system are examples of products that integrate advanced digital capabilities to enhance consumer engagement and brand relevance. These connected products and services enable companies to create new value spaces that did not exist before, providing tailored solutions that resonate with consumers on a more intimate level [1].

Digital tools also facilitate frequent and meaningful consumer interactions, allowing companies to gather granular insights and drive closed-loop product development. By sharing targeted information about product provenance, authenticity, and tailored offerings, companies can build trusted relationships and deliver winning consumer experiences.

TABLE 4.1 shows an overview of how digital technologies reshape manufacturing processes, focusing on product development.

Industries transitioning to the Industry 5.0 era are adopting smart manufacturing approaches that utilize multi-scale dynamic modeling, simulation, and intelligent decision-making to enhance production capabilities [31]. Digital Twins (DT) technology plays a key role in smart manufacturing by creating virtual representations of physical systems. Within the broader framework of Cyber-Physical Systems (CPS), the DT paradigm aligns seamlessly with a life cycle-centered perspective. Consequently, DT technology is becoming a focal point for enhancing and evolving global manufacturing processes [79].

The cyber-physical revolution necessitates integrating new digital technologies into everyday practices. Companies now have access to advanced technologies that enable process monitoring and decision-making, as shown in TABLE 4.1, seamlessly integrated with digital systems. Although the generation of visual representations of concepts in product design often relies on analog tools [81], the advent of digital technologies and the integration of LCE principles offer significant opportunities to modernize this process. The ability to sensor, control, and gather information throughout the design, development, and manufacturing phases of a product creates a valuable database for enhancing production efficiency. This data foundation supports a dynamic Life Cycle Inventory, facilitating continual improvement and optimization [42].

Integrating digital technologies in manufacturing processes, especially in product development, leads to a paradigm shift in how products are designed, tested, and brought to market. This digital transformation enhances agility, reduces time-to-market, and fosters innovation by leveraging advanced tools and analytics throughout the product development life cycle.

Innovation alone is not sufficient; it must be combined with cost-efficiency to ensure profitability. Many new products fail to succeed in the market, which underscores the need for a strategic approach to maximize margins. The concept of "design to margin" involves revisiting product portfolios with a focus on profitable innovation, cost reduction, and efficient resource utilization.

An integrated approach to design to margin requires aligning priorities across various functions—marketing, supply chain, design, manufacturing, and more. Collaboration with external partners and suppliers is essential to create a coherent, data-driven strategy that optimizes profitability. Successful companies integrate these capabilities to respond efficiently to market changes, experiment with new offerings, and reduce time to market.

TABLE 4.1
Main Digital Technologies that are Reshaping Manufacturing Processes

Digital Technology	Concept	Application
Digital Twins	Digital twin technology involves creating a virtual replica (digital twin) of a physical product or system.	In product development, digital twins allow real-time monitoring, simulation, and analysis, enabling engineers to optimize designs and predict performance before physical prototypes are built.
Additive Manufacturing (3D Printing)	Additive manufacturing builds objects layer by layer from digital models.	Accelerates prototyping and enables the production of complex geometries. Reduces material waste and allows for on-demand and localized manufacturing.
Advanced Simulation and Modeling	Utilizing sophisticated simulation and modeling tools for virtual testing and validation.	Predictive modeling and simulation help optimize designs, assess product performance, and identify potential issues early in the development process.
Artificial Intelligence (AI) and Machine Learning (ML)	AI and ML algorithms enable machines to learn from data and make intelligent decisions.	Enhances product design through generative design algorithms, improves quality control, predicts maintenance needs, and optimizes manufacturing processes.
Internet of Things (IoT)	Connecting physical devices and objects to the internet for data exchange and communication.	In product development, IoT sensors provide real-time data on product usage, enabling a better understanding of customer behavior and feedback for iterative improvements.
Big Data Analytics	Analyzing large datasets to extract meaningful insights and patterns.	Helps to make data-driven decisions, optimize supply chain logistics, and identify trends in consumer preferences for more targeted product development.
Augmented Reality (AR) and Virtual Reality (VR)	AR overlay digital information onto the real-world environment, while VR creates immersive virtual environments.	Facilitates collaborative design reviews, training, and prototyping by providing a visual and interactive experience.

Digital Technology	Concept	Application
Collaborative Robotics (Cobots)	Collaborative Robots are designed to work alongside humans.	Improves efficiency and safety in the design process, including the first assembly of prototypes and testing, by collaborating with human workers.
Cloud Computing	Using remote servers to store and process data.	Facilitates real-time collaboration, data storage, and access to powerful computing resources, particularly beneficial for large-scale simulations and analyses.

4.2 Product as 5S Systems

Developing products using the 5S system, encompassing Social, Sustainable, Sensing, Smart, and Safe principles ensures products are efficient, ethically produced, environmentally responsible, and designed with user safety and societal well-being in mind. Applying the 5S framework to product development ensures that the product meets high standards of functionality and efficiency and addresses broader social and environmental concerns (see FIGURE 4.1).

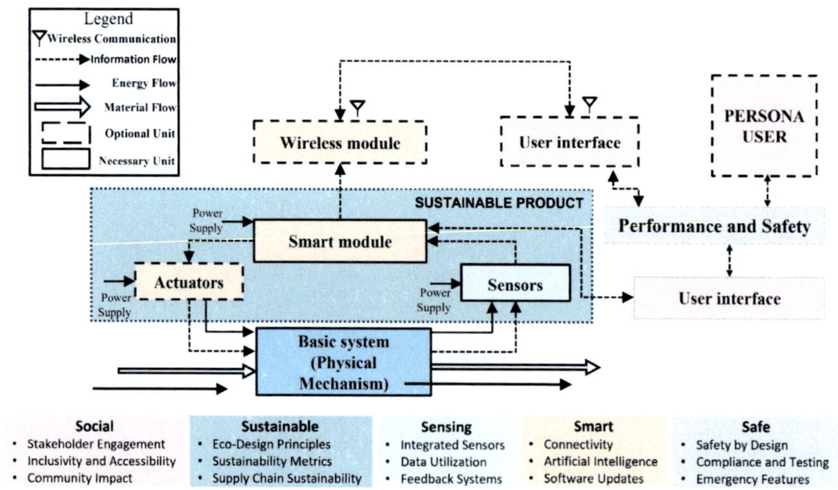

FIGURE 4.1
5S products (social, sustainable, sensing, smart and safe).

Here is how to integrate each principle into the product development process:

- **Social:**
  - ***Stakeholder Engagement:*** Engage with various stakeholders, including potential users, community groups, and industry experts, to gather diverse perspectives and ensure the product meets various social needs and does not inadvertently exacerbate social issues.
  - ***Inclusivity and Accessibility:*** Design the product to be accessible to people with different abilities, ages, and cultural backgrounds, ensuring it is user-friendly and inclusive.
  - ***Community Impact:*** Assess the potential impacts of the product on communities and design initiatives that contribute positively, such as local sourcing, community development programs, or educational outreach.
- **Sustainable:**
  - ***Eco-Design Principles:*** Implement eco-design principles to minimize environmental impacts. This includes using renewable or recycled materials, designing for energy efficiency, and facilitating easy disassembly for recycling at the product's end of life.
  - ***Sustainability Metrics:*** Define and track sustainability metrics such as carbon footprint, water usage, and life cycle impact to continuously assess and improve the product's environmental performance.
  - ***Supply Chain Sustainability:*** Ensure that the supply chain adheres to sustainability standards, from raw material sourcing to manufacturing and distribution.
- **Sensing:**
  - ***Integrated Sensors:*** Depending on the product, integrate sensors that can provide useful data to the user or help in the maintenance and operation of the product, such as sensors for monitoring energy consumption, wear and tear, or environmental conditions.
  - ***Data Utilization:*** Design the product to collect and analyze data to offer improved usability, performance enhancements, and personalized user experiences.
  - ***Feedback Systems:*** Implement systems that allow the product to adapt based on feedback from its environment or direct user input, enhancing functionality and user satisfaction.
- **Smart:**
  - ***Connectivity:*** Incorporate connectivity features that enable the product to interact with other smart devices and systems, enhancing

user experience and creating integrated solutions within the user's ecosystem.

  - ***Artificial Intelligence:*** Use AI to add functionalities such as predictive analytics, automation, and adaptive learning, which can optimize performance and make the product more intuitive and responsive to user needs.
  - ***Software Updates:*** Ensure the product can receive over-the-air software updates to improve its functionality and security over time, extending its lifespan and adaptability.

- **Safe:**
  - ***Safety by Design:*** Integrate safety into the product design, ensuring it is safe under all expected conditions and usage scenarios. This includes physical safety, such as robust construction and ergonomic design, and cybersecurity measures if the product is connected.
  - ***Compliance and Testing:*** Rigorously test the product against all relevant safety standards and regulations before it goes to market. Continually monitor product safety post-launch and address any issues promptly.
  - ***Emergency Features:*** Include features that enhance the safety of users in case of product failure or during unexpected situations, such as automatic shut-offs, fail-safes, or alert systems.

By integrating the 5S framework into product development, the process becomes a dynamic model that focuses on creating a high-quality product and prioritizes ethical practices, environmental responsibility, user engagement, and safety. This approach enhances the product's market appeal, fosters customer loyalty, and aligns with global trends towards sustainability and smart technology.

4.3 Life Cycle Engineering Description

From the selection of raw, new, or treated materials to end-of-life disposal, aiming for sustainability, cost-effectiveness, and high performance, LCE considers all phases in product development. The integrative approach searches to optimize the entire life cycle of products. When LCE principles are applied, the environmental and societal impacts are minimized throughout a product's lifespan, and processing methods that reduce waste and conserve energy are created.

Integrating environmental aspects in the product development process makes it possible to balance ecological and economic requirements with established requirements. To do so, eco-design and eco-innovation approaches must be considered. On the one hand, according to the United Nations Environment Programme (UNEP), eco-design considers environmental aspects at all stages of the product development process, striving to achieve products that generate the lowest possible environmental impact throughout the entire product life cycle [139]. Complementing this, eco-innovation promotes sustainability throughout a product's whole life cycle, helping small- and medium-sized enterprises stay ahead of regulations and standards mostly related to the environment [141]. For example, in some disposable products, sustainable vision can make a difference in all the industrial sectors. Consider a disposable toothbrush. In the redesign of the toothbrush, the substitution of the materials it is made of, or a change in the technical management of materials, manufacturing, as well as the usability, comfort, energy consumption, and water consumption, made a huge difference in the product life cycle [142].

Considering the user's behaviors and the difficulty in integrating standards in product design, the eco-guidelines can be easily implemented. Another important aspect where these environmental aspects can be started is, for example, the packaging design. Packaging designs tend to consider the lowering life cycle environmental impact, but the repercussions that design interventions on the packaging can have on the life cycle of the contained product are sometimes neglected [133]. This aspect is an implication when going back to the example of the disposable toothbrush, the packaging is used only once, so small actions can go a long way in greening up the supply chain, reducing the packaging size, knowing the product's carbon footprint, using renewable materials, and choosing recycled materials over virgin plastics. Strategies for sustainable product development include using sustainable materials, designing for durability, and minimizing packaging.

Achieving these changes to have all the benefits is not as easy as described. Adopting LCE practices requires a significant investment. Implementation can be resource-intensive, from detailed data collection and analysis to using experts. With this in mind, digital technologies can help reduce costs when implementing LCE practices. Digital technologies further enhance LCE by streamlining the development of sustainable materials, improving product performance, reducing waste in prototyping, and optimizing manufacturing processes. The use of AI and machine learning reduces time-to-market, promotes predictive modeling and real-time feedback loops, and enhances efficiency, which are vital for the competitiveness of various industries. Aside from making sure that your product designs are market-ready and that LCE has been considered in every stage, employing the right technologies can speed up your product development life cycle and is, therefore, also of utmost importance. From the use of tools like Computer-Aided Manufacturing (CAM) to cloud technology, which helps in the scalability of your developer's works, these digital technologies can conserve resources and accelerate development.

4.3.1 Benefits and Limitations of Applying Life Cycle Engineering Principles in Product Development

Applying LCE principles in product development offers numerous benefits that can significantly enhance a company's performance. Businesses can leverage state-of-the-art technology and interconnected processes to improve efficiency, flexibility, and collaboration across the product life cycle. Digital tools integrate Social, Mobile, Analytics, and Cloud technologies, providing valuable insights by aggregating information from multiple sources, which drives innovation in products and services. Companies can also create new revenue streams through smart products that provide insights into customer preferences and usage patterns. By transforming operations to be more consumer-centered, agile, and responsive, businesses can enhance customer satisfaction and loyalty while also promoting sustainability and trust through transparent value chains [1, 25].

LCE principles are at the forefront of leveraging digital technologies like the Internet of Things, Big Data, Artificial Intelligence, and Blockchain to promote a more sustainable and circular economy. LCE supports product development through techniques such as Design for Excellence (DFX), which emphasizes the proactive consideration of diverse factors early in the design process. By integrating these principles, DFX minimizes the need for costly modifications in later stages, fostering the creation of superior products that excel throughout their life cycles. The universal emergence of this concept underscores its critical importance in modern engineering methodologies.

Design for X (DFX) and LCE collectively enhance product development by adopting a holistic and systematic approach from concept generation to final delivery. DFX emphasizes good practices and guidelines to ensure efficient design and manufacturing processes, minimizing the need for modifications later. Compared to traditional engineering design, which often involves costly iterations and corrections, DFX aims for early defect correction, reduced product iterations, and a broader scope, focusing on manufacturability, assembly, quality, and supply chain considerations from the outset.

Key characteristics of DFX include early defect correction through virtual designs and simulations, reducing physical iterations and tool requirements, and promoting efficiency. An inclusive design team that encourages collaboration between designers, suppliers, and manufacturers can further enhance the design process.

Specific applications of DFX within LCE include [33]:

- **Design for Manufacturing (DfM):** Focuses on ease of manufacturing to create better products at lower costs. Incorporating LCE gives a holistic perspective, considering environmental, social, and economic impacts throughout the product's life cycle, facilitating recycling and reducing waste.

- **Design for Assembly (DfA):** Simplifies assembly processes by reducing the number of parts and complexity. LCE integrates sustainable practices into DFA, such as using renewable energy sources and minimizing hazardous substances.
- **Design for Manufacturing and Assembly (DfMA):** Combines DFM and DFA to create efficient, cost-effective products. Incorporating LCE ensures products are designed with manufacturability and assembly in mind from the beginning, enhancing overall efficiency.
- **Design for Reliability (DfR):** Builds reliability into products from the earliest design stages, emphasizing robust design practices, high-quality materials, and reliable manufacturing processes to ensure product longevity.
- **Design for Quality (DfQ):** Integrates quality checks throughout the production system, achieving high-quality standards while being eco-friendly.
- **Design for Supply Chain (DfSC):** Incorporates logistics considerations early in the design process.
- **Design for Testing (DfT):** Builds testing methods into the product design for economical and effective quality assurance.
- **Design for Maintenance (DfMt):** Ensures products are easy to maintain with modular designs and accessible components.
- **Design to Cost (DtC):** Controls development and manufacturing costs by considering cost as a design parameter.
- **Design for Sustainability (DfS):** Focuses on reducing environmental impact through recyclable materials and green manufacturing practices.
- **Design for Product Life Cycle (DfPLC):** Enhances profitability and ease of changes throughout the product's life cycle.
- **Design for Environment (DfE):** Focuses on minimizing environmental impacts through sustainable design choices.
- **Design for Disassembly (DfD):** Ensures products are easy to disassemble at the end of their useful life, facilitating recycling and reuse.
- **Design for Recycling (DfRc):** Emphasizes using materials and design practices that make recycling more efficient and effective.

Implementing LCE principles enhances the sustainability of products by addressing environmental impacts throughout their life cycle. This proactive approach allows for the early identification and mitigation of negative environmental effects, resulting in more eco-friendly products. Additionally, LCE supports improved decision-making through tools that evaluate sustainability performance across various design options, ensuring comprehensive environmental, economic, and social considerations. Adopting LCE principles leads to greater resource efficiency by optimizing material use, enhancing product durability, and facilitating recycling and reuse, ultimately contributing to cost savings and a reduced environmental footprint.

However, there are limitations to applying LCE principles in product development. Implementing a digital product framework requires significant changes in existing processes and systems, which can be complex and resource-intensive. The initial investment needed for adopting advanced digital technologies and integrating various systems can be substantial, involving time, money, and human resources. Organizational resistance to change can also hinder successful implementation, as employees and stakeholders may be reluctant to adopt new processes and technologies. Furthermore, managing and ensuring the accuracy, consistency, and security of aggregated data from multiple sources can be challenging. The reliance on digital technologies means that any disruption, such as cyber-attacks or system failures, can significantly impact the entire product development process. Additionally, scaling the digital PLM framework across different regions and product lines can be difficult, with each having unique requirements that must be addressed separately. The rapid pace of technological advancement necessitates continuous upgrades of systems and processes, requiring ongoing investment in training and development [4, 33].

Another major challenge is the availability and management of life cycle data. Companies often struggle to obtain comprehensive data on their products' environmental and social impacts, particularly at the end-of-life (EOL) stage. This data gap can hinder effective decision-making and the implementation of sustainable practices. Additionally, integrating LCE principles into existing product development processes can be complex. Many organizations face difficulties aligning their established workflows and organizational culture with the requirements of LCE, which can lead to resistance and implementation challenges. Furthermore, while LCE can result in long-term cost savings, the initial investment in sustainable technologies and practices can be substantial. Market forces and consumer preferences can also undermine the benefits of LCE, especially when recycled materials are undervalued compared to new materials.

Despite these challenges, the strategic application of Life Cycle Engineering principles is crucial for companies aiming to innovate and compete effectively in the modern market. The benefits often outweigh the limitations, as companies implementing these principles can achieve improved business performance, greater efficiency, and enhanced customer satisfaction. Moreover, by embracing these principles, businesses can better navigate the complexities of today's market, drive sustainable growth, and build stronger, more transparent relationships with their customers. Adopting Life Cycle Engineering principles ultimately positions companies to respond more effectively to market demands, regulatory requirements, and competitive pressures, ensuring long-term success and resilience in an increasingly dynamic and complex business environment.

Therefore, while implementing LCE principles presents certain challenges, the overall benefits to sustainability, efficiency, and innovation in product development make it a worthwhile investment for forward-thinking companies.

4.3.2 How Life Cycle Engineering Contributes to Sustainable Manufacturing in Product Development

LCE plays a pivotal role in fostering sustainable manufacturing in product development by integrating environmental, economic, and social considerations throughout a product's entire life cycle. This holistic approach ensures that sustainability is embedded in every stage of product development, from initial design to end-of-life disposal. By adopting principles like Design for Excellence (DFX), which includes Design for Environment (DfE), products are designed with their entire life cycle in mind. This means considering the environmental impact of materials, manufacturing processes, and product use from the outset. For example, selecting recyclable materials and designing for easy disassembly can significantly reduce a product's environmental footprint.

One of the primary goals of LCE is to enhance resource efficiency. This involves optimizing the use of materials and energy throughout the product life cycle. Techniques such as DfM and DfA help minimize waste and reduce the required raw materials. Additionally, advanced digital tools and simulations allow for the testing and refinement of designs before physical production, further conserving resources. LCE promotes the reduction of energy consumption across all stages of product development. Companies can lower their carbon footprint by incorporating energy-efficient practices in the manufacturing process and utilizing renewable energy sources where possible. For instance, using energy-efficient machinery and processes during production can significantly reduce greenhouse gas emissions.

LCE's focus on waste minimization greatly supports sustainable manufacturing. Through principles like DfR and DfD, products are designed to be easily broken down at the end of their life cycle, facilitating recycling and reducing landfill waste. This not only conserves natural resources but also minimizes environmental pollution. LCE enhances product durability and longevity by emphasizing robust design practices and high-quality materials. Design for Reliability ensures that products are built to last, reducing the need for frequent replacements and thereby conserving resources. Predictive maintenance, enabled by technologies such as the IoT, helps identify potential failures before they occur, extending product life and reducing waste.

LCE supports the transition to a circular economy by encouraging practices that close the loop on product life cycles. This involves designing products that can be refurbished, remanufactured, or recycled at the end of their useful life. Companies can reclaim materials from old products and reuse them in new ones, significantly reducing the demand for virgin resources and minimizing environmental impact. LCE promotes sustainability throughout the supply chain by incorporating logistics and supplier considerations into the design process. DfSC ensures that products are designed with the entire supply chain in mind, optimizing transportation efficiency and reducing the environmental impact of logistics. Additionally, collaborating with suppliers to adopt sustainable practices and materials further enhances overall sustainability.

Sustainable manufacturing through LCE benefits the environment and offers economic advantages. Companies can achieve significant cost savings by reducing waste, improving energy efficiency, and extending product life. Furthermore, sustainable products often meet consumer demand for environmentally friendly options, enhancing brand reputation and customer loyalty. LCE helps companies comply with increasingly stringent environmental regulations and standards. By proactively addressing environmental impacts throughout the product life cycle, businesses can avoid potential fines and penalties associated with non-compliance. Moreover, adopting sustainable practices can mitigate risks related to resource scarcity and supply chain disruptions, ensuring long-term operational resilience.

Integrating LCE principles into product development fosters innovation by encouraging the exploration of new materials, processes, and technologies. Companies that lead in sustainable manufacturing can differentiate themselves in the market, gaining a competitive advantage. Sustainable products that offer superior performance and environmental benefits are increasingly valued by consumers, driving market growth and profitability. Life Cycle Engineering is integral to sustainable manufacturing in product development. By embedding sustainability into every product life cycle stage, LCE helps companies create environmentally friendly, resource-efficient, and economically viable products. This supports the transition to a more sustainable and circular economy and offers significant business benefits, including cost savings, regulatory compliance, and competitive advantage. Through LCE, businesses can contribute to a more sustainable future while achieving long-term success.

LCE's contributions to product development center around design optimization, helping in material selection, managing energy efficiency, and reducing waste. These aspects directly impact supply chain management, as LCE emphasizes supply chain transparency, responsible sourcing, and ethical manufacturing practices. By collaborating with suppliers to ensure compliance with environmental and social standards, companies can minimize their supply chains' environmental and social impact and promote sustainable manufacturing practices. Using life cycle assessment techniques to evaluate the environmental impacts of products throughout their life cycle, companies can identify areas of high environmental impact and prioritize improvements to minimize their overall environmental footprint. In turn, this enables regulatory compliance. LCE helps companies comply with environmental regulations and meet sustainability standards. By staying informed about relevant regulations and proactively addressing environmental and social issues, companies can reduce compliance risks and build trust with customers and stakeholders.

By encouraging the use of renewable, recyclable, and environmentally friendly materials and selecting materials with lower environmental impact and toxicity levels, companies can minimize their environmental footprint and promote sustainable manufacturing practices. LCE focuses on optimizing energy efficiency throughout the product life cycle, from manufacturing and transportation to product use and disposal. Companies can reduce energy consumption, lower greenhouse gas emissions, and decrease their carbon footprint by adopting energy-efficient technologies and processes. LCE promotes continuous improvement and innovation in sustainable manufacturing practices by monitoring environmental performance, gathering feedback from stakeholders, and implementing best practices. Companies can identify opportunities for improvement and drive ongoing progress toward sustainability. LCE aims to minimize waste generation and promote waste reduction strategies such as recycling, reuse, and remanufacturing. By designing products for disassembly and implementing closed-loop manufacturing processes, companies can reduce waste sent to landfills and conserve resources.

Overall, LCE contributes to sustainable manufacturing in product development by integrating environmental, social, and economic considerations into product design, production, and use. By adopting LCE principles, companies can create more sustainable products that minimize their environmental footprint, promote resource efficiency, and contribute to a more sustainable future.

4.4 Life Cycle Diagram Flow

Product: Ideation, Conceptualization, Engineering and Prototyping

The most important factor management professionals use to decide when it is appropriate to make changes in the process of strategizing ways to maintain a product continuously is the life cycle of it. Life Cycle refers to the time from the product creation until it is removed from the sales. According to [56], a six-step theoretical framework can be used as a point of reference for discussing the development of systemic approaches to avoid the gaps and weaknesses of product life cycle management.

FIGURE 4.2 presents the life cycle diagram flow under the 5S framework, where product development can follow the described stages. Life cycle engineering applied to product development is a systematic approach that considers the entire life cycle of a product, from its ideation, conceptualization, and design to prototyping, manufacturing, use, and, ultimately, until the end of its useful life. Combining several methodologies under the LCE principles, the product design can be sustainable in all its processes. At the same time, when including the 5S framework, product development becomes Social, Sustainable, Sensing, Smart, and Safe.

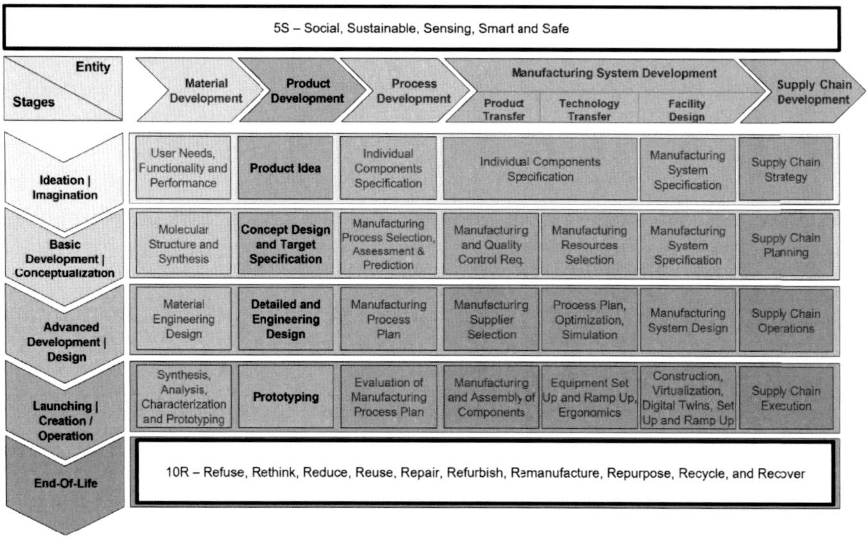

FIGURE 4.2
Life cycle diagram flow: product development and its stages.

Here is an overview of how to use digital technologies in the different stages of the product development process:

- **Design Phase:** Conduct a LCA to understand environmental impacts. Implement DfE principles. Utilize digital twin technology for virtual prototyping and simulation. Optimize material selection for sustainability and recyclability. Explore generative design for innovative and efficient product concepts.
- **Manufacturing Phase:** Implement energy-efficient manufacturing processes. Utilize advanced manufacturing technologies, such as 3D printing. Adopt lean manufacturing principles to reduce waste. Integrate IoT sensors for real-time monitoring and optimization. Ensure supply chain transparency and responsible sourcing.
- **Product Use Phase:** Design for energy efficiency and durability. Implement smart technologies for energy monitoring and optimization. Provide user manuals and information on sustainable product use. Encourage repairability and modular design for upgrades. Develop predictive maintenance using IoT and AI technologies.

- **End-of-Life Phase:** Design for disassembly to facilitate recycling. Implement take-back programs for responsible product disposal. Explore remanufacturing options to extend product life. Utilize advanced recycling technologies. Comply with regulations for environmentally friendly disposal.
- **Continuous Improvement:** Establish feedback loops for gathering data from each life cycle phase. Analyze data to identify areas for improvement. Innovate in response to emerging technologies and sustainability standards. Collaborate with stakeholders for ongoing optimization.
- **Documentation and Reporting:** Create comprehensive documentation of the life cycle assessment and sustainability measures. Generate reports on environmental performance and improvements. Communicate achievements to stakeholders, customers, and regulatory bodies.

Applying Life cycle Engineering throughout these phases ensures that environmental, social, and economic considerations are integrated into every product development and use aspect. It promotes a circular economy mindset, aiming for sustainable practices contributing to long-term ecological balance and societal well-being. As mentioned in Section 2.2.2, integrating these 5S principles into product development and management leads to products that are functional, innovative but also responsible, sustainable, and beneficial for society and the environment.

4.4.1 Entity Description: Product

According to [137], a company sells a product to its customers. To go deeper into the definition, here we consider a product, everything manufactured under prior planning, as a result of a production process, and whose purpose is to satisfy a need. Conversely, product development begins with the perception of a market opportunity and ends in producing, selling, and delivering a product [137].

LCE practices are being implemented when considering the life cycle from ideation to launch during product development. Integrating LCE with digital technologies redefines and enhances the product development landscape, aligning it with the principles of Industry 5.0.

Starting with the design phase, incorporating eco-design principles and eco-innovation is essential to minimize environmental impacts. This includes considering durability, recyclability, and minimizing the use of hazardous substances. Making products energy-efficient and educating consumers about sustainable usage are critical steps. Optimizing the stages to reduce waste, energy, and resource consumption also positively impacts manufacturing. Finally, developing and promoting recycling and disposal programs for end-of-life products is crucial.

Designers and manufacturers can mitigate environmental impacts and enhance product value by deploying these technologies within a structured

LCE framework. Digital technologies significantly enhance the speed and efficiency of product development, enabling rapid prototyping, enhanced collaboration, and improved design accuracy. These tools and techniques allow for implementing LCE practices at every product development stage, as described in subsection 4.4.2.

4.4.2 Stages

The LCE framework outlines a structured approach to product development, ensuring that each phase is carefully considered to promote sustainability, efficiency, and innovation. The stages under this framework encompass the entire life cycle of a product, from the initial ideation to the final disposal. Each stage addresses specific aspects of the product development process, incorporating eco-design principles, customer needs, technological advancements, and economic viability. The following subsections detail the stages involved:

4.4.2.1 Idea—Imagination

The initial stage of the product development process under LCE method begins with generating ideas for the new product. The ideation stage consists of brainstorming product concepts based on the problem to solve, the customer needs, the pricing, and the market research [30]. This idea generator can lead to the product definition, which is an analysis of the product concept improvement. Product ideas are generated and screened for their technical and economic viability. It is a brainstorming phase to conceptualize what products could be developed and to outline the project's scope and plan.

In the product concept development phase, the target market's needs are identified, alternative product concepts are generated and evaluated, and one or more concepts are selected for further development and testing. A concept is a description of the form, function, and features of a product usually accompanied by a set of specifications, an analysis of competitive products, and an economic justification of the project [137].

A systematic search for the selection and development of promising product ideas leads to the product development entity. It obtains both technically and economically viable applications. The scope of the project and the project plan are defined in this stage, and even with the help of artificial intelligence or big data analytics, they must be done considering the sustainability of the final product and the use of eco-design principles.

- **Activity Inputs:** Market research data, customer needs and feedback, emerging technology trends, brainstorming sessions.
- **Activity Outputs:** Generated product ideas, preliminary project scope and plan, product definition, selected product concepts for further development.

4.4.2.2 Basic Development—Conceptualization

The conceptual design and target specification involve collecting information about constraints and customer requirements to be embodied in the solution. A conceptual design is formulated during this phase, considering constraints and customer requirements. The significance lies in translating customer needs into a tangible design that can be further developed. This conceptualization under the idea creation can be complemented using, for example, simulation software and even augmented reality [66], [107].

- **Activity Inputs:** Selected product concepts, detailed customer requirements, technological constraints, simulation software and augmented reality insights.
- **Activity Outputs:** Conceptual designs, target specifications, architectural design documents, feasibility studies.

4.4.2.3 Advanced Development—Design

Detailed and engineering design develops all individual parts' arrangement, form, dimensions, and surface properties. Drawings and manufacturing documents are produced. Detailed and engineering designs of all parts are completed. This includes exact arrangements, dimensions, and surface properties, all necessary for manufacturing. Design CAD tools or other advanced simulation software contributes to the digital representation of the final product designs. Engineers and designers can create detailed and accurate virtual models of products and systems by leveraging digital tools such as computer-aided design (CAD), virtual prototyping, and simulation software. These digital representations enable stakeholders to visualize and analyze product designs more effectively, facilitating collaboration, iteration, and decision-making throughout the product development life cycle. One of the biggest points in product design these days is using data to optimize the design. When using the information taken from a sensor, new technologies, and consumer insights, it is possible to design for durability, repairability, and end-of-life considerations.

- **Activity Inputs:** Approved conceptual designs, engineering specifications, CAD tools and simulation software, data from technology and consumer insights.
- **Activity Outputs:** Detailed design documents, engineering drawings, virtual prototypes, design for sustainability considerations (durability, recyclability, minimal hazardous substances).

4.4.2.4 Launching—Creation/Operation

This stage considers the prototyping. Tests are carried out to check the functionality and potential design modifications through verification of any remaining problems in the product. For the product transfer, the individual component specifications are made under a life-cycle economic and sustainable analysis [143]. Product information is collected and a bill of the materials (BOM) is carried out to identify materials, standard components, quality requirements, and delivery times. Product prototypes are created and tested, which is a crucial phase for identifying and rectifying any issues before mass production. It is natural to think of Additive Manufacturing to accelerate the prototyping and visualize the product. Also, helps to reduce material waste and helps with the product presentation to validate with the final users before starting with the manufacturing line. The use of reverse engineering for product design [150] or the consideration of LCA of 3D printed products [20] for the production scheme changes the material supply chains and helps to relocate the energy consumption hotspots around the life cycle of a product.

It is possible to implement smart features that can extend the product's lifespan when the maintenance can be predicted or the software can be updated. Or simply giving the users information about the proper disposal or resources to recycle the product at the end of its life.

- **Activity Inputs:** Approved conceptual designs, engineering specifications, CAD tools and simulation software, data from technology and consumer insights.
- **Activity Outputs:** Detailed design documents, engineering drawings, virtual prototypes, design for sustainability considerations (durability, recyclability, minimal hazardous substances).

4.4.2.5 End of Life: 5S—Social, Sustainable, Sensing, Smart, Safe

The final stage allows us to develop a safe and environmentally friendly disposal or recycling plan for the product. With knowledge about how to handle and dispose of materials, including hazardous ones, under regulations, we can provide instructions to consumers on how to disassemble or recycle the product safely. This stage is very important because it encompasses the disposal, recycling, and repurposing of a product once it reaches the end of its useful life.

- **Social:** This aspect ensures that the product development process addresses broader societal needs and does not exacerbate social issues. Engaging with stakeholders, promoting inclusivity, and assessing community impacts are crucial. This stage also involves educating consumers about responsible disposal and recycling, engaging them in sustainable practices, and ensuring that products contribute positively to communities. Providing consumers with information on proper disposal methods and recycling options empowers them to make environmentally responsible choices. Clear labeling and instructions guide disassembling, recycling, or returning products for end-of-life management.

  - ***Activity Inputs:*** Consumer feedback on usability and end-of-life concerns, community impact assessments, partnerships with community organizations.
  - ***Activity Outputs:*** Programs and initiatives for community engagement and support, reports on product impacts and improvements on community well-being, educational materials on responsible product disposal.

- **Sustainable:** This involves using eco-design principles, such as utilizing renewable or recycled materials, designing for energy efficiency, and ensuring easy disassembly. It also includes tracking sustainability metrics and ensuring that the supply chain adheres to sustainability standards. The goal is to enhance the product's environmental performance throughout its life cycle and to develop recycling and disposal plans that comply with regulations. Designing products for durability, repairability, and remanufacturing extends their lifespan and reduces the frequency of disposal. Utilizing materials that are easily detachable or have high recycling rates enhances the feasibility of material recovery. Modular construction and standardized interfaces facilitate disassembly and recycling.
  - ***Activity Inputs:*** Data on material life cycles and recyclability, sustainability assessments, regulatory requirements for environmental compliance.
  - ***Activity Outputs:*** Recycling and reuse strategies, material recovery plans, compliance with environmental regulations and sustainability certifications.

- **Sensing:** This incorporates integrating sensors and data utilization to enhance product functionality and user experience. Sensors can monitor various factors like energy consumption, wear and tear, or environmental conditions, providing valuable feedback for continuous improvement and adaptation of the product.
  - ***Activity Inputs:*** Data from product usage sensors and analytics, feedback mechanisms integrated within the product, environmental impact data during product use.
  - ***Activity Outputs:*** Insights on product usage patterns and environmental impacts, data-driven recommendations for product life cycle improvements, strategies to minimize environmental impact based on usage data.

- **Smart:** This aspect integrates advanced technologies such as connectivity features and artificial intelligence to optimize product performance and user interaction. Products are designed to be part of a connected ecosystem, capable of receiving updates that improve functionality and security over time, thus extending their lifespan and adaptability.

  - ***Activity Inputs:*** Smart technology integration for product life cycle management, software tools for monitoring product health and usage, user interaction data through IoT and connected devices.
  - ***Activity Outputs:*** Predictive maintenance schedules and updates, smart decommissioning processes, firmware or software updates to extend product life or repurpose usage.

- **Safe:** This ensures that products are designed with safety as a priority. This includes physical safety through robust construction, ergonomic design, and cybersecurity measures for connected products. Products are rigorously tested against safety standards and regulations to ensure they pose no harm to users. Features that enhance user safety in case of product failure or unexpected situations are also incorporated. Planning for safe and responsible disposal, including the identification of hazardous materials, ensures they can be properly managed at the end-of-life of the product.
  - ***Activity Inputs:*** Safety standards and guidelines, hazardous material identification, end-of-life handling procedures for safety risks.
  - ***Activity Outputs:*** Safety plans for product disassembly and disposal, guidelines and kits for safe recycling or disposal, training programs for safe handling and decommissioning of products.

These activities ensure that each aspect of the 5S framework is addressed specifically, allowing for focused strategies that enhance the overall life cycle management of products. This ensures that they are socially responsible, environmentally sustainable, enhanced through smart technologies, safely decommissioned, and their impacts sensibly monitored.

4.4.3 Techniques and Methods Required During Product Development

As previously discussed, manufacturing companies are increasingly interested in adopting a circular economy model. Implementing this model provides a framework that aligns organizational goals with the SDGs. Product development is a complex and multifaceted process that necessitates integrating various techniques and methodologies to create sustainable, efficient, and marketable products.

According to [33], when a corporation adopts various value-retention options (R-strategies), it can create, preserve, and recover the value of assets and products. Main R-strategies identified by [117] include refuse, reduce, resell/reuse, repair, refurbish, remanufacture, repurpose, recycle, recover, and re-mine. Each of these strategies contributes to creating, preserving, and recovering input value in innovative ways, from composting food waste to repurposing plastics. Approximately 80% of a product's total environmental impact is determined by the R-strategies employed in its design and development [33].

Moreover, some of the techniques and methods are crucial during different stages of product development.

- **Project Management and Agile Methodologies:**
  - ***Scrum and Kanban:*** Utilize agile frameworks to manage and streamline workflows, enhance team collaboration, and adapt to changes quickly and efficiently.
  - ***Critical Path Method (CPM):*** Employ CPM to identify the longest stretch of dependent activities and schedule them effectively to predict project duration.
- **Design and Prototyping:**
  - ***Computer-Aided Design (CAD):*** Implement CAD software to create detailed 3D models and drawings for visualizing the product design before physical production.
  - ***Rapid Prototyping:*** Use technologies like 3D printing to quickly produce a prototype for testing and validation of design concepts.
- **User-Centric Design:**
  - ***User Experience (UX) Design:*** Focus on the usability, accessibility, and pleasure provided in the interaction with the product.
  - ***Design Thinking:*** Apply a solution-oriented approach that requires designers to engage with users, challenge assumptions, and redefine problems in an attempt to identify alternative strategies and solutions.
- **Sustainability Analysis:**
  - ***Eco-design:*** Integrate environmental aspects into product design to minimize negative impacts throughout the product's life cycle.
- **Materials and Supply Chain Management:**
  - ***Supplier Relationship Management:*** Develop strategic partnerships with suppliers to ensure high-quality materials and sustainable procurement practices.
  - ***Inventory Management Techniques:*** Implement systems like Just-In-Time (JIT) to reduce inventory costs and increase the efficiency of the production systems.
- **Quality Assurance and Testing:**
  - ***Statistical Quality Control:*** Utilize statistical methods for monitoring and controlling the quality during the production process.

- *Testing Standards and Protocols:* Apply rigorous testing procedures to ensure that the product meets the required specifications and quality standards.

- **Marketing and Market Analysis:**
  - *Market Research:* Conduct comprehensive research to understand market needs, competitor analysis, and potential customer base.
  - *Product Positioning and Branding:* Strategically position the product in the market to maximize visibility and appeal to the target audience.

- **Regulatory Compliance:**
  - *Compliance with Standards:* Ensure all products comply with national and international standards and regulations relevant to the product.
  - *Safety Certifications:* Obtain necessary certifications that affirm the product's safety and reliability.

The successful integration of these techniques and methods enhances product development, enabling organizations to deliver products that meet or exceed customer expectations while adhering to regulatory requirements and sustainability principles. This holistic approach ensures product quality and compliance and contributes to the organization's reputation and bottom line. In addition, implementing these strategies within the circular economy model reduces the materials and resources used in essential product components, thereby minimizing carbon emissions throughout the product's life cycle. The goal is to ensure these components and materials find new uses rather than being discarded.

For instance, eco-design provides a well-established framework for environmental management systems. Eco-design involves specifying product functions and conducting environmental assessments to identify the stages or processes with the highest environmental impact [100]. These steps can be aligned with LCE principles, starting with product function specification in the initial Life Cycle stage and ending with technical solutions in the final prototyping stage. Integrating eco-design with optimization processes and digital technologies fosters innovation throughout product development, offering a competitive advantage.

To accelerate product development, leveraging advanced technologies is crucial. Section 4.1.2 discusses the use of digital technologies in product development. These applications enhance the sustainability and optimization of the product launch process. The following technologies can be employed to expedite product development:

- **Product Idea Stage:** AI and ML enable machines to learn from data and make intelligent decisions, enhancing generative design, quality control, demand forecasting, and process optimization. Generative design, which uses algorithms to generate multiple design iterations based on specified parameters, explores numerous possibilities and optimizes for performance, weight, and other criteria.

- **Concept Design and Target Specification:** Advanced Materials Science explores innovative and sustainable materials, leveraging materials with enhanced properties, durability, and environmental sustainability. Simulation and Modeling Tools enable virtual testing and validation, predicting performance, identifying issues early, and reducing the need for physical prototypes.
- **Detailed and Engineering Design:** IoT sensors connect physical devices to the internet for data collection and communication, providing real-time data on product usage for iterative improvements and maintenance predictions. Cobots enhance efficiency in manufacturing processes, including assembly and testing, by working alongside human workers. Cloud Computing facilitates collaboration, provides access to powerful computing resources for simulations, and ensures seamless communication among distributed teams.
- **Prototyping:** Rapid prototyping tools enable the quick creation of physical prototypes for testing and iteration, allowing designers and engineers to validate and refine concepts cost-effectively. Additive Manufacturing (3D printing) builds objects layer by layer from digital models, accelerating prototyping, enabling rapid iteration, and allowing for the production of complex geometries. AR and VR facilitate collaborative design reviews, training, and prototyping by providing visual and interactive experiences. Digital twins allow the manipulation of virtual prototypes, providing insights without needing a manufacturing process.

Complementing these technologies, blockchain technology provides a decentralized and secure ledger for recording and verifying transactions, ensuring transparency and traceability in the supply chain. Supply chain digitization integrates digital technologies to enhance visibility and efficiency, speed up material procurement, reduce lead times, and improve overall supply chain responsiveness.

Integrating these technologies can significantly accelerate product development, reduce time to market, and foster innovation. The choice of technologies should align with the specific needs and goals of the product development project.

Considering that a product can be hardware or software, FIGURE 4.3 illustrates the differences between stages in the life cycle diagram flow. While the main stages remain the same, the methods for analyzing and executing each stage differ. Software products, like hardware products, have environmental implications that must be considered. [129] provides an analysis of the environmental impact of the software life cycle.

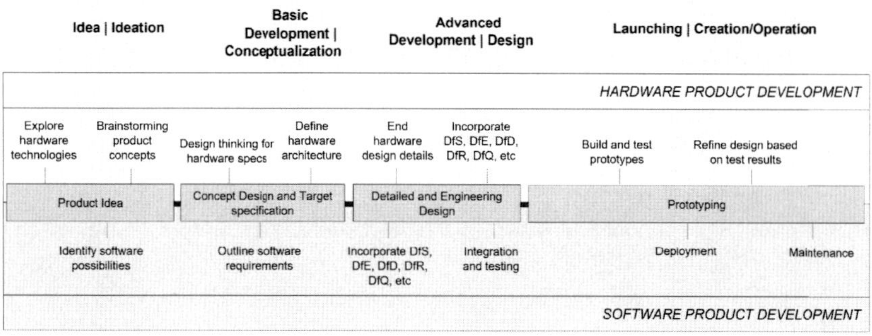

FIGURE 4.3
Differences in life cycle diagram for software or hardware product development.

While the fundamental principles of LCE apply to software and hardware product development, key differences exist in their application. FIGURE 4.3 highlights these differences, from physical components, prototyping, and testing to supply chain management, manufacturing and production, and physical distribution and logistics in hardware development. Software development includes codebase development, prototyping and simulation, the software development life cycle, continuous integration and deployment, and digital distribution.

4.4.4 The Ten Core Technologies Applied to the Entity

Adopting cutting-edge technologies in product development can significantly accelerate the process, reduce costs, and improve product quality and sustainability. This section outlines key technologies instrumental in modernizing and speeding up product development stages, as depicted in FIGURE 4.4.

- **AI and ML:**
  - ***Generative Design:*** AI and ML algorithms can generate multiple design alternatives based on predefined criteria. This helps explore numerous possible iterations and optimize designs for performance, weight, cost, and other factors.
  - ***Predictive Analytics:*** Employ AI to forecast trends and behaviors, enhancing decision-making in market analysis, material selection, and production planning.
- **Simulation Tools:**
  - ***Feasibility Testing:*** Simulate concepts during Ideation to assess their technical and economic viability.
  - ***Stress and Functional Analysis:*** Validate designs in conceptualization and advanced development stages.

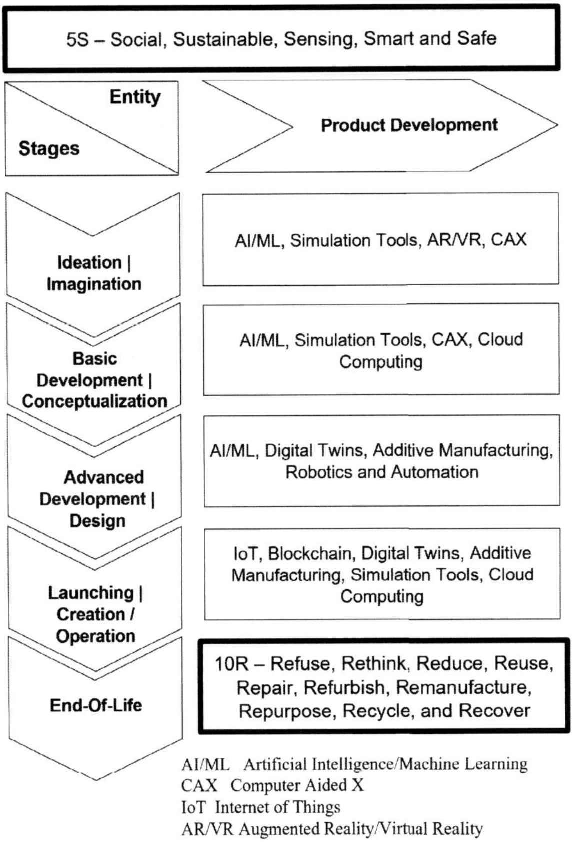

FIGURE 4.4
Core technologies applied to this entity.

- **IoT:**
  - ***Smart Manufacturing:*** IoT devices can monitor and collect data from manufacturing equipment, providing real-time insights into production processes. This enables predictive maintenance, reduces downtime, and enhances manufacturing efficiency.
  - ***Product Use Data Collection:*** Embedding IoT sensors in products to gather data on how they are used, which can feed back into the design and manufacturing processes for continuous improvement.
- **Digital Twins:**
  - ***Virtual Prototyping:*** Use digital twins to simulate and analyze the performance of a product in a virtual environment. This helps identify potential issues early in the development process, reducing the need for physical prototypes.

- *Life cycle Management:* Digital twins can be utilized throughout the product life cycle for ongoing optimization and maintenance, extending the product's useful life.

- **Additive Manufacturing (3D Printing):**
  - *Rapid Prototyping:* 3D printing allows for quick production of prototypes, enabling physical testing and refinement of designs at a much faster rate than traditional methods.
  - *Custom Manufacturing:* Enables the production of complex geometries and customized parts without the need for expensive tooling, reducing the time and cost associated with small production runs or custom products.

- **Computer-Aided Tools (CAX):**
  - *Integrated Design Tools:* CAD and CAE tools streamline conceptual and detailed design, enabling engineers to efficiently define form, dimensions, and surface properties.
  - *Advanced Engineering:* CAX systems with AI integration provide enhanced simulations and analyses.

- **Big Data Analytics and Blockchain:**
  - *Secure Traceability:* Blockchain technology provides a transparent and secure method for tracking the provenance of materials and components throughout the supply chain, enhancing compliance and sustainability practices.

- **Cloud Computing:**
  - *Scalable Resources:* Utilize cloud-based platforms for high-performance computing tasks such as simulations, data analysis, and storage. This provides the scalability needed to handle large datasets and complex calculations without significant upfront investment in IT infrastructure.
  - *Collaboration Tools:* Cloud services can enhance collaboration among distributed teams by providing access to shared tools and platforms, streamlining communication, and project management.

- **AR and VR:**
  - *Collaborative Design and Training:* AR and VR can facilitate immersive design reviews and collaborative sessions, allowing teams to work together effectively from different locations.

  - ***Interactive Prototyping:*** Enables designers and engineers to interact with their designs in a realistic 3D environment, improving the design process and enhancing user experience testing.

- **Robotics and Automation:**
  - ***Automated Testing:*** Robotics systems streamline prototype testing, reducing errors and enhancing scalability.
  - ***Production Scalability:*** Automation ensures efficient, repeatable production processes, especially in high-volume manufacturing.

By integrating these technologies into the product development process, companies can achieve a faster time-to-market, enhanced product quality, and increased innovation. This technological approach supports rapid development and deployment and aligns with sustainability goals by optimizing resource use and reducing waste.

4.5 Case Study Applying the Life Cycle Engineering of this Entity: Mio-Tracker

Electromyography (EMG) is a diagnostic technique that measures and records the electrical activity produced by skeletal muscles. By detecting the electrical potential generated by muscle cells when they are electrically or neurologically activated, an EMG device provides essential data that can be used to assess the health and function of muscles and the nerve cells that control them.

The application of electromyography is particularly valuable in both therapeutic and sports environments. In physical therapy, EMG is used to diagnose muscle disorders, guide rehabilitation progress, and tailor treatment plans according to the specific needs of patients. It allows therapists to objectively measure improvements in muscle function over time, making it an indispensable tool for recovery from injuries or managing chronic conditions. In the realm of sports, electromyography helps athletes optimize their training based on precise muscle activity data, enhancing performance and preventing overuse injuries by identifying imbalances and areas of strain.

The development of a single-channel electromyograph tailored for these purposes represents a significant advancement. This device simplifies the complex technology of EMG into a more accessible tool for physical therapists and sports professionals alike, enabling them to conduct on-the-spot assessments of muscular tone and condition. This report outlines the journey of creating a single-channel electromyograph, focusing on its design, functionality, and the impact it aims to have in enhancing therapeutic outcomes and sports performance through detailed muscular analysis.

Mio-Tracker is a medical device designed to evaluate the muscular activity of people in a rehabilitation process, high-performance sportsmen, seniors, or anyone who wants to know the health state of their muscles. This device consists of two parts: (1) the electronic equipment and (2) the digital platform. The first is the non-invasive device with an automatic adjustment algorithm. It consists of a portable electronic apparatus with an amplifier. The concept behind the latest iteration of our single-channel electromyograph was driven by feedback and data gathered from previous versions. Each iteration has sought to better meet the needs of physical therapists and sports professionals, who require precise, easy-to-use, and efficient tools for evaluating muscular tone. This new version of the electromyograph was specifically designed to address key areas for improvement identified in earlier models: usability, data management, and device portability.

The second is the digital platform to visualize the muscular response of each user/patient. It is designed for both the user/patient and the medical specialist. The interface was designed under the LCE and 5S frameworks, involving and integrating the principles of sustainability, social responsibility, sensing, smart technology, and safety throughout the product life cycle.

From the beginning, the Mio-Tracker was intended as a portable device for clinics, physiotherapists, and other medical specialists. Every part of Mio-Tracker has been designed according to LCE principles.

FIGURE 4.5 depicts all the principles and concepts required to be addressed under LCE while developing a 5S product applied to Mio-Tracker. This case study reflects a comprehensive understanding of products' role in shaping a sustainable future, emphasizing the need for thoughtful design, ethical production, and continuous improvement.

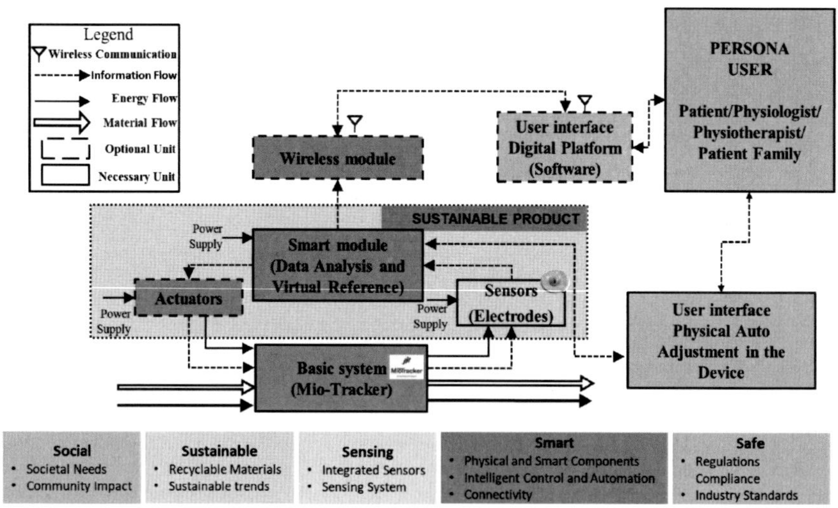

FIGURE 4.5
Mio-Tracker study case applying 5S products (social, sustainable, sensing, smart and safe).

4.5.1 Walkthrough of the Entire Life Cycle Process: From Idea to Launching – Mio-Tracker: The Device

For all product design stages of Mio-Tracker, it is considered the LCE principles and the 5S framework:

- **Mio-Tracker:** Ideation (see FIGURE 4.6)
  - ***Social:*** Brainstorm product ideas that address societal needs and contribute positively to communities, in the case of Mio-Tracker, to patients.

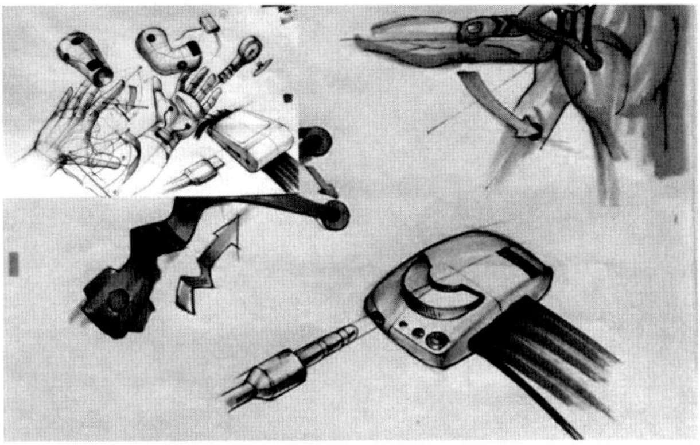

FIGURE 4.6
Mio-Tracker: Ideation and sketch.

  - ***Sustainable:*** Consider the environmental impact of potential products and prioritize designs that minimize resource consumption and waste generation. For this, Mio-Tracker had had several iterations to reduce the size of the device and use less material to manufacture it.
  - ***Sensing:*** The use of market research and consumer insights allows us to understand social and environmental trends, ensuring that product ideas align with evolving preferences and sustainability goals. For this, in Mio-Tracker, the materials are recyclable, and the consumables are reusable.
  - ***Smart:*** Mio-Tracker is an innovative technology with intellectual property [5, 6], and features that enhance the product's social value and environmental sustainability, such as energy efficiency or accessibility features. This is why, Mio-Tracker's new design can be used with one hand, and the battery ensures portability and long operational life.
  - ***Safe:*** Since Mio-Tracker is a medical device, each regulation has been considered in its design. With this product, it is ensured that product concepts prioritize consumer safety and well-being, adhering to relevant regulations and industry standards.

- **Mio-Tracker:** Conceptualization
  - ***Social:*** Engage stakeholders, including customers and community members, in the conceptualization process to gather diverse perspectives and address social needs effectively. For this, interviews with the users and the community were conducted to understand the concept of Mio-Tracker, which covers global necessity.
  - ***Sustainable:*** Evaluate materials, manufacturing processes, and supply chain practices to minimize environmental impact and promote sustainability throughout the product's life cycle. With the help of manufacturers, recycled materials were proposed, low maintenance and the use of less material in the construction of Mio-Tracker was considered.
  - ***Sensing:*** Incorporate feedback mechanisms and data collection processes to gather insights on user preferences, environmental performance, and social impact, informing design decisions. The interviews work for this, too, but also the digital platform where feedback is required to improve the device and follow the impact.
  - ***Smart:*** Integrate smart technologies and features that enhance the product's functionality, efficiency, and user experience while promoting social inclusivity and environmental responsibility. In itself, the Mio-Tracker is a high-tech device that uses AI to process information and follow up on every patient's session.
  - ***Safe:*** Design products with built-in safety features and considerations to protect users from potential hazards and ensure product reliability and longevity. Medical regulations are considered in the electrical, physical, and software design, and the medical device norms are accomplished in Mio-Tracker.
- **Mio-Tracker:** Design
  - ***Social:*** Design products that foster positive social interactions and contribute to community well-being, considering factors such as accessibility, inclusivity, and cultural sensitivity. In the third interaction with users, Mio-Tracker was modified to be used with one hand; this makes it accessible and inclusive. Also, the device size considers the on-field work some users do, so it is comfortable to use.
  - ***Sustainable:*** The last Mio-Tracker interaction optimizes its design for resource efficiency, durability, and recyclability, minimizing environmental impact and promoting sustainable consumption patterns.
  - ***Sensing:*** Mio-Tracker incorporates sensors and data analytics capabilities in the digital platform to monitor performance, usage patterns, and environmental impacts, enabling continuous improvement and optimization.

  - ***Smart:*** Mio-Tracker implements smart features and connectivity options that enhance product functionality, customization, and user experience while promoting energy efficiency and environmental sustainability.
  - ***Safe:*** Ensure product safety through rigorous testing, compliance with safety standards, and user-friendly design features that mitigate potential risks and hazards. Several tests and certifications to cover the standards have been made to Mio-Tracker, also, it is designed for ease of use and patient familiarity.

- **Mio-Tracker:** Operation (see FIGURE 4.7)
  - ***Social:*** Provide support resources, user communities, and engagement platforms to foster positive social interactions and user empowerment, promoting product adoption and loyalty.
  - ***Sustainable:*** Encourage responsible product use and maintenance practices that extend product lifespan, reduce resource consumption, and minimize waste generation.
  - ***Sensing:*** Collect and analyze data on product usage, performance, and environmental impact to optimize operations, identify opportunities for improvement, and enhance user satisfaction.
  - ***Smart:*** Leverage smart technologies and automation to streamline product operation, improve efficiency, and enhance user convenience while minimizing energy consumption and environmental footprint.
  - ***Safe:*** Implement safety protocols, user training, and maintenance procedures to ensure safe and reliable product operation, reducing the risk of accidents, injuries, and product failures.

- **Mio-Tracker:** End-of-life
  - ***Social:*** Develop end-of-life management strategies that prioritize social responsibility, including options for product reuse, recycling, or responsible disposal that minimize negative impacts on communities. This point applies to every medical device, gadget, and electronic equipment.
  - ***Sustainable:*** Design products for easy disassembly, material recovery, and recycling, promoting a circular economy and reducing waste generation at the end of the product life cycle.
  - ***Sensing:*** Incorporate sensors and tracking systems to monitor product disposition and end-of-life processes, enabling traceability and accountability for environmental and social impacts.
  - ***Smart:*** Implement smart recycling and waste management solutions that optimize resource recovery, reduce landfilling, and promote sustainable material reuse and repurposing.

- ***Safe:*** Ensure that end-of-life processes prioritize worker safety, environmental protection, and regulatory compliance, minimizing risks associated with product disposal and recycling activities.

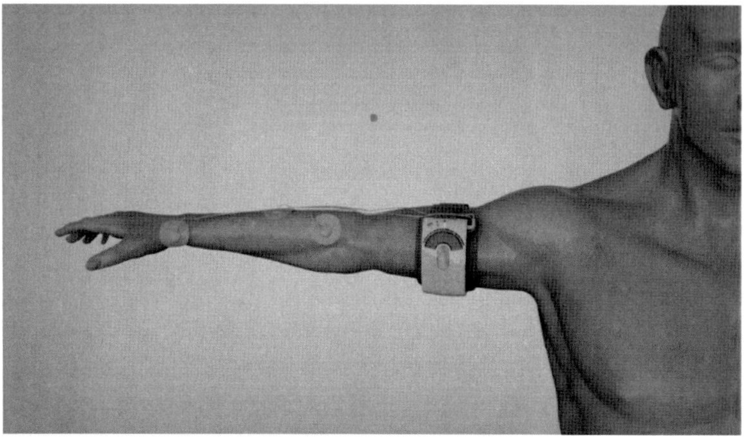

FIGURE 4.7
Last version of Mio-Tracker, a portable EMG device.

The design is made to have a large life, which implies that the product can be used in more than one life epoch. Also, it means, that the product can have different usages besides the primary use. The design also respects the environment under the philosophy of Design for Environment. Mio-Tracker, considers the whole life cycle, going beyond just the use of recycled materials or proper packaging or disposal.

4.5.1.1 Mio-Tracker: The Software Interface

The software interface of Mio-Tracker prioritizes user-centric design, ensuring accessibility and inclusivity for users with varying levels of technical expertise and physical abilities. This digital platform incorporates features that facilitate collaboration and communication among healthcare professionals, researchers, and patients, promoting knowledge sharing and patient engagement. Provide educational resources and user support within the software interface to empower users with information on EMG technology, data interpretation, and best practices for usage. All these, make Mio-Tracker socially responsible.

It was challenging to design the software interface with energy efficiency in mind to minimize the device's power consumption and environmental impact. To do so, the platform implements data optimization and storage management features to reduce the software's resource footprint and enable efficient use of storage resources. Mio-Tracker is working on extending the software's lifespan and supporting long-term usability and accessibility by considering its compatibility with different operating systems and devices in the actual scaling stage.

For every iteration, Mio-Tracker integrates real-time data visualization and analysis tools into the software interface to provide users with immediate feedback on EMG signals and muscle activity. Incorporate features for data filtering, signal processing, and noise reduction to enhance the accuracy and reliability of EMG measurements and analysis. Enable connectivity with external sensors and devices to expand the software's sensing capabilities and support interdisciplinary research and applications.

Leverage smart algorithms and machine learning techniques within the software interface to automate repetitive tasks, improve user workflow, and enhance diagnostic capabilities. Also, to follow every patient's sessions to make better decisions, and help them to improve their therapy. Implement customizable user profiles and preferences to personalize the software experience and adapt to individual user needs and preferences. For this, the Mio-Tracker digital platform incorporates games and other features for every user. Enable remote monitoring and telehealth capabilities to facilitate virtual consultations, remote patient monitoring, and data sharing among healthcare providers and patients, making it a smart device available for everyday use.

Mio-Tracker prioritizes data security and privacy within the software interface by implementing robust authentication, encryption, and access control mechanisms to protect sensitive patient information according to regulations. Incorporate safety features and alerts to mitigate risks associated with EMG testing, such as electrode placement errors, signal artifacts, and patient discomfort. It provides comprehensive user training and documentation within the software interface to ensure safe and proper EMG device and software platform usage.

4.6 The Future of Sustainable Manufacturing for Product Development

The future of sustainable manufacturing for product development, especially in the context of technological advancements and evolving market dynamics, indicates a transformative shift towards integration, agility, and customer-centricity. Here are several key aspects to consider:

- **Offering Management Over Product Management:** Traditional product management is being outpaced by a broader, more integrative approach known as "offering management." This model emphasizes full life cycle accountability—from validating market needs through design, implementation, and customer feedback loops. This approach facilitates a closer alignment with customer needs and promotes faster adaptation to market changes, which is crucial for sustainable manufacturing [18].

- **Customer-Centric and Data-Driven Approaches:** Modern product development is increasingly influenced by a wealth of data, allowing product managers to assume a "mini-CEO" role, leveraging insights to drive decisions. This role transformation is significant for sustainable manufacturing, where understanding broad, evolving consumer and environmental needs is crucial. It enables manufacturers to adapt products for better sustainability outcomes [85].
- **Agile and Responsive Product Development:** The documents emphasize the importance of agile methodologies, which allow for rapid testing and adaptation of features based on real-world customer feedback and market dynamics. This agility benefits sustainable manufacturing, allowing for the swift integration of environmentally friendly practices and technologies [18].
- **Technological Integration:** As technology sectors like AI, IoT, and digital platforms disrupt traditional manufacturing and product development landscapes, there is a growing need to integrate these technologies into manufacturing processes. This integration can lead to more efficient use of resources, reduced waste, and better life cycle management of products, all key to sustainability [18].
- **Comprehensive Skill Development for Product Managers:** The emerging profile of product managers includes technical and market knowledge and a deep understanding of sustainability. This knowledge is essential for driving the development of products that are not only efficient and marketable but also environmentally sound [86].

These trends suggest that the future of sustainable manufacturing in product development will be deeply intertwined with technological advancements, a shift towards comprehensive offering management, and a strong focus on environmental impact and customer needs. Furthermore, the future of sustainable manufacturing, especially when considering LCE and the 5S principles, points towards an integrated and holistic approach that deeply embeds these principles into every aspect of product development.

LCE principles will guide the design phase, considering the entire life cycle from raw material extraction through to end-of-life disposal. This comprehensive focus helps reduce environmental impacts at every stage. Moreover, advanced tools and software for LCE will be used more frequently to measure and optimize the environmental footprint of products throughout their life cycle. In addition, the following is depicted how the 5S principles might shape the future:

- **Social:**
  - ***Community and Stakeholder Engagement:*** Future product development will increasingly involve stakeholders, including local

communities and end-users, in the design process. This inclusivity ensures that products are well-tailored to real needs and can enhance social acceptance and market adoption.

  - ***Equitable Design:*** Emphasis will be placed on creating accessible products to all segments of society, including vulnerable and underserved populations, thus promoting social equity.

- **Sustainable:**
  - ***Resource Efficiency:*** Utilizing advanced materials and technologies that minimize environmental impact will be crucial. This includes using recycled materials, biodegradable components, and renewable energy sources during production.
  - ***Circular Economy Models:*** Manufacturing processes will evolve to support circular economy models, with products designed for easier disassembly, reuse, and recycling, thereby reducing waste and extending product life cycles.
- **Sensing:**
  - ***IoT and Sensors:*** Products will be increasingly equipped with sensors that provide continuous feedback on their use and environmental impact. This data will be crucial for optimizing product design and operation to minimize ecological footprints.
  - ***Predictive Maintenance:*** Sensor data will also facilitate predictive maintenance, reducing downtime and extending the lifespan of products, which is a key aspect of sustainability.
- **Smart:**
  - ***AI and Machine Learning:*** AI will play a pivotal role in product development, enhancing the ability to analyze sensor data and make intelligent real-time adjustments to product design and manufacturing processes.
  - ***Integration of Digital Twins:*** Digital twins will become more prevalent, allowing for the simulation and testing of products in virtual environments to optimize performance and sustainability before physical models are built.
- **Safe:**
  - ***Safety by Design:*** Safety will be an integral part of the design process, ensuring that products are safe to use and contribute to the environment's safety through reduced emissions and safer materials.
  - ***Compliance and Standards:*** Regulatory compliance will drive the adoption of safer practices in manufacturing, with stringent adherence to global standards on emissions, waste management, and labor safety.

By integrating these principles, the future of sustainable manufacturing for product development entities will not only focus on minimizing negative impacts but will actively contribute to a regenerative industrial economy. Technological advancements and a growing regulatory and consumer demand for sustainable, smart, and safe products will likely support this approach.

4.7 Conclusion

Sustainable product development is essential for companies striving for long-term success while addressing environmental, economic, and social responsibilities. Integrating LCE principles throughout the product life cycle minimizes environmental impact from material selection to end-of-life disposal. Companies can create economically viable, environmentally friendly, socially responsible products by embracing strategies like Design for Environment, Design for Disassembly, and Design for Recycling.

The 5S principles—Social, Sustainable, Sensing, Smart, and Safe—further enhance sustainable product development by embedding comprehensive and holistic considerations into every product life cycle stage. Engaging with stakeholders and communities ensures that products meet diverse needs, promote inclusivity, and contribute positively to society. Utilizing eco-design principles minimizes environmental impacts, promotes resource efficiency, and supports circular economy models. Incorporating IoT and sensor technologies monitors product performance and environmental impacts, enabling continuous improvement and predictive maintenance. Leveraging AI, machine learning, and digital twins enhances product functionality, customization, and user experience while optimizing sustainability. Ensuring product safety through rigorous testing, compliance with safety standards, and user-friendly design features protect both users and the environment.

Modern technologies such as Digital Twins, IoT, AI, and Big Data Analytics are transforming product development processes. They enable companies to design, test, and bring products to market more efficiently and sustainably. These technologies facilitate a comprehensive approach, considering the entire product life cycle and integrating sustainability at every stage. Companies should set clearly defined goals for sustainable product development. This should include the desired outcomes regarding environmental, economic, and social impacts and should be linked to the company's overall sustainability objectives.

The future of sustainable manufacturing will likely see a shift towards offering management over traditional product management, focusing on customer-centric, data-driven approaches. Agile methodologies, technological integration, and comprehensive skill development for product managers will be crucial in driving sustainable innovation and meeting evolving consumer and regulatory demands.

Ultimately, sustainable product development and manufacturing, guided by LCE and 5S principles, contribute to a more sustainable future. By adopting these principles and leveraging modern technologies, businesses can achieve significant environmental, economic, and social benefits, positioning themselves as leaders in the sustainable economy. This holistic approach ensures that products are efficient, marketable, environmentally sound, and socially beneficial, fostering a more sustainable and equitable world.

5

Manufacturing Process: Selection, Development and Planning

This chapter explores the integration of Life Cycle Engineering (LCE) with the 5S Manufacturing Process Development framework—Social, Sustainable, Sensing, Smart, and Safe—across various stages of product life cycle management. It outlines how manufacturers can create economically viable, environmentally friendly, and socially responsible systems by optimizing key phases such as ideation, development, operation, and end-of-life. Using LCE principles, the chapter emphasizes early design considerations, sustainable material selection, and efficient manufacturing processes to minimize environmental impact while enhancing economic and social benefits.

The chapter introduces advanced tools and methodologies, including Group Technology, PRIMA matrices, and Computer-Aided Process Planning (CAPP) systems (Variant, Generative, Hybrid). It also highlights the role of digital technologies such as IoT, AI, machine learning, and digital twins in optimizing manufacturing processes. Furthermore, applying the 10R framework (Refuse, Rethink, Reduce, Reuse, Repair, Refurbish, Remanufacture, Repurpose, Recycle, and Recover) demonstrates how manufacturers can transition toward circular economy principles during the decommissioning phase.

Through detailed examples and using rigid coupling as a case study, this chapter underscores the importance of aligning manufacturing processes with sustainability goals while leveraging advanced technologies for efficiency, adaptability, and innovation. By adopting this holistic approach, manufacturers can ensure resilient, future-ready systems that balance social, environmental, and economic dimensions.

Future trends in sustainable manufacturing are explored, including artificial intelligence-driven decision-making, green energy adoption, and innovative material science developments. These advancements position LCE and the 5S framework as essential tools for addressing modern manufacturing challenges, ensuring competitive, sustainable, and socially responsible systems.

DOI: 10.1201/9781003501619-5

5.1 Introduction

A manufacturing process must combine raw materials, a manufacturing transformation process of the material, manufacturing resources (machine, equipment, energy, water, consumables), operators, and a control system to control, monitor, and execute the process to produce a product/component/part. It must also be connected and integrated to manufacturing systems (facility/production lines) and the supply chain to obtain products under a 5S approach (social, sustainable, sensing, smart, and safe). The 5S approach includes value efficiency, productivity, ethical practices, environmental responsibility, safety, and technology integration.

In an era marked by significant environmental challenges, sustainable manufacturing is a critical response to the global need for more environmentally conscious industrial practices. Industries are increasingly driven to adopt sustainable manufacturing by evolving environmental regulations, consumer preferences for eco-friendly products, and the global commitment to sustainable development goals. These pressures encourage industries to innovate and transform traditional manufacturing processes to meet modern demands for sustainability and efficiency.

International initiatives such as the United Nations Sustainable Development Goals (SDGs), particularly Goals 9 (Industry, Innovation, and Infrastructure) and 12 (Responsible Consumption and Production), underline the importance of integrating sustainability into manufacturing processes. These goals advocate for building resilient infrastructure, promoting inclusive and sustainable industrialization, and fostering innovation. Moreover, businesses face an evolving regulatory landscape where compliance with environmental standards is not just about good stewardship but is crucial for market access and competitiveness.

The selection, development, and planning process in manufacturing ensures that products are designed, produced, and launched efficiently and sustainably. This process begins with ideation that feedback on product design and material selection, with attention to customer needs and environmental impact. The development phase involves selecting appropriate manufacturing processes and planning production in detail, considering resource efficiency and cost management factors. The advanced development focuses on fine-tuning these plans to achieve quality and optimize time, cost, volume, and flexibility. Finally, the launching phase tests these plans against real-world criteria to ensure that production is viable at the desired scale. Each stage is augmented by digital technologies, which provide critical data and analytics to refine and optimize the manufacturing process.

It is essential to mention that the successful application of the 5S framework to manufacturing processes depends on applying the 5S approach to the other components of the manufacturing environment. For instance, the correct selection of raw materials will define the sustainability of the manufacturing process. If the raw materials have been made following a 5S approach, the manufacturing process can meet the criteria of 5S.

The development of a manufacturing process must be done using Life cycle Engineering; the details of this process are discussed in Chapter 2. In this sense, the stages of manufacturing process development can be seen in the following figure (FIGURE 5.1):

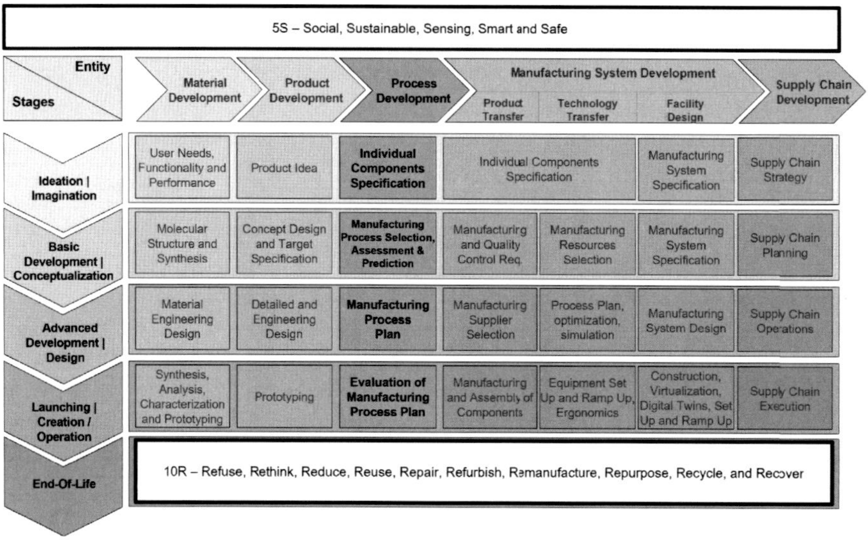

FIGURE 5.1
The 5S framework focuses on manufacturing process development using life cycle engineering.

The characteristics and concepts for a manufacturing process to meet the 5S framework are explained in the next sections. The application of Life cycle Engineering is then described to achieve a 5S manufacturing process. Finally, digital enablers related to each stage of the manufacturing process development will be discussed.

5.2 Manufacturing Process as 5S System

Developing a manufacturing process using the 5S framework—focusing on Social, Sustainable, Sensing, Smart, and Safe—requires an integrated approach ensuring the manufacturing operations are efficient, innovative but also responsible, and sustainable. The following list explains how to embed each of these principles into manufacturing process development:

- **Social:**
  - ***Community Engagement:*** Actively engage with the communities where manufacturing facilities are located. Explain to the community the behavior of the manufacturing process in terms of energy, water, CO_2 emissions, and waste produced. This might include listening to community concerns about these issues and how the manufacturing process (facility location) contributes to local development projects. Ensure that the manufacturing process provides social benefits such as job creation.
  - ***Ethical Labor Practices:*** Ensure fair labor practices across the manufacturing process. This includes fair wages, reasonable working hours, safe working conditions, and respect for workers' rights.
  - ***Diversity and Inclusion:*** Promote a diverse workforce and create inclusive work environments. This enhances social equity and can drive innovation by incorporating a wide range of perspectives and ideas.
- **Sustainable:**
  - ***Resource Efficiency:*** Optimize the use of resources such as materials, water, energy, and consumables to reduce waste and emissions. Implement techniques like lean manufacturing to minimize waste and reduce costs. The usage of green technologies can enhance efficiency. The concept of Circular Economy is a must in the design of manufacturing processes.
  - ***Sustainable and Eco-Friendly Materials:*** Select materials that are sustainable, recyclable, or biodegradable to minimize environmental impact.
  - ***Life cycle Assessment:*** Conduct life cycle assessments to identify and mitigate the environmental impacts of products from design through disposal.
  - ***Recycling and Reuse:*** Feedback the design of products for disassembly and recycling, enabling materials to be reused and reducing landfill waste.
  - ***Waste Management:*** Implement strategies to reduce, reuse, and recycle waste generated during manufacturing processes.
  - ***Renewable Energy:*** Utilize renewable energy sources, such as solar and wind, to power manufacturing operations and reduce carbon footprint.
  - ***Pollution Prevention:*** Implement clean technologies and processes that reduce pollution and minimize environmental impact.

- ***Emission Reduction:*** Aim to reduce emissions through cleaner production techniques and the adoption of technologies that limit environmental impact, such as filtration systems and renewable energy sources.

- **Sensing:**
  - ***Data Collection:*** Incorporate sensors throughout the manufacturing process to collect data on machine efficiency, energy consumption, and production quality. This data is vital for monitoring performance and driving continuous improvement.
  - ***Process Optimization:*** Continuously analyze data from sensors to optimize the manufacturing process, adjusting parameters in real time to improve quality and efficiency.
  - ***Predictive Analytics:*** Use data analytics and machine learning to predict maintenance needs and optimize production schedules, reducing downtime and costs.
  - ***Environmental Sensors:*** Monitor emissions, energy use, and waste to ensure compliance with environmental regulations and improve sustainability metrics.
  - ***IoT and Real-Time Monitoring:*** Deploy IoT sensors to monitor equipment performance, environmental conditions, and product quality in real-time.
  - ***Predictive Maintenance:*** Use sensor data to predict when machines and equipment will need maintenance before breakdowns occur, reducing downtime and increasing productivity.

- **Smart:**
  - ***Automation and Robotics:*** Implement automation technologies and robotics to enhance precision, reduce errors, and increase efficiency in manufacturing processes. Robotics and automated systems can handle repetitive tasks, improve production rates, and reduce human error.
  - ***Artificial Intelligence (AI):*** Utilize AI to analyze data, optimize production, and improve decision-making processes. Leverage AI to analyze data from the manufacturing process to make informed decisions, enhance quality control, and predict trends and outcomes.
  - ***Integration:*** Ensure that smart technologies are well integrated, allowing for seamless communication between different stages of the manufacturing process and among various departments and areas of the manufacturing facility.
  - ***Digital Twins:*** Create digital twins of manufacturing processes to simulate and optimize operations, allowing for proactive problem-solving and innovation.

- ***Safety:***
  - ***Safety Protocols:*** Develop and enforce comprehensive safety protocols to protect workers and reduce the risk of accidents. Use technology to enhance safety measures, such as automatic machine shut-offs and real-time monitoring systems for hazardous conditions.
  - ***Risk Management:*** Regularly conduct risk assessments to identify potential safety hazards. Implement proactive measures to mitigate these risks.
  - ***Real-Time Hazard Detection:*** Use sensors and IoT devices to detect potential hazards and trigger alerts for immediate action.
  - ***Emergency Preparedness:*** Establish and practice emergency procedures, including evacuation plans, safety drills, and the proper use of safety equipment.
  - ***Training and Education:*** Provide employees with regular training on safety practices and the use of advanced technologies to ensure a safe working environment.
  - ***Safe Disposal:*** Implement strategies for the safe disposal of non-recyclable components to prevent environmental harm.

To effectively integrate LCE with social sustainability, sensing technologies, smart systems, and safety, manufacturers should adopt a holistic approach that focuses on continuous improvement and stakeholder engagement:

- **Holistic Product Development:** Incorporate sustainability, social responsi-
  bility, and safety considerations into every phase of product development and manufacturing.
- **Stakeholder Collaboration:** Engage stakeholders, including employees, suppliers, customers, and communities, in the development and implementation of sustainable practices.
- **Continuous Improvement:** Use data-driven insights and feedback loops to continuously refine processes, adapt to changing conditions, and drive innovation.

The benefits for undertaking this approach for the development of a 5S Manufacturing Process are:

- **Environmental Sustainability:** Reduce waste and emissions through efficient resource use and sustainable practices.
- **Economic Efficiency:** Optimize processes to reduce costs and increase productivity, leading to enhanced competitiveness.

- **Enhanced Safety and Social Responsibility:** Improve worker safety and contribute positively to society, building trust and reputation.
- **Innovation and Adaptability:** Foster innovation through the integration of advanced technologies, enabling adaptability to market changes and future challenges.

By systematically applying the 5S principles to manufacturing process development, businesses can ensure that their operations are not only efficient and innovative but also align with broader social, environmental, and ethical standards. This holistic approach enhances the sustainability and safety of manufacturing processes, improves social impact, and leverages modern technology to meet contemporary challenges. A 5S manufacturing process is conceived and depicted in FIGURE 5.2:

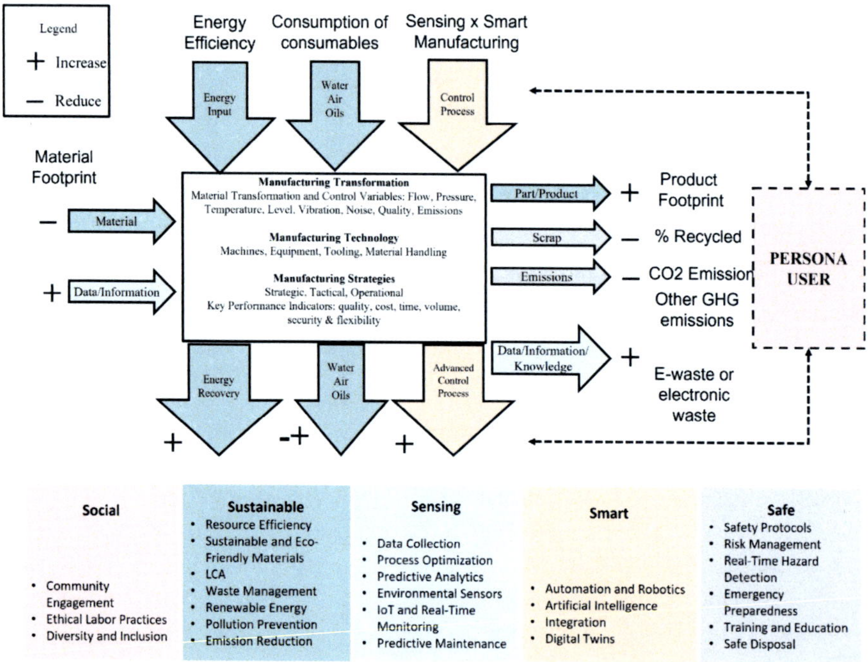

FIGURE 5.2
5S process development (social, sustainable, sensing, smart and safe).

1. A materials transformation process that is controlled by different variables: flow, pressure, temperature, levels and volumes, vibration, noise, quality specifications and emissions.
2. The materials is transformed into products using manufacturing technologies (machines, equipment, tools, and material handling equipment) and operators.

3. In the transformation process scrap is produced, emissions are generated, and data and information are required and generated to control the process. The objective of the process is to reduce product footprint, reduce waste, CO_2 and GHG emissions, and electronic waste (e-waste). It is also important to increase the capacity to recycle and reuse materials and scrap.
4. The manufacturing process also requires resources (energy, water, air, oils, consumables) and data/information/knowledge to control and execute the process. Ideally the process must reduce and recover energy, reduce the consumption of water, air, oils, and consumables. The control process should be able to achieve a sensing and smart manufacturing process that is optimal, effective, and efficient.
5. The manufacturing process should be centered in the person (operator) that monitors and controls the execution of the process. The process should be harmless, safe and secure for the operator.

5.3 Life Cycle Engineering Applied to Develop 5S Manufacturing Process Development

Manufacturers can create economically viable, environmentally friendly, socially responsible systems by applying a life cycle engineering approach to developing social, sustainable, sensing, smart, and safe manufacturing processes. This comprehensive strategy ensures long-term success and resilience in a rapidly changing world. Life Cycle Engineering optimizes each product's life cycle phase to minimize environmental impact and enhance social and economic benefits. The key phases include ideation, basic development, advanced development, launching and operation, and end-of-life. Let's review each phase of the 5S Manufacturing Process development in more detail using Life cycle Engineering.

5.3.1 Ideation: Individual Component Specifications

During the Ideation or Imagination stage, the initial design and specifications of each product/component/part are critically examined through the lens of 5S framework and Life Cycle Engineering. This early integration of LCE principles ensures that sustainability is a core consideration from the outset, influencing material selection and overall life cycle impacts. Some important considerations are described below:

- **Alignment of Customer Requirements with Sustainability Goals:** Specifications related to geometry, material, and production rates are optimized to reduce environmental impact while meeting customer needs.

- **Life cycle Analysis for Each Component:** Evaluate the environmental impacts of materials and processes intended for each component to ensure sustainable choices are made early in the design phase.
- **Sustainable Design:** Incorporate eco-design principles to define how the different products/components can be produced using an energy-efficient process, zero water usage, and reduced waste and emissions. This includes feedback on selecting sustainable materials and considering the entire product life cycle.
- **Socially Responsible Design:** Feedback on product design that considers user accessibility and safety, ensuring they meet diverse needs and improve the quality of life.
- **Sustainable Project Planning:** The project scope and plan are developed with clear sustainability targets, considering the entire life cycle of the product.

The inputs and outputs of this stage are described in TABLE 5.1. Furthermore, a visual explanation is depicted in FIGURE 5.3:

TABLE 5.1
Inputs and Outputs During the Ideation—Imagination Stage

	Inputs		Outputs
1.	Initial Design Concepts: A digital prototype from product development is required to evaluate its manufacturability.	1.	Individual Component Specifications: Detailed specifications for each product component derived from the initial designs.
2.	Product/Component specification: Detailed information about geometry, tolerances, surface finish, morphology, material, and product quality.	2.	BOM (Bill of Materials): BOM of the product to be manufactured with the specifications of geometry, tolerances, surface finish, materials, volume, quality and costs.
3.	CAD product models & 2D drawings	3.	Material Selection Decisions: Decisions on the types of materials to be used that align with sustainability criteria.
		4.	Project Scope and Plan: A defined project scope and a comprehensive plan that outlines the sustainability targets, timelines, and resources required.

5.3.2 Basic Development: Manufacturing Process Selection

In the Basic Development (Manufacturing process selection), decisions are made regarding the most appropriate manufacturing processes for each

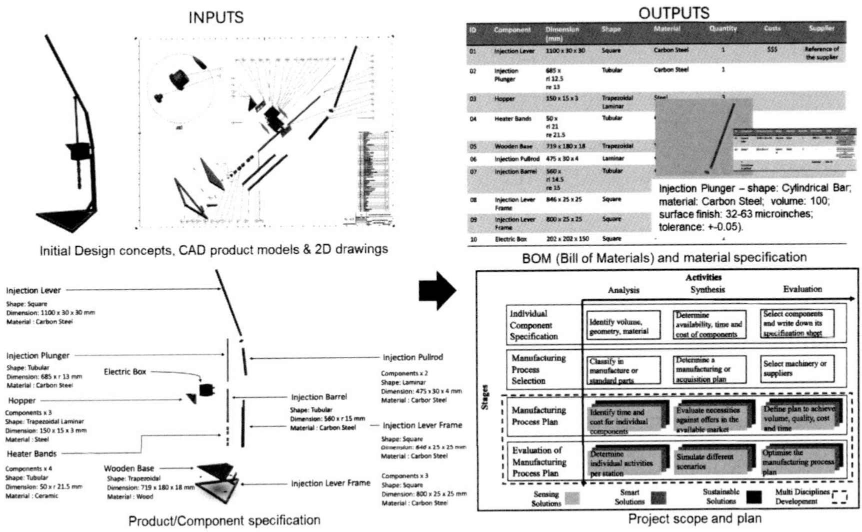

FIGURE 5.3

Inputs and outputs of ideation: Individual component specifications.

product/component/part. This stage is crucial for selecting process that align with LCE objectives, thereby optimizing resource use and minimizing waste and emissions. This stage must be done simultaneously/concurrently with the material and product development stage from a life cycle process. The following are major considerations for this stage:

- **Sustainable or Eco-friendly Materials:** Prioritize the use of sustainable, reusable, or recyclable materials to enhance sustainability.
- **Choosing Eco-friendly Manufacturing Processes:** Select processes that minimize energy consumption and reduce waste production, integrating technologies that support resource conservation.
- **Component Classification Based on Life cycle Impacts:** Components are classified by their function and environmental footprint, guiding decisions on whether they should be standardized or custom-made.

The inputs and outputs of this stage are described in TABLE 5.2. Furthermore, a visual explanation is depicted in FIGURE 5.4.

In practice, engineers might use a combination of Group Technology, PRIMA matrices, and process selection matrices or software to optimize the manufacturing process selection:

1. Group Technology helps organize the manufacturing process by categorizing parts into families, reducing complexity.
2. The PRIMA Matrix maps product design requirements to potential manufacturing processes, offering a clear visual representation.

TABLE 5.2
Inputs and Outputs during the Basic Development/Conceptualization Stage

Inputs	Outputs
1. Individual Component Specifications: Detailed specifications for each product component derived from the initial designs.	1. Classification and coding of components in BOM using Group Technology (e.g., OPITZ coding system, FIGURE 5.5)
2. BOM (Bill of Materials): BOM of the product to be manufactured with the specifications of geometry, tolerances, surface finish, materials, volume, quality and costs.	2. PRIMA Matrix for manufacturing process selection: Decisions on the types of manufacturing process to be used for each component in BOM that align with the materials and quantities using the PRIMA selection matrix (FIGURE 5.6).
	3. Matrices for manufacturing process selection: Decisions on the types of manufacturing process to be used that align with type of materials, shape classification, mass (weight), thickness, tolerances, surface roughness, and economic batch size (FIGURE 5.7).
	4. Software for manufacturing process selection: Decisions on the types of manufacturing process to be used that align with type of materials, shape, surface finish, tolerance, maximum wall thickness, quantity, and lead time (https://www.custompartnet.com/process-selector) (FIGURE 5.8).
	5. Final manufacturing process selection by defining preparation process, shaping process and secondary process (FIGURE 5.9).

3. Process Selection Matrix refines the selection by evaluating specific criteria for each potential process. This matrix considers various factors, including shape, material properties, cost, lead time, and process capabilities. Final process selection determines the preparation process (if needed), shaping process, and secondary process (if needed because of tolerances or roughness) with feature recognition.
4. Process Selector Software streamlines the entire process, allowing for quicker and more informed decision-making. Some popular process selection software includes CES Selector by Granta Design, PRO-SELECTOR, and other custom-built decision support systems used in the industry.

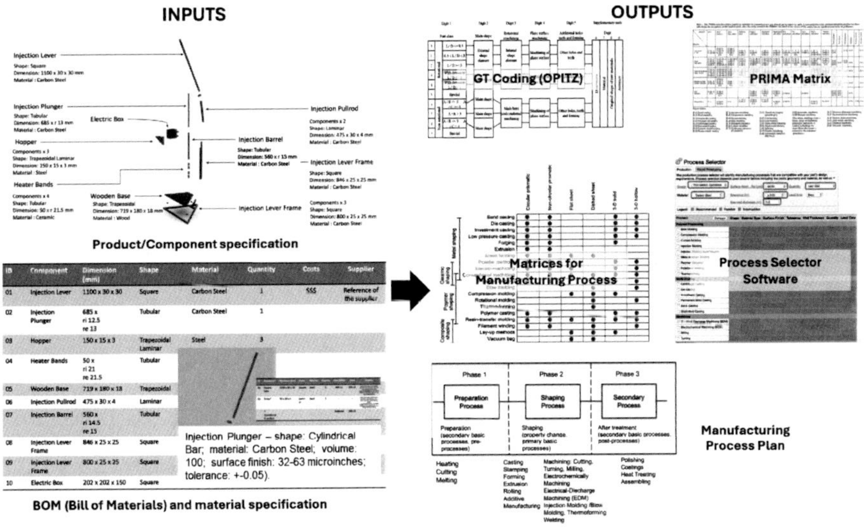

FIGURE 5.4
Inputs and outputs of basic development: Manufacturing process selection.

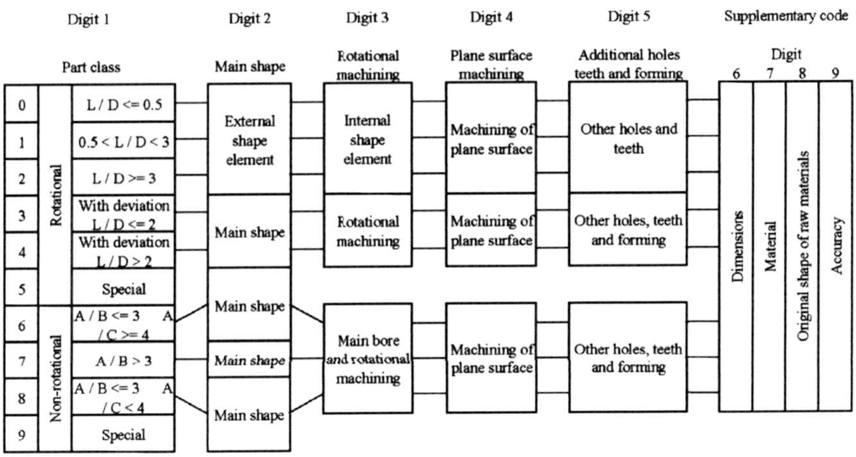

FIGURE 5.5
Opitz coding and classification system.

By leveraging these tools, manufacturers can achieve a more efficient, cost-effective, and streamlined production process that aligns with their design and business objectives. The selection of a manufacturing process is very important because it defines the fundamental process of transformation of the material and determines manufacturing technology (equipment/machines, tooling, auxiliary equipment) that must be used to achieve a sustainable and optimal production process (FIGURE 5.9).

Note - The PRIMA selection matrix cannot be regarded as comprehensive and should not be taken as such. It represents the main common industrial practice but there will always be exceptions at this level of detail. Also, the order in which the PRIMAs are listed in the nodes of the matrix has no significance in terms of preference.

Material / Quantity	Irons	Steel (carbon)	Steel (tool, alloy)	Stainlees steel	Copper and alloys	Aluminium and alloys	Magnesium and alloys	Zinc and alloys	Tin and alloys	Lead and alloys	Nickel and alloys	Titanium and alloys	Thermoplastics	Thermosets	FR composites	Ceramics	Reactive metals	Precious metals
Very small 1- 100	[1.5][1.6] [1.7][4.M]	[1.5][1.7] [3.6][4.M] [5.1][5.5] [5.6]	[1.1][1.7] [3.6][4.M] [5.1][5.5] [5.6]	[1.7][3.6] [4.M][5.1] [5.5][5.6]	[1.5][1.7] [3.6][4.M] [5.1]	[1.5][1.7] [3.6][4.M] [5.1][5.5]	[1.6][1.7] [3.6][4.M] [5.1][5.5]	[1.1][1.7] [3.6][4.M] [5.5]	[1.1][1.7] [3.6][4.M] [5.5]	[1.1][3.6] [4.M][5.5]	[1.5][1.7] [3.6][4.M] [5.1][5.5][5.6]	[1.1][1.6] [4.M][5.1] [5.5][5.6]	[2.3] [2.5]		[2.6]	[5.6]		[5.5]
Small 100- 1000	[1.2][1.3] [1.5][1.6] [1.7][4.M] [5.3][5.4]	[1.2][1.3][1.5] [1.7][3.6] [4.M][5.1] [5.3][5.4][5.5]	[1.1][1.7][3.6] [4.M][5.1] [5.3][5.4] [5.5][5.6]	[1.2][1.7] [3.6][4.M] [5.1][5.3] [5.4][5.5]	[1.2][1.3][1.5] [1.7][1.8][3.3] [3.6][4.M][5.1] [5.3][5.4]	[1.2][1.3][1.5] [1.7][1.8][3.6] [4.M][5.3] [5.4][5.5]	[1.3][1.6] [1.7][1.8] [3.6][4.M] [5.5]	[1.1][1.3] [1.7][1.8] [3.6][4.M] [5.5]	[1.1][1.3] [1.7][1.8] [3.6][4.M] [5.5]	[1.1][1.3] [1.8][3.6] [4.M][5.5]	[1.2][1.3][1.5] [1.7][3.6] [4.M][5.1] [5.3][5.4][5.5]	[1.1][1.6] [4.M][5.3] [5.4][5.5] [5.6]	[2.2] [2.3] [2.5]	[2.2]	[2.2] [2.6]	[5.6]		[5.5]
Small to medium 1000- 10000	[1.2][1.3] [1.5][1.6] [1.7][3.7] [4.A][5.2]	[1.2][1.3][1.5] [1.7][3.1][3.2] [3.7][4.A][5.2] [5.3][5.4][5.5]	[1.2][1.7][3.1] [3.2][3.7][4.A] [5.2][5.3] [5.4][5.5]	[1.2][1.7][3.1] [3.2][3.7][4.A] [5.2][5.3] [5.4][5.5]	[1.2][1.3][1.4] [1.5][1.8][3.1] [3.2][3.7][4.A] [5.2][5.3][5.4]	[1.2][1.3][1.4] [1.5][1.8][3.1] [3.2][3.7][4.A] [5.3][5.4][5.5]	[1.3][1.4] [1.6][1.8] [3.1][3.2] [4.A][5.5]	[1.3][1.4] [1.8][3.2] [4.A][5.5]	[1.3][1.4] [3.2]	[1.3][1.4] [3.2]	[1.2][1.3][1.5] [1.7][3.1][3.2] [3.7][4.A][5.2] [5.3][5.4][5.5]	[3.1][3.7] [4.A][5.2] [5.3][5.4] [5.5]	[2.1] [2.2] [2.3] [2.4]	[2.2]	[2.2]	[5.2]		[5.5]
Medium to high 10000- 100000	[1.2][1.3] [3.7][4.A]	[3.1][3.2] [3.3][3.7] [3.8][4.A] [5.5]	[3.2][3.3] [3.8][4.A] [5.2]	[3.1][3.2] [3.3][3.7] [3.8][4.A]	[1.2][1.4] [3.1][3.2] [3.3][3.7] [3.8][4.A]	[1.2][1.3][1.4] [3.1][3.2][3.3] [3.7][3.8] [4.A][5.5]	[1.3][1.4] [3.1][3.2] [3.3][3.8] [4.A]	[1.4][3.2] [3.8][4.A]	[1.4][3.2] [3.8]	[1.4][3.2] [3.3][3.8] [4.A]	[3.3][3.7] [3.8][4.A] [5.2][5.5]	[3.7][3.8] [4.A][5.2] [5.5]	[2.1] [2.3] [2.4] [2.7]	[2.1] [2.2] [2.7]	[2.2]	[3.7]		[3.3]
High 100000+	[1.2][1.3] [3.7]	[3.1][3.2] [3.3][3.8] [4.A]			[1.2][3.2] [3.3][3.7] [3.8][4.A]	[1.2][1.3] [1.4][3.2] [3.3][3.8] [4.A]	[1.3][1.4] [3.1][3.8] [4.A]	[1.4][3.2]		[1.4][3.2]			[2.1] [2.4] [2.7]	[2.1] [2.2] [2.7]		[3.7]		
All quantities	[1.1]	[1.1][1.6] [3.4][3.5]	[1.6]	[1.1][1.6] [3.4][3.5]	[1.1][1.6] [3.4][3.5] [5.5]	[1.1][1.6] [3.4][3.5]	[1.1][3.4] [3.5]	[3.4][3.5]			[1.1][1.6] [3.4][3.5]	[3.4] [3.5]				[5.5]	[1.6]	[1.6]

Key to matrix:

[1.1] Sand casting
[1.2] Shell moulding
[1.3] Gravity die casting
[1.4] Pressure die casting
[1.5] Centrifugal casting
[1.6] Investment casting
[1.7] Ceramic mould casting
[1.8] Plaster mould casting

[2.1] Injection moulding
[2.2] Compression moulding
[2.3] Vacuum forming
[2.4] Blow moulding
[2.5] Rotational moulding
[2.6] Contact moulding
[2.7] Continuous extrusion (PLASTICS)

[3.1] Closed die forging/ upset forging
[3.2] Cold forming
[3.3] Cold heading
[3.4] Sheet metal shearing
[3.5] Sheet metal forming
[3.6] Spinning
[3.7] Powder metallurgy
[3.8] Continuous extrusion (METALS)

[4.A] Automatic machining
[4.M] Manual machining
(The above headings cover a broad range of machining processes and levels of control technology. For more detail, the reader is referred to the individual processes.)

[5.1] Electrical discharge machining
[5.2] Electrochemical machining
[5.3] Electron beam machining
[5.4] Laser beam machining
[5.5] Chemical machining
[5.6] Ultrasonic machining

FIGURE 5.6
The PRIMA selection matrix [134].

Digital technologies play a pivotal role in modernizing and enhancing the manufacturing process selection when using methodologies like Group Technology (GT), the PRIMA matrix, process selection matrices, and process selector software. Here is how various digital technologies are integrated into these methods:

1. **Digital Technologies in Group Technology (GT)**
   (a) ***Data Analytics and Machine Learning:***
   - *Automated Part Classification:* Advanced data analytics and machine learning algorithms can analyze large datasets of parts and automatically classify them into families based on similarities in design, material, and manufacturing processes. These algorithms can process vast amounts of data more quickly and accurately than manual methods.
   - *Pattern Recognition:* Digital tools equipped with pattern recognition capabilities can identify common features among parts, such as geometric shapes or material properties, that may not be easily noticeable through manual inspection.

   (b) ***CAD/CAM Integration:***
   - *Feature Recognition:* Integration with CAD (Computer-Aided Design) and CAM (Computer-Aided Manufacturing) systems allows for automatic extraction of part

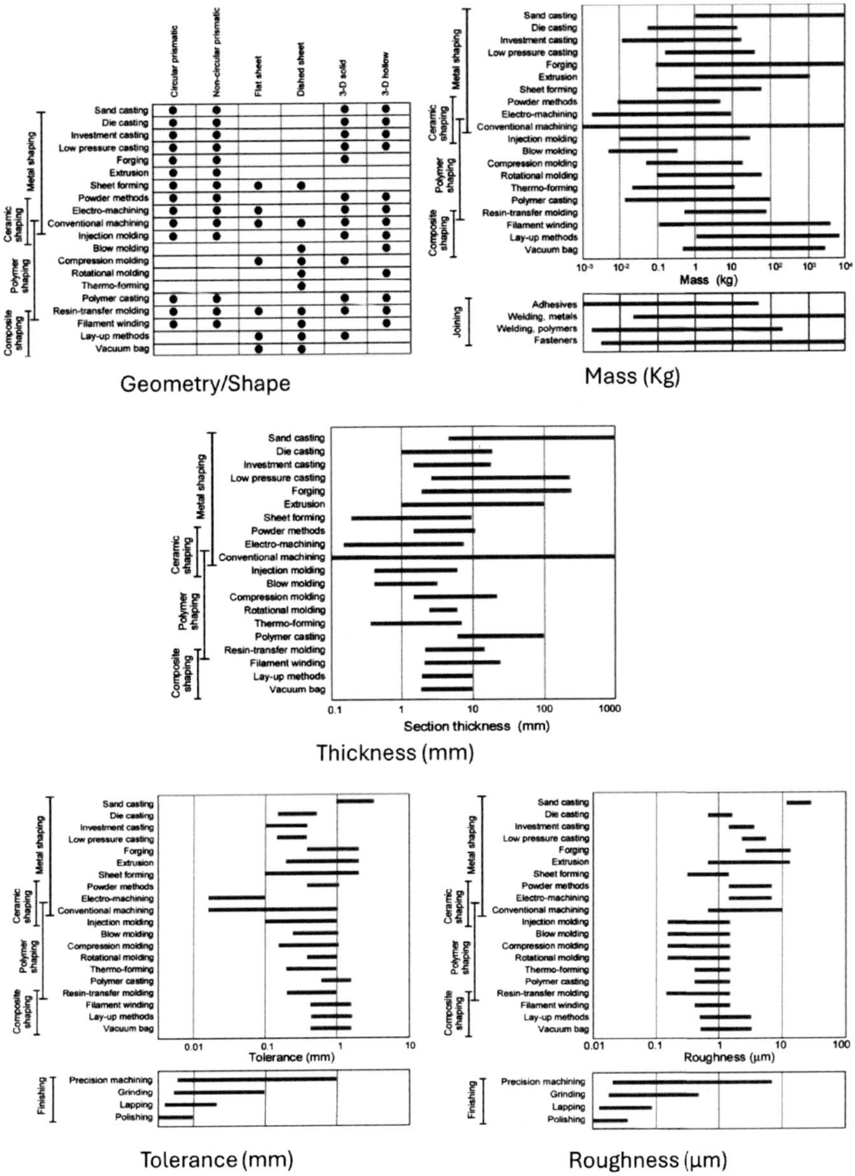

FIGURE 5.7
Manufacturing process selection using matrices. A red dot or line indicates that the process is compatible (Adapted from [10]).

features directly from digital models, facilitating the grouping of parts in GT. This integration also supports automated process planning based on the recognized features.

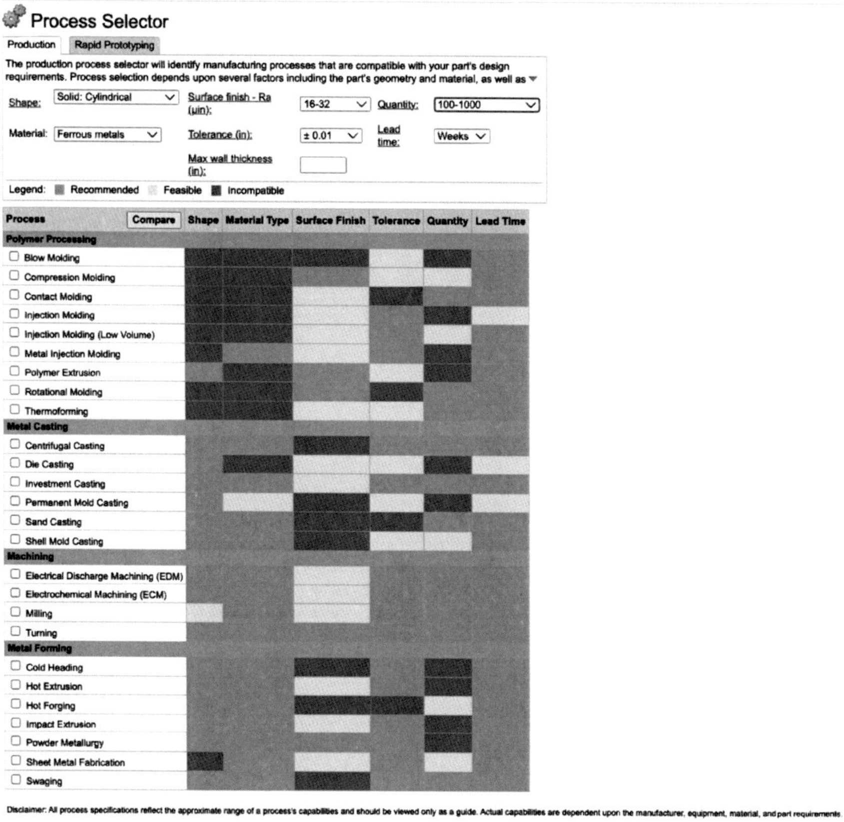

FIGURE 5.8
An example of the manufacturing process selection using process selector software (https://www.custompartnet.com/process-selector).

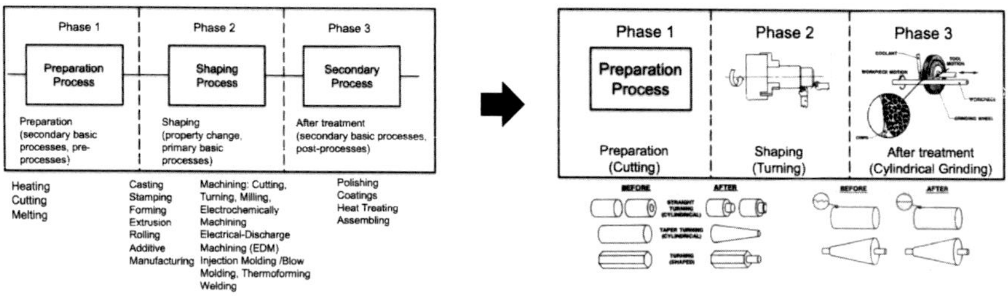

FIGURE 5.9
Final manufacturing process selection describing preparation process, shaping process and secondary process for a rotational component (Injection Plunger).

(c) ***Digital Twin Technology:***

- *Simulation of Manufacturing Cells:* Digital twins of manufacturing cells or entire production lines can be used to simulate the production process for different part families, optimizing cell design and operation before physical implementation.

2. **Digital Technologies in PRIMA Matrix (Process Information Map)**

   (a) ***Digital Mapping Tools:***

   - *Interactive PRIMA Matrix:* Digital platforms allow for the creation of interactive PRIMA matrices where design requirements and manufacturing processes can be mapped in real-time. These tools often include drag-and-drop interfaces, making it easier to update and visualize the matrix.

   (b) ***AI and Machine Learning:***

   - *Automated Mapping:* AI-driven tools can automate the mapping process by suggesting optimal manufacturing processes based on the input design requirements. These tools learn from historical data to continuously improve their recommendations.
   - *Data-Driven Decision Making:* Machine learning models can analyze past process performance data to predict how well different manufacturing processes will meet the specified design requirements, enabling more informed decision-making.

   (c) ***Cloud Computing:***

   - *Collaborative Tools:* Cloud-based PRIMA matrix tools enable teams across different locations to collaborate in real-time, ensuring that everyone has access to the most up-to-date information and can contribute to the process selection.

3. **Digital Technologies in Process Selection Matrix**

   (a) ***Decision Support Systems (DSS):***

   - *Multi-Criteria Analysis Software:* Digital decision support systems incorporate advanced multi-criteria decision-making algorithms (such as Analytic Hierarchy Process, AHP) to evaluate and rank manufacturing processes based on various factors like cost, time, and material compatibility.

   (b) ***Simulation Software:***

   - *Process Simulation:* Before finalizing the process selection, simulation software can be used to model and predict the outcomes of different manufacturing processes. These tools help in understanding the implications of each process choice on factors like production time, quality, and cost.

(c) ***IoT and Real-Time Data Integration:***

- Dynamic *Matrices:* IoT (Internet of Things) sensors and devices can provide real-time data from the manufacturing floor, which can be integrated into the process selection matrix. This enables dynamic adjustments to the process selection criteria based on actual performance data.

4. **Digital Technologies in Process Selector Software**

(a) ***AI-Driven Process Selection:***

- *Intelligent Process Selector:* AI algorithms embedded in process selector software can analyze extensive datasets, including material properties, manufacturing process capabilities, and historical performance data, to recommend the most suitable processes. These recommendations can adapt over time as more data is collected.

(b) ***Cloud-Based Platforms:***

- *Scalability and Access:* Cloud-based process selector software offers scalability and easy access from anywhere, allowing manufacturers to utilize powerful computational resources without the need for significant on-premises infrastructure.
- *Data Sharing and Collaboration:* Cloud platforms enable seamless data sharing and collaboration among teams, ensuring that process selection is based on the most comprehensive and up-to-date information available.

(c) ***Digital Twins and Virtual Prototyping:***

- *Pr ocess Simulation and Optimization:* Digital twins of manufacturing processes allow for virtual prototyping, where different process scenarios can be tested and optimized in a virtual environment before physical implementation. This reduces the risk and cost associated with process selection.

(d) ***Integration with PLM and ERP Systems:***

- *End-to-End Integration:* Process selector software can be integrated with Product Life cycle Management (PLM) and Enterprise Resource Planning (ERP) systems to ensure that process selection is aligned with broader business goals and that decisions are consistent across the product development life cycle.

AI can significantly enhance the manufacturing process selection by integrating with methodologies such as Group Technology (GT), the PRIMA matrix, process selection matrices, and specialized software. Here's how AI can be utilized in each of these areas:

1. **AI in GT**

   (a) ***Classification and Clustering:***

   - *Automated Part Family Identification:* AI, particularly machine learning algorithms, can be used to analyze large datasets of parts and components to automatically classify them into families based on design and manufacturing similarities. This can include using unsupervised learning techniques like clustering to identify patterns in the data that might not be immediately apparent.
   - *Pattern Recognition:* AI can be trained to recognize patterns in part features (e.g., geometry, material properties) that are indicative of specific manufacturing processes, thereby streamlining the GT process.

   (b) ***Optimization:***

   - *Production Flow Optimization:* AI can optimize the layout and sequencing of manufacturing operations by analyzing the grouped parts and suggesting the most efficient manufacturing cells, reducing lead times and improving resource utilization.
   - *Predictive Analytics:* AI-driven predictive models can forecast the performance of different part families within specific manufacturing cells, helping in the optimal arrangement of parts and processes.

2. **AI in PRIMA Matrix**

   (a) ***Data Analysis and Mapping:***

   - *Automated Mapping:* AI can automate the creation of the PRIMA matrix by analyzing design requirements and matching them with the most suitable manufacturing processes based on historical data and industry standards. This reduces the time and effort required to manually fill out the matrix.
   - *Natural Language Processing (NLP):* NLP techniques can be used to analyze design requirements expressed in natural language and map them to suitable manufacturing processes. This is particularly useful for integrating design documents with process selection tools.
   - *Recommendation Systems:* AI-powered recommendation engines can suggest optimal processes by learning from

previous decisions and outcomes, continuously improving the accuracy and relevance of the process suggestions.

(b) ***Predictive Analytics:***

- *Predictive Capabilities:* AI can predict the performance of various manufacturing processes against specific design requirements by leveraging predictive modeling and simulations. This can include predicting the impact of process variables on product quality, cost, and production time.
- *Regression Models:* AI can use regression models to predict the outcomes of different manufacturing processes based on input parameters like material type, design complexity, and desired tolerances. These predictions help in filling out the PRIMA matrix with quantitative data.
- *Simulation-Based Learning:* Machine learning models can be trained on data generated from process simulations, enhancing the accuracy of the PRIMA matrix by providing insights into how different processes will perform under various conditions.

(c) ***Cloud Computing:***

3. **AI in Process Selection Matrix**

(a) ***Multi-Criteria Decision Making:***

- *Automated Scoring:* AI can automate the scoring process within a selection matrix by evaluating multiple criteria such as cost, material compatibility, and production volume. It can apply advanced decision-making algorithms (e.g., AHP, TOPSIS) to balance conflicting criteria and identify the best process.
- *Real-time Data Integration:* AI can integrate real-time data from manufacturing operations, market trends, and material availability to adjust the scores dynamically, ensuring that the process selection is always based on the most current information.
- *Dynamic Matrices:* AI systems can continuously update process selection matrices in real-time by integrating data from IoT devices, production monitoring systems, and supply chain data. This ensures that the selection process is always based on the most current information, adapting to changes in production conditions, material availability, and cost factors.

(b) ***Optimization Algorithms:***

- *Optimization of Process Selection:* AI-driven optimization algorithms (like genetic algorithms or particle swarm optimization) can be employed to search through the space of possible manufacturing processes, finding the most efficient and cost-effective solutions based on the specified criteria.

4. **AI in Process Selector Software**

(a) ***Intelligent Process Recommendation:***

- *AI-Driven Process Selector:* AI can enhance process selector software by incorporating machine learning models that are trained on vast amounts of data, including material properties, process capabilities, and historical production outcomes. This allows the software to provide more accurate and customized process recommendations.
- *Adaptive Learning:* AI can enable the software to learn from each decision, improving its recommendations over time by continuously refining the models based on feedback and new data.

(b) ***Integration with CAD/CAM Systems:***

- *Seamless Integration:* AI can facilitate the integration of process selection tools with CAD/CAM systems, allowing for real-time adjustments to the design and manufacturing process based on AI-generated insights. This integration ensures that the design is optimized for manufacturability from the outset.

(c) ***Simulation and Digital Twins:***

- *Process Simulation:* AI can be used in conjunction with digital twins to simulate the performance of selected manufacturing processes in a virtual environment. This allows engineers to test and refine their choices before actual production, reducing risks and costs.
- *Predictive Maintenance:* AI can also predict potential failures or inefficiencies in the selected manufacturing processes, suggesting proactive measures to avoid down-time and maintain production efficiency.
- *Digital Twin Integration:* Machine learning models can be embedded within digital twins of manufacturing processes to provide real-time feedback and predictive insights. This integration allows for continuous optimization of the manufacturing process selection, adapting to changes in production variables and environmental conditions.

(d) ***User Interaction and Customization:***

- *AI-Powered User Interfaces:* AI can enhance user interaction with process selector software through intelligent, user-friendly interfaces. For example, AI can use natural language processing to allow users to query the system using conversational language, making the software more accessible to non-experts.
- *Customization and Personalization:* Machine learning algorithms can personalize the process selection experience by adapting to the user's preferences and past decisions, ensuring that the recommendations align with the user's specific needs and objectives.

Digital Technologies and AI can significantly enhance the manufacturing process selection by automating and optimizing various stages, from classifying parts using Group Technology to refining process selection with advanced decision-making algorithms. By integrating AI with tools like the PRIMA matrix, process selection matrices, and process selector software, manufacturers can make more informed, efficient, and cost-effective decisions, ultimately improving product quality and production efficiency.

5.3.3 Advanced Development: Manufacturing Process Plan and Evaluation

Manufacturing process planning is a critical phase in the production life cycle where the sequence of operations and resources needed to manufacture a product are determined. The goal is to optimize efficiency, reduce costs, and ensure the product meets quality standards. The manufacturing process plan includes the following activities (FIGURE 5.10):

- **Input and Data Collection:** Import CAD model, design specifications and BOM (Bill of Materials), use Group Technology to understand the part/component/product, and use the manufacturing process selected to begin the planning of individual operations of the production process. Identify geometric features and manufacturing requirements.
- **Machine and Resource Assignment:** Assign machines, tools, and fixtures and manufacturing strategies (FIGURE 5.11)
- **Manufacturing Process Simulation and Optimization:** Simulate, analyze, and optimize the material transformation, generally using a CAM system.
- **Generation of Process Documentation:** Create detailed operation sheets, tool lists, and NC code.
- **Process Planning:** Planning using the transformation process to describe inputs and outputs of the process, with factors of production (materials, workforce, means, and information), type of production process

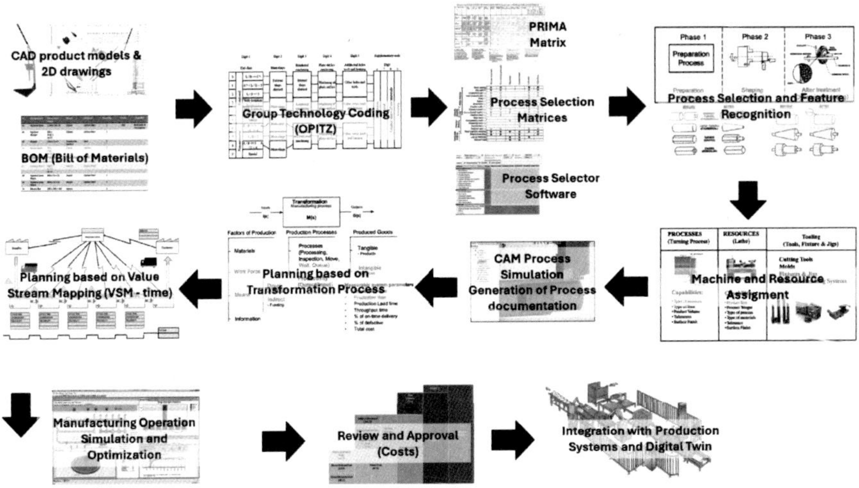

FIGURE 5.10
Workflow of manufacturing process planning and evaluation.

(transformation, inspection, moving, waiting, and queuing), production of goods (tangible and non-tangible) and performance measures (time, quality, volume, cost). See FIGURE 5.12.

- **Operation Planning:** Involves defining and sequencing manufacturing operations by applying Value Stream Mapping (VSM) to identify and streamline key metrics (processing time, production time, and lead time) to improve overall efficiency. See FIGURE 5.13.
- **Manufacturing Operation Simulation and Optimization:** Simulate, analyze, and optimize the plan.
- **Review and Approval:** Engineer review and stakeholder approval of the process plan (costs and time). The workflow from FIGURE 5.14 ensures that process planning is thorough, efficient, and aligned with design intent and production capabilities. FIGURE 5.15 shows a calculation of the cost proposal for the manufacturing process.
- **Integration with Production Systems:** Implement the plan in ERP/MES systems and Digital Twin.
- **Continuous Improvement and Maintenance:** Collect feedback and refine the process.
- **Reporting and Documentation:** Finalize documentation and generate reports.

The inputs and outputs from TABLE 5.3 depict the manufacturing process planning.

This process can be automated and integrating various stages using CAPP systems. CAPP systems significantly reduce the time and effort

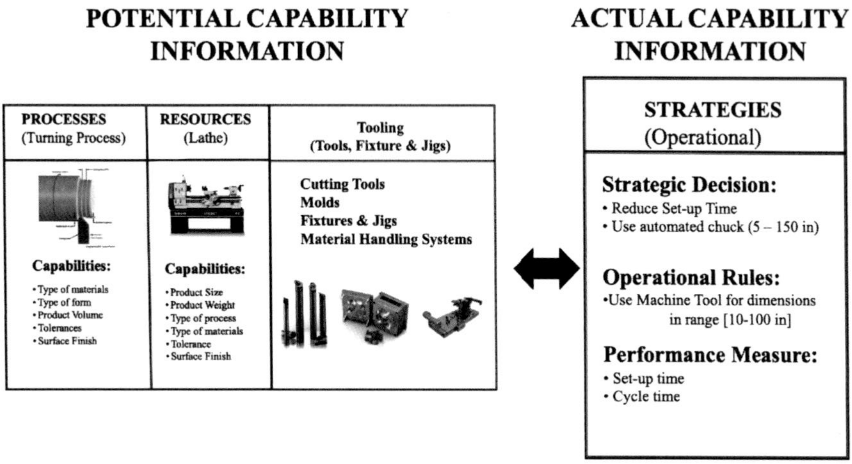

FIGURE 5.11
Machine, resource assignment, and manufacturing strategies.

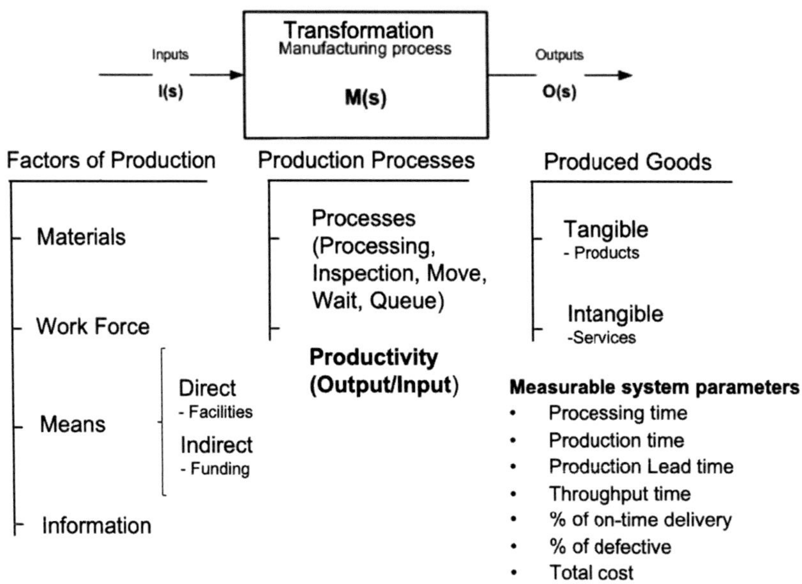

FIGURE 5.12
Planning using the transformation process (adapted from [60]).

required to develop high-quality manufacturing process plans. Computer-Aided Process Planning (CAPP) is classified into three primary types based on the methodologies and approaches used to generate process plans:

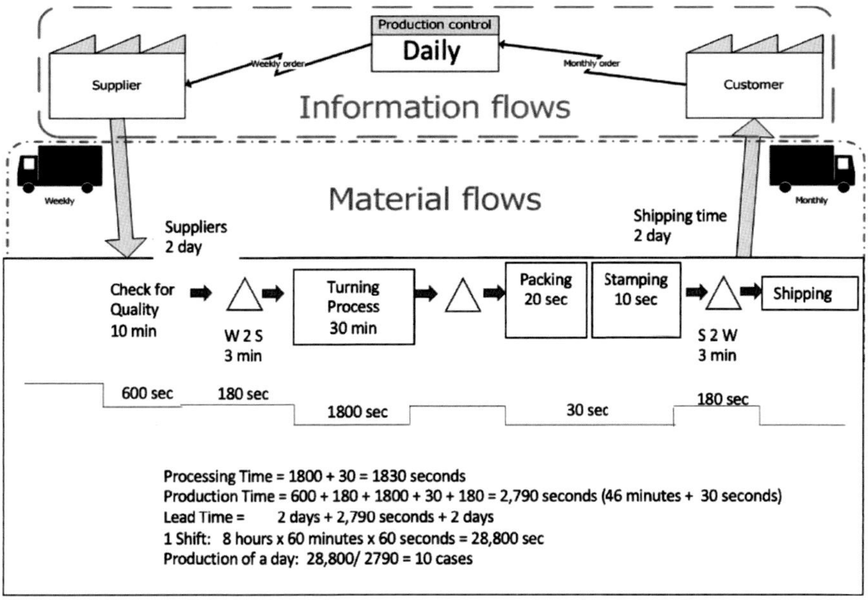

FIGURE 5.13
Operation planning using VSM (value stream mapping) to define operational times.

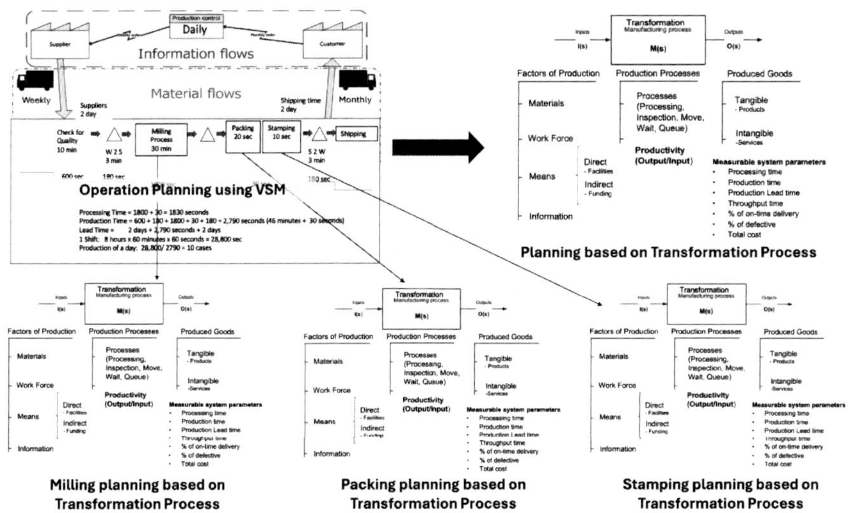

FIGURE 5.14
The complete operation plan has defined each operation with the specification to comply with the total manufacturing process.

1. **Variant CAPP:** Template-based, efficient for repetitive production, less flexible.

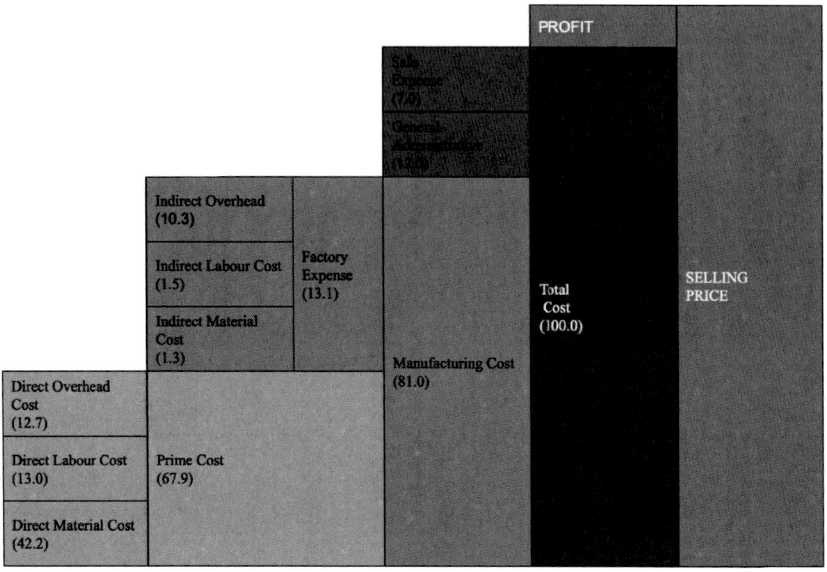

FIGURE 5.15
A proposal for the calculation of costs for the manufacturing process (adapted from [60]).

TABLE 5.3
Inputs and Outputs during the Advanced Development/Design Stage

Inputs	Outputs
• CAD Data (3D models, 2D drawings) • Material Information (type, standards) • Bill of Materials (material requirements, consumables) • Manufacturing Resources (machines, tools, fixtures) • Production Requirements (quantity, delivery time, quality) • Process Constraints and Rules (design, manufacturing, regulatory) • Historical Data (past process plans, performance data) • Knowledge Base (manufacturing knowledge, databases)	• Process Plan Documentation (VSM – Value Stream Mapping, operation sheets, routing sheets) • Machine, Tooling and Fixture Lists • NC Code (CNC machine programs) • Process Simulation Results (simulation reports, virtual prototyping data) • Cost and Time Estimates (cost, time) • Quality Control Plans (inspection procedures, testing requirements) • Production Schedules (job schedules, resource allocation) • Feedback and Improvement Reports (process performance, continuous improvement)

2. **Generative CAPP:** Rule-based, highly flexible, and automated, suitable for complex parts.
3. **Hybrid CAPP:** Combines the best of both Variant and Generative approaches, offering a balance between standardization and customization.

Each type of CAPP has its own use cases depending on the complexity of the manufacturing environment, the variety of parts produced, and the level of automation desired. Each type has its own strengths and is suitable for different manufacturing systems and requirements. CAPP systems efficiently convert design and manufacturing data into actionable process plans, ensuring that production is carried out effectively, within budget, and to the desired quality standards. Hybrid CAPP represents a modern approach that leverages the strengths of both Variant and Generative CAPP systems to meet the needs of diverse manufacturing operations.

5.3.3.1 Variant CAPP

Variant CAPP is based on the Group Technology (GT) principle, where parts with similar features and manufacturing processes are grouped into families. A standard process plan is developed for each family, which can be retrieved and adapted for new parts within the same family. The following are the key characteristics of a Variant CAPP:

- **Template-Based:** Process plans are created using predefined templates that correspond to specific part families. When a new part is introduced, the closest matching template is selected and modified to fit the new part's requirements.
- **Reliance on Human Expertise:** Engineers significantly modify the retrieved process plans to account for differences between parts.
- **Standardization:** Promotes standardization of process plans within part families, leading to consistent manufacturing practices.

The following is the workflow used in this type of systems:

- **Part Classification:** Parts are classified into families based on their geometric and manufacturing attributes using a coding and classification system.
- **Plan Retrieval:** A standard process plan for the relevant family is retrieved from the system.
- **Plan Modification:** Engineers modify the retrieved plan to meet the specific requirements of the new part.
- **Plan Documentation:** The modified process plan is documented and implemented in production.

The advantages of the variant approach are:

- **Efficiency in Repetitive Production:** Suitable for environments where similar parts are produced regularly, reducing the time and effort required to develop new process plans.
- **Consistency:** Ensures consistency in manufacturing practices across similar parts.

But there are also limitations:

- **Limited Flexibility:** Less effective for unique or highly complex parts that do not fit neatly into existing families.
- **Manual Intervention:** Significant manual intervention is required to modify process plans for new parts.

5.3.3.2 Generative CAPP

Generative CAPP systems automatically generate process plans from scratch using rules, algorithms, and a comprehensive knowledge base. This approach is more flexible and can handle a wide variety of parts and processes without relying on pre-existing templates. The most important characteristics of this type of CAPP system are:

- **Rule-Based Planning:** The system uses decision logic and rules to select processes, sequence operations, and assign resources based on the part's design and manufacturing requirements.
- **Knowledge-Based Systems:** Relies on a knowledge base containing information on materials, machines, tools, and processes.
- **Automation:** Minimizes the need for human intervention, allowing for the automatic generation of process plans.

The following are the core functions of the generative CAPP:

- **Feature Recognition:** The system automatically recognizes features from the CAD model.
- **Process Selection:** The system selects appropriate manufacturing processes/operations using predefined rules and algorithms based on recognized features.
- **Operation Sequencing:** The system determines the optimal sequence of operations, considering factors like tool availability, machine capabilities, and production constraints.
- **Plan Generation:** The system generates the complete process plan, including tool paths, machine setups, and work instructions.

The major advantages of this type of CAPP system are:

- **Flexibility:** It can handle various parts and processes, making it suitable for complex or customized manufacturing environments.

- **Reduced Lead Time:** Automatically generates process plans, significantly reducing the time required to move from design to production.
- **Consistency and Quality:** Ensures consistent process planning and reduces the likelihood of human errors.

However, the limitations must be taking into consideration when selecting generative CAPP system:

- **Complex Implementation:** Requires a comprehensive and well-maintained knowledge base, which can be time-consuming and costly to develop.
- **High Initial Cost:** The setup and maintenance of a generative CAPP system can be expensive.

5.3.3.3 Hybrid CAPP

Hybrid CAPP systems combine the strengths of both Variant and Generative approaches. They provide the flexibility of Generative CAPP while leveraging the efficiency and standardization of Variant CAPP. Hybrid systems allow for the retrieval of standard process plans (like in Variant CAPP) but also enable automatic generation or modification of plans using rule-based logic (like in Generative CAPP). The key characteristics of this type of system can be defined as:

- **Combines Variant and Generative:** Hybrid CAPP systems can retrieve standard plans from a library and then use generative techniques to modify or extend these plans based on the specific requirements of a new part.
- **Customization and Flexibility:** Offers the ability to adapt standard plans while also supporting the creation of new plans for parts that do not fit existing templates.
- **Scalability:** Can be scaled to fit different levels of process complexity and production volume.

The main functionalities of this system are:

- **Part Classification and Retrieval:** Similar to Variant CAPP, the system classifies the part and retrieves a standard process plan if one exists.
- **Generative Modification:** The retrieved plan is then automatically modified using generative techniques, or a new plan is generated if no suitable template exists.
- **Plan Finalization:** The system finalizes the process plan, which may involve further adjustments by engineers before it is implemented.

Key advantages can be declared as:

- **Versatility:** Suitable for various manufacturing environments, from highly standardized production to more flexible, custom manufacturing.

- **Optimized Efficiency:** Combines the quick retrieval of standard plans with the flexibility of generative methods, optimizing both speed and customization.
- **Balanced Automation:** Offers a balance between automation and manual control, allowing for efficient yet adaptable process planning.

Even this type of system represents the state of the art in CAPP systems there are still some limitations:

- **Complexity:** Integrating variant and generative approaches can make the system more complex to develop and maintain.
- **Potential for Over-Reliance:** If not managed carefully, there is a risk of over-reliance on the variant approach, which could limit the generative component's flexibility.

5.3.3.4 Digital Technologies for Computer-Aided Process Planning (CAPP)

Digital technologies are transforming Computer-Aided Process Planning (CAPP) systems by making them more intelligent, flexible, and integrated. These technologies enhance the ability of CAPP systems to automate, optimize, and manage complex manufacturing processes more effectively. TABLE 5.4 describes some key digital technologies used in CAPP systems.

Integrating these digital technologies into CAPP systems revolutionizes the way process planning is conducted, making it more dynamic, efficient, and adaptable to the evolving needs of modern manufacturing.

TABLE 5.4
Some Digital Technologies Employed during the Process Selection

Digital Technology	Application
Digital Twins	**Virtual Replicas:** Digital twins are virtual replicas of physical manufacturing processes, products, or systems. In CAPP, digital twins can simulate the entire manufacturing process in real-time, allowing for testing and optimization before production begins. Real-Time Monitoring: Digital twins enable real-time monitoring and adjustment of manufacturing processes. Any deviations or issues can be quickly identified and corrected by mirroring the physical process in a digital environment. **Predictive Maintenance and Optimization:** With the help of data from sensors and IoT devices, digital twins can predict potential failures or inefficiencies in the manufacturing process, enabling preemptive adjustments to the process plan.

Digital Technology	Application
	Automated Decision Making: AI and machine learning algorithms enhance CAPP by automating complex decision-making processes. They can analyze historical data, recognize patterns, and suggest optimized process plans with minimal human intervention.
	Predictive Analytics: Based on historical data, machine learning models can predict outcomes such as tool wear, machine failure, or quality issues. These predictions help in creating more robust and reliable process plans.
	Continuous Improvement: AI systems in CAPP can learn from past process plans and improve their accuracy and effectiveness over time. This leads to increasingly optimized manufacturing processes.
Industrial Internet of Things (IIoT)	**Data Collection and Analysis:** IIoT involves using interconnected sensors, devices, and systems that collect vast amounts of data from the manufacturing environment. This data is invaluable in CAPP for understanding machine performance, tool wear, environmental conditions, and other factors influencing process planning.
	Real-Time Feedback: IIoT enables real-time feedback loops where data from the production floor is continuously fed back into the CAPP system. This allows for immediate adjustments to the process plan, improving efficiency and reducing downtime.
	Enhanced Decision Making: Integrating IIoT with CAPP systems provides planners with detailed, real-time insights into the manufacturing process, leading to more informed and effective decision-making.
Cloud Computing	**Scalable Computing Power:** Cloud computing provides scalable computing resources to handle modern CAPP systems' complex calculations and data storage needs. This is particularly useful for running simulations, processing large datasets, and performing complex optimization tasks.
	Collaboration and Accessibility: Cloud-based CAPP systems enable collaboration across different teams and locations. Engineers and

Digital Technology	Application
	planners can access and modify process plans from anywhere, facilitating more efficient and coordinated planning. **Integration with Other Systems:** Cloud platforms facilitate the integration of CAPP with other enterprise systems such as ERP (Enterprise Resource Planning), MES (Manufacturing Execution Systems), and SCM (Supply Chain Management), ensuring seamless data flow and process coordination across the organization.
Virtual Reality (VR), Mixed Reality (MR), Extended Reality (XR)	**Immersive Process Visualization:** Planners can use XR technologies to experience a fully interactive, virtual replica of the manufacturing floor. VR allows users to "walk through" the factory, assess setups, and identify potential challenges, while MR combines virtual and physical elements for a more grounded view of process stages. This immersive visualization enables proactive adjustments to be made before production begins.
Augmented Reality (AR), Mixed Reality (MR)	**Training and Simulation:** AR and MR offer hands-on training for operators and engineers by overlaying digital information directly onto physical equipment. This allows learners to interact with real and virtual components, improving their understanding and shortening the learning curve for complex processes. SC and XR further enhance training by creating fully immersive simulations of equipment and process environments.
Spatial computing (SC)	**Real-Time, Context-Aware Guidance:** Through AR and SC, operators receive real-time, step-by-step guidance during the production process, with instructions and alerts displayed directly within their line of sight. This personalized support ensures precise adherence to process plans and immediate response to critical alerts, minimizing errors and improving overall process efficiency.
Big Data Analytics	**Data-Driven Process Planning:** Big data analytics enables the analysis of large datasets collected from various sources, such as production history, machine logs, and quality reports. This analysis helps identify trends, correlations, and insights that inform better process planning decisions.

Digital Technology	Application
	Optimization of Processes: By analyzing production data, big data tools can identify inefficiencies and areas for improvement in the manufacturing process. These insights can then optimize the process plan for better performance.
	Quality Control: Big data analytics can also monitor quality across the manufacturing process. It helps identify patterns that lead to defects, allowing adjustments to the process plan to reduce scrap and rework.
Additive Manufacturing (3D Printing)	**Process Planning for 3D Printing:** In CAPP systems, additive manufacturing requires specific planning tools. These tools must address unique considerations such as layer-by-layer construction, material usage, and post-processing steps like support removal and surface finishing.
	Customization and Prototyping: Additive manufacturing allows rapid prototyping and customization, and CAPP systems must accommodate these capabilities. Digital technologies in CAPP help streamline the process planning for custom and prototype parts, ensuring efficient production.
Blockchain Technology	**Supply Chain Transparency:** Blockchain can be integrated into CAPP systems to provide a secure and transparent record of every step in the manufacturing process. This is particularly important for industries with stringent regulatory requirements or where traceability is critical.
	Smart Contracts: Blockchain technology can automate contract execution between different stakeholders in the manufacturing process, ensuring that conditions such as quality standards and delivery timelines are met before payments are processed.
	Data Security: Blockchain provides a secure method for storing and sharing process planning data, ensuring that sensitive information is protected from unauthorized access or tampering.
Advanced Simulation Tools	**Finite Element Analysis (FEA):** FEA allows for the simulation of a product's behavior under various physical conditions (stress, heat, vibration). Integrating FEA into CAPP enables

Digital Technology	Application
	planners to anticipate potential issues and refine the process plan accordingly.
	Discrete Event Simulation: This type of simulation models the sequence of events in a manufacturing process, allowing planners to test different scenarios, optimize workflow, and identify bottlenecks before production begins.
	Process Optimization: Advanced simulation tools allow for the virtual testing of different process parameters, helping to identify the most efficient and cost-effective manufacturing strategies.
Cyber-Physical Systems (CPS)	**Integration of Physical and Digital Worlds:** CPS integrates physical manufacturing systems with digital networks, enabling real-time monitoring, control, and optimization of the manufacturing process. In CAPP, CPS allows for the dynamic adjustment of process plans based on real-time data from the shop floor.
	Smart Manufacturing: CPS facilitates smart manufacturing by enabling machines and systems to communicate, share data, and make decisions autonomously. This leads to more responsive and adaptable manufacturing processes.
	Enhanced Automation: CPS increases the level of automation in process planning and execution, reducing the need for manual intervention and enabling more complex and precise manufacturing operations.

5.3.3.5 AI and Machine Learning for CAPP

Artificial Intelligence (AI) and Machine Learning (ML) are increasingly integrated into CAPP systems to enhance their capabilities, making them more intelligent, adaptive, and efficient. The use of AI and ML in CAPP helps automate complex decision-making, optimize processes, and continuously improve process plans based on data-driven insights. Here's a detailed overview of how AI and ML are used in CAPP systems:

1. **Automated Feature Recognition**
   - ***Machine Learning Algorithms:*** AI-driven algorithms can automatically recognize and classify geometric features from CAD models, such as holes, slots, pockets, and contours. This feature recognition is crucial for determining the appropriate manufacturing processes and tools.

- ***Deep Learning for Complex Features:*** Deep learning models, particularly convolutional neural networks (CNNs), can handle complex and non-standard features that traditional methods may not easily recognize, enhancing the flexibility and robustness of the CAPP system.

2. **Intelligent Process Selection**
   - ***Expert Systems:*** AI-based expert systems use knowledge from human experts encoded into rules and logic to select the most suitable manufacturing processes for a given part. These systems consider material properties, feature geometry, tolerances, and production volume.
   - ***Adaptive Learning:*** AI can learn from previous process plans and outcomes to improve future process selection. For example, reinforcement learning algorithms optimize process selection by learning from past successes and failures.
   - ***Fuzzy Logic Systems:*** Fuzzy logic allows for more flexible decision-making in CAPP by handling uncertainties and imprecise data. This is particularly useful in selecting processes when exact data is not available or when there are multiple conflicting criteria.

3. **Optimizing Operation Sequencing**
   - ***Genetic Algorithms (GA):*** AI uses genetic algorithms to find the optimal sequence of manufacturing operations that minimize cycle time, tool changes, and machine setups. GAs simulates natural selection, iteratively improving the sequence to find the most efficient solution.
   - ***Simulated Annealing:*** This AI technique searches for the optimal operation sequence by probabilistically exploring different sequences and gradually converging on the best one. It's effective in avoiding local minima and finding a global optimum.

4. **Tool Selection and Optimization**
   - ***AI-Driven Tool Selection:*** AI can assist in selecting the optimal tools for each operation based on factors like material, required precision, and tool life. By analyzing data from previous operations, AI recommends tools that minimize wear, maximize cutting efficiency, and reduce overall machining time.
   - ***Neural Networks for Predictive Tool Performance:*** AI, particularly neural networks, can predict tool performance and wear rates by analyzing historical data on similar operations and tools. This predictive capability helps select

tools to maintain performance throughout the process, reducing downtime and the need for tool replacements.

- ***Multi-Criteria Decision Making (MCDM):*** AI can apply MCDM techniques to evaluate and rank different tools based on various criteria, such as cost, availability, machine compatibility, and expected tool life. This helps in making balanced decisions that consider all relevant factors.

5. **Real-Time Process Monitoring and Adjustment**
   - ***Adaptive Control:*** AI systems can monitor manufacturing processes in real-time and dynamically adjust parameters like feed rate, cutting speed, and tool path. This adaptive control is driven by AI algorithms that respond to variations in material properties, tool wear, and machine performance to maintain optimal conditions throughout the operation.
   - ***IoT Integration for Real-Time Data:*** AI can leverage data from IoT sensors embedded in machines to monitor variables such as vibration, temperature, and force during manufacturing. By analyzing this data, AI can detect anomalies or predict failures before they occur, allowing real-time adjustments to the process plan.

6. **Knowledge-Based Systems for Process Planning**
   - ***Expert Systems:*** These AI systems encode the expertise of human process planners into a knowledge base, which is then used to make decisions about process planning. Expert systems can suggest manufacturing processes, tools, and operation sequences based on predefined rules and heuristics. They can also handle exceptions and unusual cases by referring to similar instances in their knowledge base.
   - ***Case-Based Reasoning (CBR):*** AI can use CBR to solve new process planning problems by referring to past cases. The system retrieves the most similar past cases, adapts their solutions to the current problem, and applies them to generate a new process plan. This approach is particularly useful when dealing with customized or one-off parts.

7. **Simulation and Virtual Testing**
   - ***AI-Driven Simulation:*** AI can simulate various manufacturing scenarios to predict outcomes and optimize process plans. These simulations include machine settings, material behavior, and tool wear. AI-driven simulations help identify potential issues and refine the process plan before production.
   - ***Digital Twins:*** AI-powered digital twins create virtual replicas of physical manufacturing processes. These digital twins

allow for real-time monitoring, simulation, and optimization, enabling continuous improvement of the process plan.

8. **Enhanced Decision Support Systems**
   - ***Decision Trees and Bayesian Networks:*** AI can utilize decision trees and Bayesian networks to support decision-making in process planning. These models help navigate complex decision spaces by breaking down decisions into manageable steps and assessing the probabilities and outcomes of different choices.
   - ***AI-Assisted Human Decision Making:*** In scenarios where fully automated decision-making is not feasible or desired, AI can assist human planners. AI enhances the decision-making process without replacing human expertise by providing recommendations, highlighting potential risks, and offering insights derived from data.

9. **Integration with Supply Chain and ERP Systems**
   - ***Supply Chain Optimization:*** AI can integrate process planning with supply chain management by ensuring that the selected processes, tools, and materials are aligned with supply chain constraints and opportunities. For example, AI can suggest alternative materials or processes if there are supply chain disruptions, or it can optimize the plan based on the availability of resources.
   - ***ERP System Integration:*** AI-enhanced CAPP systems can be integrated with Enterprise Resource Planning (ERP) systems to ensure the process plan synchronizes with broader production schedules, inventory levels, and financial constraints. This integration helps in creating more realistic and executable process plans.

10. **Continuous Learning and Improvement**
    - ***Machine Learning for Continuous Improvement:*** AI systems can continuously learn from the outcomes of executed process plans, identifying patterns and areas for improvement. Over time, the AI system becomes more effective at generating optimal process plans as it accumulates more data and refines its algorithms.
    - ***Feedback Loops:*** AI can establish feedback loops where the performance of the manufacturing process is constantly monitored, and the data is fed back into the CAPP system. This ensures that future process plans are informed by the latest performance data, leading to incremental efficiency, quality, and cost improvements.

- ***Decision Support Systems:*** AI enhances human decision-making with decision trees, Bayesian networks, and advanced analytics.

The integration of AI and Machine Learning into CAPP systems significantly enhances their ability to handle complex manufacturing processes, reduce errors, and optimize production, leading to more efficient and cost-effective manufacturing operations.

5.3.4 Launching: Evaluation of Manufacturing Process Plan

Evaluating a manufacturing process plan using the dimensions of social, sustainability, sensing, smart, and safety involves a holistic approach that considers the manufacturing process's technical and economic aspects and its broader impact on society, the environment, and the workplace. The following is depicted how each S dimension can be assessed:

1. **Social Dimension**
    - ***Purpose:*** This dimension focuses on the impact of the manufacturing process on workers, communities, and society at large.
    - ***Key Evaluation Criteria:***
        - *Community Impact:* Consider the impact on local communities, including job creation, infrastructure development, and potential disruptions (e.g., noise, pollution, safety).
        - *Ethical Considerations:* Ensure that the process plan aligns with ethical standards, including fair trade practices and human rights.
        - *Labor Practices:* Assess working conditions, fair wages, and adherence to labor laws. Evaluate how the process plan affects employment levels, job quality, and employee well-being.
    - ***Example:*** A process plan that automates a significant portion of the production line should include retraining programs for displaced workers to maintain employment levels and support community stability.
2. **Sustainability Dimension**
    - ***Purpose:*** This dimension evaluates the environmental impact of the manufacturing process, focusing on resource efficiency, waste reduction, and long-term environmental stewardship.
    - ***Key Evaluation Criteria:***
        - *Resource Efficiency:* Assess raw materials, energy, and water use. Evaluate the process's ability to minimize waste and optimize resource usage.

  - *Waste Management:* Consider how the process handles waste, including solid, liquid, and gaseous emissions. Evaluate the recycling and reuse strategies.
  - *Carbon Footprint:* Measure the greenhouse gas emissions associated with the process and explore opportunities for reducing the carbon footprint.
  - *Life Cycle Impact:* Evaluate the environmental impact of the product throughout its life cycle, from raw material extraction to disposal.
- ***Example:*** Implementing energy-efficient machinery and using renewable energy sources can reduce the overall carbon footprint of the manufacturing process.

3. **Sensing Dimension**
   - ***Purpose:*** This dimension involves the integration of sensors and data collection systems to monitor the manufacturing process in real-time.
   - ***Key Evaluation Criteria:***
     - *Real-time Monitoring:* Evaluate using sensors to monitor key process parameters (e.g., temperature, pressure, humidity) and detect anomalies.
     - *Data Collection and Analysis:* Assess the ability to collect, store, and analyze data for process optimization and predictive maintenance.
     - *Feedback Loops:* Consider integrating sensing systems with control systems to enable real-time adjustments based on sensor data.
   - ***Example:*** A manufacturing process plan with IoT-enabled sensors to monitor machine performance can reduce downtime through predictive maintenance and improve product quality by maintaining optimal process conditions.

4. **Smart Dimension**
   - ***Purpose***: This dimension uses advanced technologies, such as AI, IoT, and automation, to make the manufacturing process more efficient, flexible, and responsive.
   - ***Key Evaluation Criteria:***
     - *Automation and Robotics:* Evaluate the use of robotics and automation to increase efficiency and reduce manual labor. Consider the flexibility of the process to adapt to changes in production requirements.
     - *AI and Machine Learning:* Assess the application of AI and ML for process optimization, quality control, and predictive analytics.

    - *Connectivity and Integration:* Consider the level of integration between machines, systems, and processes. Evaluate the use of IoT for real-time data sharing and decision-making.
- ***Example:*** A smart manufacturing process plan might use AI to optimize production schedules based on real-time demand data, reducing lead times and inventory costs.

5. **Safety Dimension**
    - ***Purpose:*** This dimension assesses the safety of the manufacturing process for workers, equipment, and the environment.
    - ***Key Evaluation Criteria:***
        - *Worker Safety:* Evaluate the potential hazards associated with the manufacturing process and the measures to protect workers. This includes using protective equipment, safety training, and emergency response plans.
        - *Process Safety:* Assess the safety of the process itself, including the risk of accidents, spills, or other incidents that could harm people or the environment.
        - *Compliance with Regulations:* Ensure the process plan complies with all relevant health, safety, and environmental regulations.
    - ***Example:*** A process plan that includes automated safety monitoring systems can detect and respond to potential hazards (e.g., gas leaks, equipment malfunctions) more quickly than human operators, reducing the risk of accidents.

To effectively evaluate a manufacturing process plan using these dimensions, it is essential to integrate them into a comprehensive evaluation framework:

1. **Multi-Criteria Decision Analysis (MCDA):** Use MCDA techniques to weigh and balance the different dimensions, considering the specific priorities and goals of the organization.
2. **Stakeholder Involvement:** Engage stakeholders, including employees, local communities, environmental groups, and regulators, in the evaluation process to ensure that all relevant perspectives are considered.
3. **Scenario Analysis:** Develop different scenarios to evaluate how changes in one dimension (e.g., increased automation) might impact others (e.g., social impact on employment).
4. **Continuous Improvement:** Implement feedback loops to continuously monitor and improve the process based on real-time

data and changing conditions, ensuring ongoing alignment with social, sustainability, sensing, smart, and safety goals.

By considering these dimensions, manufacturers can develop and evaluate process plans that are efficient, cost-effective, socially responsible, environmentally sustainable, technologically advanced, and safe for all stakeholders involved.

TABLE 5.5 depicts the inputs and outputs of a manufacturing process plan. The inputs for evaluating a manufacturing process plan in these dimensions generally include data, technologies, regulatory requirements, and operational practices. The outputs typically reflect improvements in process efficiency, sustainability, safety, and overall social and economic impact. The evaluation involves a holistic approach, ensuring that the process plan is not only effective and efficient but also responsible and forward-thinking in terms of its impact on people, the environment, and the overall system's intelligence and safety.

These results are crucial for decision-making, allowing the company to make informed choices that enhance its manufacturing operations' overall effectiveness, responsibility, and competitiveness.

5.3.4.1 Digital Technologies for Evaluation of a Manufacturing Process Plan

Digital technologies play a critical role in supporting the evaluation of a manufacturing process plan across social, sustainability, sensing, smart, and safety dimensions. These technologies enable real-time data collection, analysis, and decision-making, which are essential for optimizing processes and ensuring compliance with various goals and standards. Here's how digital technologies can be applied to each of these dimensions:

1. **Social**
   - ***Human Resource Management Systems (HRMS):*** Digital HRMS platforms help manage workforce data, track employee satisfaction, and monitor diversity and inclusion initiatives. These systems can provide insights into the social impact of manufacturing operations, such as employee turnover, training needs, and workforce demographics.
   - ***Social Media Analytics Tools:*** These tools can analyze public sentiment and community engagement by tracking mentions and discussions on social media platforms, providing insights into the social impact of the manufacturing process.
   - ***Employee Feedback Platforms:*** Digital platforms for collecting and analyzing employee feedback (e.g., surveys, suggestion boxes) allow for continuous monitoring of employee satisfaction and morale, helping to identify areas where the process may negatively affect the workforce.

TABLE 5.5
Inputs and Outputs during the Launching—Creation/Operation Stage

Inputs	Outputs
• Process Plan Documentation (operation sheets, routing sheets) • Machine, Tooling and Fixture Lists • NC Code (CNC machine programs) • Process Simulation Results (simulation reports, virtual prototyping data) • Cost and Time Estimates (cost, time) • Quality Control Plans (inspection procedures, testing requirements) • Production Schedules (job schedules, resource allocation) • Feedback and Improvement Reports (process performance, continuous improvement)	• **Balanced Scorecard:** The results may be presented in a balanced scorecard format, summarizing the performance across social, sustainability, sensing, smart, and safety dimensions. The evaluation of the dimensions of sensing, smart, and sustainability can be found in the taxonomy presented in Molina et al. [92] and Cortés et al. [29]. The taxonomy for the social dimension includes: (1) ethical practices, (2) labor conditions, (3) community impact, (4) social equity, and (5) human development. The taxonomy for safety dimension includes (1) hazard identification, (2) risk assessment, (3) process safety management, (4) safety standards and compliance, and (5) system environmental safety. See TABLES 5.6 and 5.7. • **Strategic Alignment:** The process plan's alignment with the company's strategic goals regarding social responsibility, sustainability, innovation, and safety would be assessed. • **Risk and Opportunity Assessment:** The results would include an assessment of risks and opportunities related to each dimension, helping guide future process improvements.

2. **Sustainability**
   - ***Energy Management Systems (EMS):*** EMS platforms monitor and optimize energy use throughout manufacturing. They provide real-time data on energy consumption, enabling adjustments that reduce waste and improve sustainability.
   - ***Life Cycle Assessment (LCA) Software:*** LCA tools analyze the environmental impact of products throughout

TABLE 5.6
The Social Dimension for a Manufacturing Process

Social Dimension	Description
• Ethical practices	• **Human Rights Compliance:** Adherence to human rights standards is a non-negotiable baseline for ethical manufacturing. • **Fair Trade and Sourcing:** Ethical sourcing of materials is important for ensuring that upstream practices align with social values. • **Corporate Social Responsibility (CSR):** Active engagement in CSR reflects a company's commitment to broader societal well-being.
• Labor conditions	• **Health and Safety:** Ensuring the safety and well-being of workers is paramount. Unsafe conditions can lead to immediate harm and long-term health issues. • **Wages and Benefits:** Fair compensation is essential for maintaining a stable and motivated workforce. • **Worker Rights:** Protecting the rights of workers, including their ability to organize and participate in decision-making, is fundamental to a fair and just work environment
• Community impact	• **Public Health:** Manufacturing processes that negatively impact public health (e.g., pollution, hazardous waste) can have severe, widespread consequences. • **Local Employment:** Creating jobs in the local community can reduce unemployment and improve economic stability. • **Infrastructure Development:** Investments that benefit the local community can enhance social welfare and contribute to long-term social stability.
• Social equity	• **Gender Equality:** Ensuring gender equity fosters an inclusive and fair work environment. • **Racial and Ethnic Equality:** Promoting racial and ethnic equity is equally important for creating a just workplace. • **Access to Opportunities:** Providing equal opportunities for all employees helps build a diverse and dynamic workforce.

Social Dimension	Description
• Human development	• **Training and Education:** Opportunities for skill development are vital for personal and professional growth, directly impacting job satisfaction and long-term employability. • **Career Advancement:** Pathways for career growth are important for retaining talent and ensuring long-term employee engagement. • **Workplace Diversity and Inclusion:** A diverse and inclusive workplace contributes to innovation and better decision-making.

their life cycle, from raw material extraction to disposal. This helps evaluate the manufacturing process's sustainability and identify improvement opportunities.

- ***Carbon Footprint Calculators:*** These digital tools help measure the carbon emissions associated with manufacturing activities. They also support the evaluation of sustainability efforts and help set reduction targets.
- ***Sustainable Supply Chain Management (SSCM) Tools:*** SSCM platforms ensure suppliers adhere to sustainability standards, enabling a more comprehensive evaluation of the manufacturing process's environmental impact.

3. **Sensing**
   - ***Industrial Internet of Things (IIoT):*** IIoT devices and platforms are central to collecting real-time data from machinery and processes. Sensors embedded in equipment can monitor parameters like temperature, pressure, and vibration, providing critical data for process evaluation and optimization.
   - ***Edge Computing:*** Edge computing enables data processing at the source of data generation (e.g., on sensors or machines), allowing for faster analysis and response times. This is crucial for real-time sensing and immediate corrective actions.
   - ***Data Analytics Platforms:*** Advanced analytics platforms (e.g., Hadoop, Apache Spark) can process and analyze vast amounts of sensor data to detect patterns, anomalies, and trends that are vital for evaluating the effectiveness of the manufacturing process.

TABLE 5.7
The Safety Dimension for a Manufacturing Process

Safety/Safe Dimension	Description
• Hazard identification	• **Chemical Hazards:** Evaluation of exposure to harmful chemicals, toxic substances, fumes, etc. • **Physical Hazards:** Assessment of risks from machinery, noise, vibration, radiation, etc. • **Biological Hazards:** Consideration of exposure to bacteria, viruses, or other biological agents. • **Ergonomic Hazards:** Identification of risks related to repetitive motion, poor posture, or inadequate workstation design.
• Risk assessment	• **Probability of Occurrence:** Estimation of how likely a hazardous event is to occur. • **Severity of Impact:** Evaluation of the potential consequences or severity if the hazard materializes. • **Exposure Frequency:** Analysis of how often workers are exposed to the hazard. • **Human Error Analysis:** Assessment of potential human errors and their impact on safety. • **Workload and Stress:** Consideration of how workload and stress levels may influence safety outcomes.
• Process safety management	• **Process Design:** Evaluation of the inherent safety in the design and layout of the manufacturing process. • **Process Controls:** Implementation of monitoring and control systems to maintain process safety. • **Emergency Response:** Planning for and implementing response protocols for potential safety incidents.
• Safety standards and compliances	• **Regulatory Compliance:** Assessment of compliance with local, national, and international safety regulations (e.g., OSHA, ISO 45001). • **Internal Safety Standards:** Review of company-specific safety guidelines and standards.
• System environmental safety	• **Waste Management:** Assessment of how hazardous waste is handled and disposed of. • **Environmental Impact:** Evaluation of the process's environmental impact, including emissions, pollution, and resource consumption.

4. **Smart**
   - ***Manufacturing Execution Systems (MES):*** MES platforms integrate real-time data across production processes, enabling better planning, scheduling, and execution of

manufacturing operations. They also help evaluate and optimize smart manufacturing practices.

- ***Digital Twins:*** Digital twins create a virtual replica of the manufacturing process, allowing for simulation and analysis of different scenarios without affecting the physical operation. They help evaluate and optimize processes before implementation.
- ***Robotic Process Automation (RPA):*** RPA automates repetitive tasks, increasing efficiency and accuracy in manufacturing processes. It supports the smart dimension by enabling more streamlined and intelligent operations.
- ***Artificial Intelligence and Machine Learning (AI/ML):*** AI/ML algorithms can optimize production processes, predict maintenance needs, and improve decision-making, contributing to the smart evaluation of the manufacturing process.

5. **Safety**
   - ***Safety Management Systems (SMS):*** SMS platforms manage and monitor safety protocols, track incidents, and ensure compliance with safety regulations. These systems are vital for evaluating the safety dimension of the manufacturing process.
   - ***Wearable Technology:*** Wearables with sensors can monitor workers' health and safety in real-time, providing alerts for potential hazards such as fatigue, exposure to harmful substances, or unsafe conditions.
   - ***Augmented Reality (AR) for Safety Training:*** AR technology can be used for immersive safety training, providing workers with realistic simulations of hazardous situations and reinforcing safety protocols. This helps evaluate and improve safety practices.
   - ***Emergency Response and Management Systems:*** Digital platforms that support emergency planning and response can be used to evaluate the readiness and effectiveness of safety measures in the manufacturing process.

To effectively evaluate a manufacturing process plan using these digital technologies, an integrated digital evaluation framework can be established:

- **Data Integration and Visualization Platforms:** Tools like dashboards and visualization software (e.g., Tableau, Power BI) can aggregate data from various digital systems and present it in a comprehensive and user-friendly manner, making it easier to evaluate performance across all dimensions.

- **Cloud-Based Platforms:** Cloud technology enables the storage, processing, and sharing of large amounts of data across multiple locations, facilitating collaborative evaluation and decision-making processes.
- **Artificial Intelligence and Machine Learning Integration:** AI/ML can be integrated into the evaluation framework to provide predictive analytics, automate data analysis, and generate insights that guide improvements across all dimensions.
- **Blockchain for Transparency and Compliance:** Blockchain technology can be used to ensure transparency and traceability in supply chains, supporting the sustainability and safety evaluations by providing immutable records of compliance with standards.

The benefits of Digital Technologies can be listed as:

- **Real-Time Monitoring and Response:** Digital technologies enable real-time monitoring and immediate corrective actions, ensuring that manufacturing processes are continuously optimized and compliant with social, sustainability, sensing, smart, and safety goals.
- **Comprehensive Data Analysis:** Advanced analytics and AI provide deep insights into process performance, helping identify areas for improvement and ensuring that all evaluation dimensions are thoroughly assessed.
- **Enhanced Collaboration:** Cloud and digital platforms support collaboration among stakeholders, enabling more comprehensive and integrated evaluation processes.

By leveraging these digital technologies, manufacturing companies can achieve a more holistic and data-driven evaluation of their process plans, ensuring they meet the highest social responsibility standards, sustainability, innovation, and safety.

5.3.4.2 AI and Machine Learning for Evaluation of a Manufacturing Process Plan

Artificial Intelligence (AI) and Machine Learning (ML) can significantly enhance the evaluation of a manufacturing process plan by providing advanced analytical capabilities, predictive insights, and automated decision-making across the dimensions of social, sustainability, sensing, smart, and safety. Here's how AI and ML can support each of these areas:

1. **Social**
    - ***Sentiment Analysis:*** AI-driven sentiment analysis can assess employee feedback, social media mentions, and community opinions to gauge the social impact of the manufacturing process. This helps in understanding workforce satisfaction, community engagement, and potential social issues.

- ***Predictive Workforce Management:*** ML models can predict workforce needs, optimizing staffing levels based on production demands, reducing turnover, and improving employee satisfaction.
- ***Diversity and Inclusion Monitoring:*** AI can analyze workforce data to ensure compliance with diversity and inclusion goals, identifying patterns or biases that need to be addressed.

2. **Sustainability**
   - ***Energy Consumption Optimization:*** AI algorithms can optimize energy use in real-time, reducing waste and lowering the carbon footprint of manufacturing operations. ML models can predict energy needs and adjust operations accordingly.
   - ***Sustainability Forecasting:*** AI can model the long-term sustainability impacts of different manufacturing strategies, predicting the effects on resource usage, emissions, and environmental impact.
   - ***Waste Reduction:*** ML algorithms can identify inefficiencies in the production process that lead to excess waste, suggesting optimizations to reduce material usage and improve recycling efforts.
3. **Sensing**
   - ***Real-Time Monitoring and Analysis:*** AI-driven systems can analyze data from IoT sensors in real-time, detecting anomalies, predicting equipment failures, and optimizing process parameters to ensure quality and efficiency.
   - ***Predictive Maintenance:*** ML models can predict when machinery is likely to fail or require maintenance, reducing downtime and extending equipment life.
   - ***Data Integration and Insights:*** AI can integrate data from various sensing devices across the manufacturing process, providing comprehensive insights and enabling more informed decision-making.
4. **Smart**
   - ***Automation and Robotics:*** AI algorithms can optimize the operation of robotics and automated systems, ensuring they adapt to changes in production requirements efficiently. This leads to increased productivity and flexibility in manufacturing.
   - ***Digital Twins:*** AI-powered digital twins simulate the manufacturing process, allowing for the testing of different scenarios and process optimizations without disrupting actual operations. This aids in refining the process plan.

- ***Supply Chain Optimization:*** ML can be used to optimize the supply chain by predicting demand, managing inventory levels, and reducing lead times, which aligns with the smart manufacturing objectives.

5. **Safety**
   - ***Hazard Detection and Prevention:*** AI systems can analyze data from safety sensors and cameras to detect potential hazards in real-time, issuing alerts or taking corrective actions to prevent accidents.
   - ***Safety Compliance Monitoring:*** AI can continuously monitor compliance with safety standards, identifying areas where procedures are not being followed and suggesting improvements.
   - ***Worker Safety Analytics:*** ML models can analyze data related to worker health and safety incidents to identify patterns, predict potential risks, and recommend preventive measures.

An AI/ML-driven evaluation system for manufacturing process plans would integrate the above functionalities into a cohesive platform:

- **Data Aggregation:** The system would collect and aggregate data from various sources, including sensors, workforce management systems, energy monitors, and safety devices.
- **AI-Powered Analytics:** Advanced AI algorithms would analyze this data in real-time, providing insights across the social, sustainability, sensing, smart, and safety dimensions.
- **Predictive Modeling:** ML models would predict future trends, such as potential safety risks, sustainability impacts, or social effects, allowing proactive management.
- **Automated Reporting:** The system could generate automated reports summarizing the evaluation results, providing a clear overview of how the manufacturing process performs across dimensions.
- **Continuous Improvement:** AI and ML would enable continuous monitoring and iterative improvements to the process plan, ensuring it remains optimal over time.

The benefits of AI/ML Integration could be:

- **Enhanced Decision-Making:** AI-driven insights lead to more informed and data-driven decisions, improving the overall effectiveness of the manufacturing process.
- **Efficiency Gains:** Automation and real-time monitoring reduce waste, energy consumption, and downtime, leading to cost savings and increased productivity.

- **Risk Mitigation:** Predictive analytics helps identify and mitigate risks before they escalate, ensuring safer and more reliable operations.
- **Sustainability and Social Responsibility:** AI supports achieving sustainability goals and social responsibility commitments by optimizing resources and improving workforce conditions.

By integrating AI and ML into the evaluation of manufacturing process plans, companies can ensure their processes are not only efficient and cost-effective but also socially responsible, sustainable, safe, and smart.

5.3.5 End-of-Life of the Manufacturing Process: Planning and Decommission

The 10R framework, often used in the context of sustainable manufacturing and circular economy, can be applied to manufacturing process planning to enhance resource efficiency, minimize waste, and optimize overall operations. The 10Rs are:

1. **Refuse:** Avoiding unnecessary use of materials and energy. In manufacturing process planning, this could mean choosing processes that inherently produce less waste or require fewer inputs. For example, opting for near-net shape manufacturing processes that minimize material removal.
2. **Rethink:** Reevaluating current processes to find more efficient alternatives. This could involve redesigning products to require less processing or considering alternative materials that are easier to work with or recycle.
3. **Reduce:** Minimizing the use of materials, energy, and time. Process planning can focus on selecting methods that reduce the number of steps, use less energy-intensive equipment, or require less material.
4. **Reuse:** Incorporating reusable components or materials in the production process. For instance, planning for processes that allow for the reuse of tooling, jigs, or fixtures can reduce overall resource consumption.
5. **Repair:** Designing processes that allow for easy repair of products or components, rather than replacing them. This could involve choosing manufacturing methods that make disassembly easier or selecting materials that can be repaired rather than discarded.
6. **Refurbish:** Planning processes that extend the life of products through refurbishment. This might involve selecting manufacturing methods that make it easier to refurbish parts or products at the end of their initial life cycle.

7. **Remanufacture:** Integrating remanufacturing into the process planning. This means choosing processes that allow for the recovery and restoration of used products to a like-new condition.
8. **Repurpose:** Planning for the repurposing of by-products or waste from the manufacturing process. For example, process planning could include methods to convert waste material into inputs for other processes or products.
9. **Recycle**: Selecting processes that facilitate recycling, either by producing materials that are easier to recycle or by ensuring that the end product can be easily recycled at the end of its life.
10. **Recover:** Extracting energy or valuable materials from waste that cannot be reused or recycled. In manufacturing, this could mean incorporating processes that allow for energy recovery through combustion of waste materials or the recovery of valuable elements using advanced extraction techniques.

For instance, the 10Rs can be applied to the process planning as follows:

1. **Process Selection:** When selecting manufacturing processes, consider those that align with the 10Rs principles. For instance, choose processes that reduce material waste (Reduce), allow for the use of recycled materials (Recycle), or facilitate easy repair or refurbishment (Repair, Refurbish), or enable the recovery of valuable resources or energy from waste materials (Recover).
2. **Material Choices:** Identify materials that can be reused, recycled, or remanufactured. This might influence the choice of a manufacturing process that is compatible with these materials.
3. **Design for Sustainability:** Incorporate the 10Rs into the design phase, directly influencing manufacturing process planning. For example, design products that are easier to disassemble for repair, recycling or resource recovery.
4. **Energy Efficiency:** Prioritize energy-efficient processes and reduce the manufacturing operation's overall carbon footprint.
5. **Waste Minimization:** Implement processes that minimize waste, either through more efficient material use (Reduce) or by repurposing waste materials into new products (Repurpose), or by recovering valuable resources or energy from any unavoidable waste (Recover).

By integrating the 10R into manufacturing process planning, companies can not only improve their environmental impact but also potentially reduce costs and improve the overall efficiency of their operations. This approach aligns well with modern trends toward sustainability and the circular economy in manufacturing.

Applying the 10R framework to decommission manufacturing processes can lead to more sustainable and cost-effective outcomes:

1. **Refuse**
   - ***Avoid Unnecessary Decommissioning:*** Before deciding to decommission a manufacturing process, evaluate whether the process can still be optimized or adapted for continued use. Refusing to decommission prematurely can conserve resources and extend the life of existing equipment.
   - ***Avoid Waste Generation:*** Refuse to generate unnecessary waste during decommissioning by carefully planning the process and identifying materials or components that can be reused or repurposed.
2. **Rethink**
   - ***Rethink the Decommissioning Process:*** Consider alternative approaches to decommissioning that minimize environmental impact. For example, instead of completely dismantling a facility, explore options for repurposing parts of it for new functions.
   - ***Rethink End-of-Life Uses:*** Evaluate whether the decommissioned equipment or materials can be used in a different context, either within the same organization or in another industry.
3. **Reduce**
   - ***Minimize Waste:*** Plan the decommissioning process to reduce waste generation, such as by deconstructing equipment and facilities to maximize material recovery.
   - ***Reduce Resource Use:*** Use resources efficiently during decommissioning, such as by minimizing energy use and reducing the need for new materials when dismantling or repurposing equipment.
4. **Reuse**
   - ***Reuse Equipment and Components:*** Identify parts of the decommissioned manufacturing process that can be reused in other operations. For example, machinery, tools, and fixtures might still be functional in other parts of the facility or in other plants.
   - ***Reuse Materials:*** Salvage and reuse materials, such as steel, concrete, or wiring, in other construction or manufacturing projects, rather than disposing of them.
5. **Repair**
   - ***Repair and Resell:*** If some equipment or components are still functional but no longer needed, repair them to restore their full functionality and resell them to other companies or use them in different areas of the business.

- ***Extend Life Through Maintenance:*** During decommissioning, identify opportunities to repair and maintain equipment, extending its life instead of replacing it entirely.

6. **Refurbish**
   - ***Refurbish Equipment:*** Consider refurbishing older equipment rather than scrapping it. Refurbished equipment can be sold, donated, or redeployed in other parts of the organization.
   - ***Upgrade for New Use:*** Some equipment can be upgraded or modified for new applications, either within the same company or by other businesses.

7. **Remanufacture**
   - ***Remanufacture Components:*** Decommissioned equipment often contains components that can be remanufactured into new parts. This can involve disassembling the equipment, cleaning and repairing parts, and then reassembling them into like-new condition.
   - ***Facilitate Future Remanufacturing:*** Design the decommissioning process to preserve key components in a condition that allows for future remanufacturing.

8. **Repurpose**
   - ***Repurpose Facilities:*** Consider repurposing the entire manufacturing facility or parts of it for new operations. For instance, a plant designed for one type of manufacturing could be converted to support a different production line.
   - ***Repurpose Equipment:*** Equipment that is no longer needed for its original purpose might be repurposed for different manufacturing processes, either within the same company or by other organizations.

9. **Recycle**
   - ***Material Recycling:*** Recycle materials that cannot be reused, repurposed, or refurbished. This could involve recycling metals, plastics, and other materials recovered during the decommissioning process.
   - ***Circular Economy:*** Integrate recycling into a broader circular economy strategy by ensuring that materials from decommissioned processes are fed back into the supply chain as raw materials for new products.

10. **Recover**
    - ***Recover Energy:*** Implement processes that recover energy from waste materials that cannot be reused, repurposed, or recycled. For example, use waste-to-energy technologies to convert waste into heat or electricity.
    - ***Recover Valuable Materials:*** Use advanced techniques to recover valuable materials from waste generated during decommissioning. For instance, recover rare metals or other critical resources from electronic components.

Furthermore, there are additional considerations for sustainable decommissioning that must be taken into consideration:

1. **Environmental Impact Assessment:** Conduct a thorough environmental impact assessment before decommissioning to identify potential environmental risks and opportunities to mitigate them.
2. **Stakeholder Engagement:** Engage with stakeholders, including employees, local communities, and regulators, to ensure that the decommissioning process is transparent and aligned with broader sustainability goals.
3. **Documentation and Reporting:** Keep detailed records of the decommissioning process, including how materials were managed, to provide transparency and support future sustainability reporting.
4. **Legal and Regulatory Compliance:** Ensure that the decommissioning process complies with all relevant environmental regulations and standards, particularly those related to waste management and recycling.
5. **Financial Optimization:** Consider the financial implications of each R strategy, such as the potential for cost savings through reuse or revenue generation from selling refurbished equipment.

By applying the 10Rs to decommissioning manufacturing processes, organizations can significantly reduce their environmental footprint, minimize waste, and extract maximum value from existing resources. This approach supports a more sustainable and circular approach to manufacturing, even at the end of a process's life cycle.

5.3.5.1 AI and Machine Learning in the 10Rs Manufacturing Process Planning

AI and Machine Learning are powerful tools that can significantly enhance the application of the 10R framework in manufacturing process planning. Here is how AI and ML can support each of the 10Rs:

1. **Refuse**
   - ***Demand Forecasting:*** AI can analyze market trends and consumer behavior to predict demand more accurately, helping manufacturers avoid overproduction and unnecessary resource use.
   - ***Material Substitution:*** ML algorithms can recommend alternatives, more sustainable materials that can be used in place of less sustainable options, reducing the need to use certain materials.
2. **Rethink**
   - ***Process Redesign:*** AI can simulate different manufacturing processes to identify more efficient and sustainable alternatives. For example, AI can propose modifications to existing processes to minimize waste or reduce energy consumption.
   - ***Product Design Optimization:*** ML models can analyze product designs to suggest changes that align with sustainability goals, such as designs that use fewer materials or facilitate easier recycling.
3. **Reduce**
   - ***Resource Optimization:*** AI can optimize the use of materials, energy, and water by analyzing real-time data from manufacturing processes. For example, it can adjust machine parameters to minimize material waste.
   - ***Predictive Maintenance:*** ML algorithms can predict when equipment is likely to fail, allowing for timely maintenance that reduces downtime and prevents resource-intensive repairs.
4. **Reuse**
   - ***Component Tracking:*** AI-powered systems can track the condition and life cycle of components, identifying those that can be reused in new products or processes.
   - ***Smart Sorting:*** ML algorithms can enhance sorting processes to ensure that reusable components are efficiently identified and separated from waste streams.
5. **Repair**
   - ***Fault Detection:*** AI can detect anomalies in product performance or manufacturing processes, indicating when repair is needed rather than replacement. This supports the repair of products and components, extending their life cycle.
   - ***Automated Repair Planning:*** ML models can develop optimized repair plans by analyzing historical data and identifying the most effective repair strategies for specific issues.

6. **Refurbish**
   - ***Quality Assessment:*** AI can assess the condition of used products and components to determine their suitability for refurbishment. This can involve visual inspection using computer vision or performance analysis using sensor data.
   - ***Refurbishment Process Optimization:*** ML can optimize refurbishment processes by identifying the most efficient methods for restoring products to like-new condition.
7. **Remanufacture**
   - ***Process Automation:*** AI can automate and optimize remanufacturing processes, such as disassembly, cleaning, and reassembly, ensuring that resources are used efficiently and waste is minimized.
   - ***End-of-Life Prediction:*** ML algorithms can predict when products are approaching the end of their useful life, allowing for timely collection and remanufacturing, thus reducing waste.
8. **Repurpose**
   - ***Material Repurposing:*** AI can identify new uses for by-products or waste materials generated during manufacturing. For example, it can analyze material properties and suggest alternative applications.
   - ***Design for Repurposing:*** ML can assist in designing products that are easier to repurpose at the end of their life cycle, ensuring that materials and components can be reused in new contexts.
9. **Recycle**
   - ***Waste Stream Analysis:*** AI can analyze waste streams to identify recyclable materials and optimize recycling processes. For example, AI can improve sorting accuracy in recycling facilities.
   - ***Material Composition Identification:*** ML algorithms can identify the composition of complex materials or products, making it easier to separate and recycle different materials effectively.
10. **Recover**
    - ***Energy Recovery:*** AI can optimize waste-to-energy processes by analyzing waste composition and predicting the most efficient methods for recovering energy from non-recyclable materials.

- ***Resource Recovery:*** ML can identify valuable materials in waste streams, such as rare metals in electronic waste, and optimize their extraction and recovery, reducing the need for virgin resource extraction.

There are additional considerations of AI and ML applications in the context of 10Rs:

1. **Supply Chain Optimization:** AI and ML can optimize supply chains to reduce resource consumption and waste. AI can suggest more sustainable sourcing options, transportation routes, and inventory management practices by analyzing supply chain data.
2. **Life cycle Assessment (LCA):** AI can automate and enhance life cycle assessments by analyzing large datasets related to the environmental impact of materials and processes. This allows manufacturers to make informed decisions that align with the 10Rs principles.
3. **Energy Management:** AI-driven energy management systems can optimize energy use across manufacturing facilities, minimizing energy consumption without compromising production efficiency.
4. **Circular Economy Modeling:** ML models can simulate different circular economy scenarios, helping manufacturers to design processes and systems that support the reuse, refurbishment, and recycling of materials and products.
5. **Sustainability Reporting:** AI can automate the collection and analysis of sustainability data, making it easier for manufacturers to report on their adherence to the 10Rs framework and identify areas for improvement.

By leveraging AI and ML, manufacturers can make more informed decisions, optimize processes for sustainability, and ultimately contribute to a more circular and resource-efficient economy. These technologies enable the effective implementation of the 10Rs framework, leading to reduced environmental impact, cost savings, and enhanced operational efficiency.

5.3.6 The Ten Core Technologies Applied to the Entity

Integrating the ten core technologies into this entity is applied strategically, as depicted in FIGURE 5.16.

1. **Ideation—Imagination:** Decomposing product design, identifying customer requirements (geometry, material, production rates), and defining project scope.

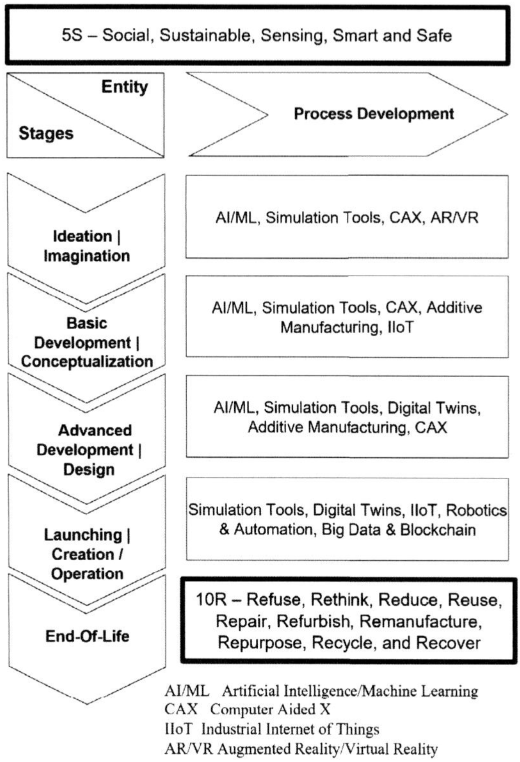

FIGURE 5.16
Core technologies applied to this entity.

(a) ***AI/ML:*** Automates analysis of customer requirements and predicts feasibility.
(b) ***Simulation Tools:*** Models initial designs and tests component feasibility virtually.
(c) ***CAX:*** CAD systems support detailed design and component specification.
(d) ***AR/VR:*** Immersive visualization aids in understanding design breakdowns and stakeholder alignment.

2. **Basic Development—Conceptualization:** Classifying components (standard vs. custom), selecting manufacturing processes, and determining whether components are transferred or manufactured.

(a) ***AI/ML***: Assists in component classification and recommends sustainable manufacturing processes.
(b) ***Simulation Tools:*** Evaluates process feasibility and simulates workflows for decision-making.

(c) ***CAX:*** Supports manufacturability analysis and integrates with CAPP for process planning.
(d) ***Additive Manufacturing:*** Produces functional prototypes and tooling to validate manufacturability.
(e) ***IIoT:*** Provides real-time data on machine capabilities to inform process selection.

3. **Advanced Development—Design:** In establishing the structure, form, and manufacturing specifications.
    (a) ***Simulation Tools:*** Models manufacturing systems and identifies bottlenecks.
    (b) ***AI/ML:*** Optimizes workflows, predicts production times, and enhances process designs.
    (c) ***Digital Twins:*** Simulates the production system dynamically to optimize performance.
    (d) ***Additive Manufacturing:*** Produces custom tooling and fixtures to refine workflows.
    (e) ***CAX:*** CAM systems provide detailed instructions for machining and production.
4. **Launching—Creation/Operation:** In the deployment and operationalization of the material system.
    (a) ***Simulation Tools:*** Validates the process plan under real-world conditions.
    (b) ***Digital Twins:*** Monitors and optimizes operational workflows dynamically.
    (c) ***IIoT:*** Ensures real-time monitoring and feedback during production.
    (d) ***Robotics and Automation:*** Streamlines assembly and ensures precision at scale.
    (e) ***Big Data Analytics and Blockchain:*** Tracks performance metrics and ensures traceability for quality assurance.

5.4 Case Study Applying the Life Cycle Engineering for Manufacturing Process Development

The case presented in this section demonstrates the application of the Life Cycle Engineering (LCE) approach for manufacturing process development is based on a rotational component used in the Micro-machine. The case study is described in TABLE 5.8 and FIGURES 5.17 and 5.18, which include the LCE stage, activities, digital technologies used, and references to similar approaches found in the literature.

TABLE 5.8
Case Study of Aluminum Rotational Component for Micro-Machine and Digital Technologies Used

Stage	Activity	Digital Technology	References
1. Ideation: Individual Component specifications	Description of Product Information	2D and 3D CAD Models	Zehtaban et al. (2016) [148].
	Generation of Bill of Materials	Databases/Spread sheet/ERP	Ashby (2011). [10]. Scallan P. (2003) [124].
2. Basic development: Manufacturing process selection	Codification of Group Technology using OPITZ	GT coding system	Zehtaban and Roller (2015) [148].
	Selection of Manufacturing Process Selection and Feature Recognition of components	Manufacturing Process selection software: https://www.custompartnet.com/process-selector	Swift and Booker (2003) [134]. Behandish et al. (2019) [16]. Mumali and Kałkowska (2024) [98].
3. Advanced development: Manufacturing process plan and evaluation	Assignment of Machine and Manufacturing Resources	Manufacturing Information Model/Object-Oriented Database	Molina and Bell (1999) [90].
	CAM Process Simulation	CAM	Hunde and Woldeyohannes (2022) [65].
	Generation of Process Documentation	CAPP	Zhou and Camba (2021) [151].
	Planning based on Transformation Process	CAPP	Soori and Asmael (2021) [132].
	Planning based on Value Stream Mapping	Value Stream Mapping software	Cortes et al. (2021) [29].
	Manufacturing Operation/Simulation and Optimization	Simulation software	Chavarría-Barrientos et al. (2017) [22].
	Review and Approval (Costs)	Costing Software/Spread sheet	Molina et al. (2021) [92].
	Integration with Production Systems and Digital Twin	Simulation software/Digital Twins	Martinez et al. (2021) [83]
4. Launching: Evaluation of Manufacturing process Plan	Determine Balance Scorecard for evaluation of 5S	Balance score card/Spread sheet	Molina et al. (2021) [92].
5. End-of-Life of the manufacturing process: Planning and Decommission	Apply 10Rs framework	Life Cycle Assessment Software	Okorie et al. (2018) [104].

This case is part of a component of the Micro-machine described in [88]. The component is a rotational component of aluminum material. The development of the manufacturing process is presented in TABLE 5.8, which describes each stage of the LCE, activity, digital technologies used, and reference to key digital developments. Integrating data, information, and knowledge is key for successfully applying digital technologies using Product, Process, and Manufacturing Resources information models [90]. The Life Cycle Assessment and principles of the 10Rs were applied during the complete life cycle process to ensure that all the aspects of sustainability are applied at each stage.

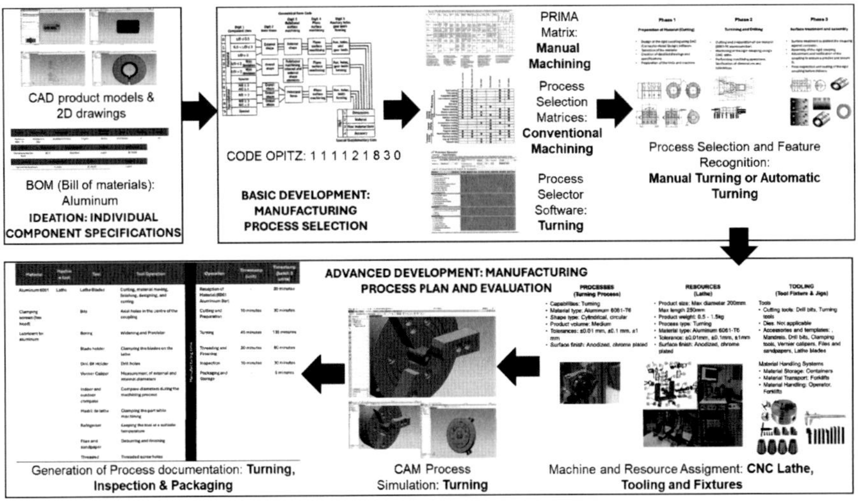

FIGURE 5.17
Case study: Stage 1—Ideation: Individual component specifications; Stage 2—Basic development: Manufacturing process selection and Stage 3—Advanced development: Manufacturing process plan and evaluation.

The rigid coupling is a critical rotational component for a micromachine, designed for standardization to reduce manufacturing costs and improve ease of assembly. This case study demonstrates how the life cycle engineering (LCE) methodology was applied to ensure the design, manufacturing, operation, and disposal of the rigid coupling align with the principles of the 5S framework, which include Social, Sustainable, Sensing, Smart, and Safe aspects.

The manufacturing process for the micromachine's rigid coupling begins with the Computer-Aided Design (CAD) stage, where 3D models and 2D technical drawings are developed. Engineers create detailed representations of the component's geometry, dimensions, and specific design features in this stage. These models serve as a blueprint for the entire manufacturing process, allowing for precise planning and reducing potential issues later in production. The CAD models are essential as they ensure the design aligns

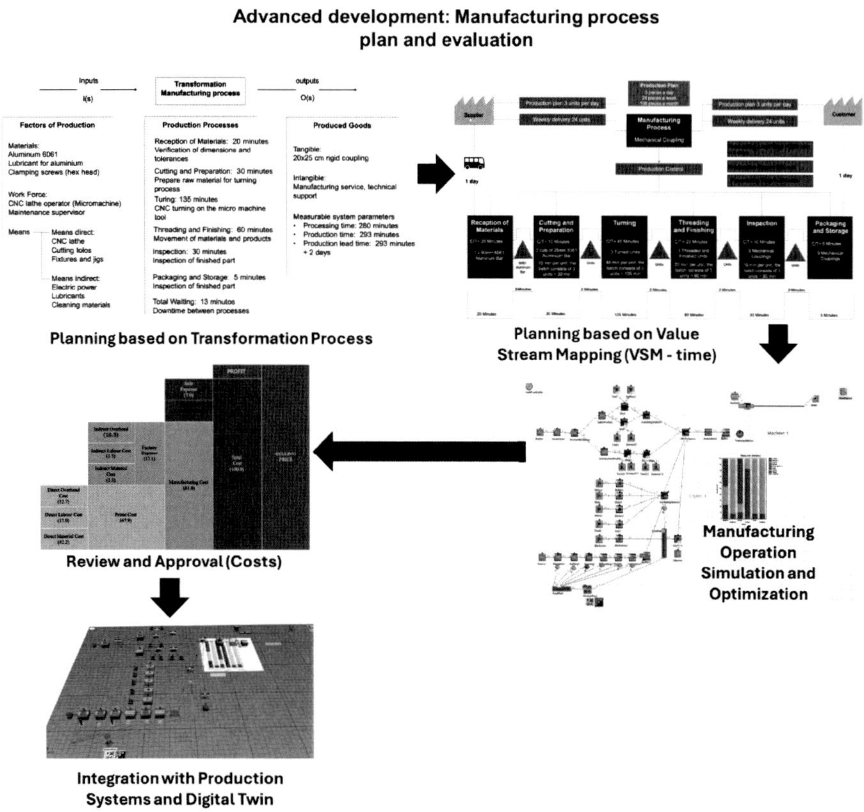

FIGURE 5.18
Case study: Stage 3 (Continuation): Advanced development: Manufacturing process plan and evaluation.

with technical requirements and provide visual and structural details necessary for each subsequent phase.

Once the CAD models are finalized, a Bill of Materials (BOM) is generated. The BOM is a comprehensive list of all materials, components, and quantities required to manufacture the rigid coupling. The BOM ensures that all resources are accounted for before the manufacturing process begins by documenting every item needed, from raw materials to fasteners. This systematic approach helps avoid production delays caused by missing components and allows for a smoother transition into the production phase by streamlining procurement and inventory management.

Group Technology Coding (OPITZ) is used to categorize the components and streamline production efficiently. This classification system organizes parts based on their geometry, material, and necessary processing methods. Coding the rigid coupling with OPITZ makes grouping it with similar components easier, thereby improving production efficiency and resource allocation. This classification is crucial for facilities that produce multiple parts, as it enables standardized processes and setups for similar items, reducing setup times and optimizing machine usage.

Following classification, Process Selector Software is employed to analyze and select the most appropriate manufacturing processes. This software evaluates different machining options based on cost, time, and equipment compatibility. Engineers can quickly identify the most efficient and cost-effective methods for each production stage through a color-coded system. Green indicators, for instance, highlight the optimal choices, while other colors may denote less suitable options. This step is instrumental in minimizing costs and ensuring the selected processes align with production goals and constraints.

In the Process Selection and Feature Recognition phase, engineers identify specific features of the rigid coupling, such as holes, threads, or grooves, and determine the processes required for each feature. This involves selecting drilling, turning, or milling methods based on the component's design specifications. By recognizing these features early on, engineers can tailor the machining steps to meet precise tolerance and quality requirements, ensuring that the final product performs as intended. This phase plays a vital role in establishing the sequence and specifications for each operation in the production line.

With processes defined, the next step is Machine and Resource Assignment. Specific machines and tools, such as CNC lathes, drills, and turning tools, are allocated to each manufacturing operation. This stage also involves organizing resources like operators and materials. Ensuring that all equipment and personnel are available and prepared minimizes downtime and maximizes production efficiency. Proper resource assignment is critical to maintaining a steady workflow and meeting production schedules, as it prevents bottlenecks caused by machine unavailability or lack of materials.

The process continues with Computer-Aided Manufacturing (CAM) Process Simulation and Documentation. CAM software simulates the entire machining process, allowing engineers to visualize the production sequence and make necessary adjustments before physical production. This simulation generates detailed process documentation, including tool paths, machine setup instructions, and safety precautions. CAM simulation ensures that the manufacturing operations are optimized, reducing the likelihood of errors and waste during actual production. By visualizing the process, engineers can fine-tune operations to achieve better precision and efficiency.

Planning based on Value Stream Mapping (VSM) is implemented for broader process optimization. VSM is a methodology that analyzes and optimizes the time taken at each production stage. By mapping out the entire workflow, engineers identify areas where time can be reduced, thus improving overall process efficiency. This phase highlights potential bottlenecks and wasteful steps in the production flow, providing a basis for strategic improvements that enhance throughput and reduce lead times, ultimately leading to a more streamlined manufacturing process.

Planning based on Transformation Processes builds on the VSM analysis, focusing on organizing each transformation (cutting, shaping, and assembly) required to produce the coupling. This stage ensures that the sequence of operations is logical, efficient, and tailored to the characteristics of the rigid coupling. It involves aligning the specific manufacturing steps with the desired output, ensuring that each transformation is performed in a way that optimizes material usage, energy consumption, and production time. This structured approach further enhances the efficiency and consistency of the manufacturing process.

Once the planning stages are completed, the workflow proceeds to Manufacturing Operation Simulation and Optimization. In this phase, a comprehensive simulation of the entire manufacturing operation is conducted, displaying how materials flow through each process step. This simulation helps identify inefficiencies and optimizes resource allocation to ensure smooth operations. By adjusting variables in the simulation, engineers can test and implement improvements before actual production, making the manufacturing process as efficient and error-free as possible. This step is vital for large-scale production, where minor adjustments can save time and cost.

Following the optimization simulation is a Review and Approval (Cost) phase. This step thoroughly reviews the projected production costs, including labor, material, and machine expenses. Engineers and managers evaluate whether the costs align with the budget and, if necessary, make adjustments to keep the project financially viable. This review ensures the production plan is technically feasible and economically sustainable, balancing quality and cost-effectiveness to meet business objectives.

Finally, the workflow concludes with Integration with the Production System, where the manufacturing plan is incorporated into the broader production environment. This step involves coordinating scheduling, inventory management, and logistics to align the coupling's production with ongoing manufacturing activities. By integrating the process with the production system, the rigid coupling is now ready for full-scale production, ensuring a smooth transition from development to manufacturing. This stage solidifies the entire process, facilitating efficient production runs and timely delivery.

5.4.1 Walkthrough of the Entire Life Cycle Process: From Idea to Launching

The rigid coupling, a critical rotational component of a micromachine, requires a structured approach to its life cycle, from design and production to end-of-life management. By applying LCE principles, we ensure that each phase of the product's life cycle—from material selection and process

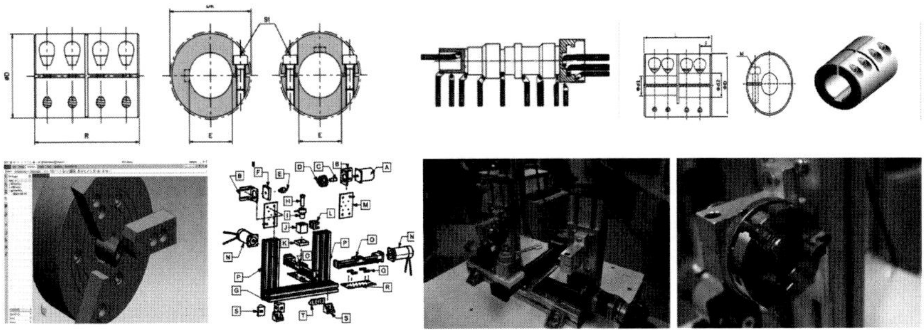

FIGURE 5.19
From design to assembly of the rigid coupling used to connect two shafts so they rotate as a single unit without allowing any flexibility or misalignment.

planning to quality control and recycling—aligns with the 5S framework (Social, Sustainable, Sensing, Smart, Safe). This systematic approach optimizes resource usage, minimizes environmental impact, and addresses community needs, balancing performance, cost, and sustainability. FIGURE 5.19 shows the development of the rigid coupling, it reflects the integration of digital and physical resources in creating a high-quality, sustainable manufacturing process for the component.

1. **Ideation Stage:** Component Specifications and Design for 5S Compliance
   (a) ***Objective:*** To design the rigid coupling with 5S principles in mind, ensuring alignment with customer needs, sustainability goals, and operational feasibility.
   (b) ***Social Responsibility:*** Consideration of fair labor practices and diversity in the workforce during the design process. Additionally, ensure that the component contributes to local job creation in assembly and machining.
   (c) ***Sustainable Design:***
   - *Material Selection:* 6061-T6 aluminum is chosen for its high recyclability and durability, minimizing environmental impact while meeting mechanical requirements.
   - *Life cycle Analysis:* Conduct a preliminary life cycle assessment to estimate environmental impacts throughout the product's life, from raw material extraction to disposal.
   - *Eco-Design:* Minimize material waste by designing a component that optimizes raw material usage and can be manufactured with minimal scrap.

(d) ***Outputs:***

- *Component Specifications:* Detailed specifications regarding geometry, tolerances, surface finish, material choice, and quality standards.
- *Bill of Materials (BOM):* Includes all components and materials for the coupling, emphasizing sustainable sources.

2. **Basic Development Stage:** Manufacturing Process Selection

(a) ***Objective:*** Select manufacturing processes that align with the 5S framework, focusing on sustainable, efficient, and safe production.

(b) ***Sustainable Manufacturing:***

- *Eco-Friendly Processes:* CNC turning is selected for its precision and lower material waste. The process is optimized to reduce energy consumption and waste production.
- *Resource Efficiency:* Implement a lean manufacturing approach to minimize waste. Optimize cutting and machining sequences to use as little material as possible and reduce energy requirements.

(c) ***Smart Process Selection:***

- *Process Optimization Tools:* Use process selection matrices and digital tools (such as CAD/CAM integration) to evaluate and select the most efficient manufacturing process. Consider factors such as material properties, cost, and production time.
- *Automation:* Program the CNC lathe with specific tolerances (± 0.01 mm, ± 0.1 mm, ± 1 mm) to achieve consistent quality and reduce human error.

(d) ***Outputs:***

- *Manufacturing Plan:* A detailed plan outlining the preparation, shaping, and secondary finishing processes required for the coupling.
- *Process Matrices:* PRIMA and Group Technology matrices are used to categorize and standardize components, ensuring efficiency and consistency in production.

3. **Advanced Development Stage:** Process Plan Evaluation and Optimization

(a) ***Objective:*** Refine the manufacturing process to ensure high quality, cost efficiency, and adherence to 5S standards before full-scale production.

(b) ***Sensing Technologies:***

- *Sensor Integration:* Install sensors on the CNC lathe to monitor parameters such as spindle speed, tool wear, and material feed rates. This data helps maintain quality and identify issues early.
- *Predictive Maintenance:* Use sensor data and predictive analytics to anticipate maintenance needs, reducing downtime and extending equipment life.

(c) ***Sustainable and Smart Process Optimization:***

- *Digital Twin:* Create a digital twin of the manufacturing process to simulate production runs and test for potential issues before physical manufacturing. This helps optimize machine parameters, reduce waste, and ensure energy efficiency.
- *Data-Driven Improvements:* Use data analytics from trial runs to adjust machining parameters and improve material efficiency, reducing resource consumption.

(d) ***Outputs:***

- *Finalized Manufacturing Process Plan:* An optimized, data-driven plan incorporating real-time monitoring and predictive analytics to maintain quality and efficiency.
- *Digital Twin Model:* A virtual representation of the manufacturing process, enabling ongoing process improvements.

4. **Launching Stage:** Full-Scale Production and Quality Control

(a) ***Objective:*** Implement the manufacturing process on a full scale while ensuring compliance with the 5S framework through continuous monitoring and quality control.

(b) ***Safety and Social Aspects:***

- *Safety Protocols:* Establish comprehensive safety guidelines for CNC lathe operations, including operator training and emergency procedures.
- *Community Engagement:* Communicate with local communities about the sustainable practices employed in manufacturing and the social benefits, such as job creation.

(c) ***Smart and Sensing Quality Control:***

- *Real-Time Monitoring:* Utilize sensors on the CNC lathe for real-time monitoring of critical parameters. This includes detecting deviations in material feed rate, temperature, or vibrations, which could affect product quality.
- *Automated Quality Inspections:* Implement automated inspections using sensors and vision systems to ensure each

coupling meets dimensional tolerances and surface finish requirements.

(d) ***Outputs:***

- *Production Report:* Detailed documentation of production metrics, quality control data, and any deviations for continuous improvement.
- *Safety and Community Engagement Documentation:* Reports outlining safety measures and community engagement activities, reinforcing the social and safety components of the 5S framework.

5. **End-of-Life Stage:** Recycling, Disposal, and Decommissioning

(a) ***Objective:*** Plan for the end of the component's life cycle, ensuring that disposal or recycling aligns with sustainability goals.

(b) ***Sustainable Disposal:***

- *Material Recovery:* The 6061-T6 aluminum used in the coupling is highly recyclable. Develop a recycling protocol to recover and reuse aluminum at the end of the component's life.
- *Waste Minimization:* Incorporate strategies for disassembling and recycling any auxiliary materials or coatings, minimizing environmental impact.

(c) ***Safe and Environmentally Friendly Disposal:***

- *Disposal Procedures:* Establish safe disposal protocols for any waste materials, including anodizing byproducts, to prevent environmental contamination.
- *Compliance with Environmental Regulations:* Ensure that all disposal and recycling practices meet relevant environmental standards and regulations.

(d) ***Outputs:***

- *Recycling Plan:* A comprehensive recycling and disposal plan details how aluminum and other materials will be recovered and reused.
- *Environmental Compliance Report:* Documentation demonstrating environmental guidelines and regulatory standards adherence.

5.5 The Future of Sustainable Manufacturing for this Entity

The future of sustainable manufacturing for the rigid coupling, as well as similar components, lies in advancing techniques and technologies that reduce resource consumption, waste, and environmental impact across the entire product life cycle. Emerging trends in sustainable manufacturing emphasize a shift towards more resilient, circular, and efficient production practices that support both environmental and economic goals.

One promising development is the increased integration of digital twins and simulation technologies. By creating a precise virtual model of the coupling's manufacturing process, manufacturers can predict and optimize resource usage, energy consumption, and production timelines in real time. This approach minimizes trial-and-error in physical production, leading to less material waste and a lower carbon footprint. Digital twins can also support predictive maintenance strategies, enhancing machine longevity and reducing the need for replacement parts.

Another area of advancement is the adoption of smart and adaptive manufacturing systems that leverage artificial intelligence (AI) and machine learning (ML). These technologies can continuously analyze sensor data during production, allowing for real-time adjustments that improve efficiency. AI can optimize parameters such as tool path, spindle speed, and coolant usage, reducing energy consumption while maintaining high quality. Moreover, ML algorithms can learn from historical production data to further refine processes, aligning with sustainable and waste-free manufacturing goals.

The implementation of circular economy principles will play a vital role in the future of sustainable manufacturing. For components like the rigid coupling, this approach focuses on designing for disassembly, ease of recycling, and material recovery. Future design iterations may consider modular or repairable features, enabling components to be refurbished or upgraded rather than discarded. Additionally, closed-loop recycling systems could be established within manufacturing facilities, allowing materials like aluminum to be recycled on-site, reducing transportation emissions and material costs.

Green energy sources and carbon-neutral manufacturing facilities are also on the horizon for sustainable manufacturing. By utilizing renewable energy sources, such as solar or wind, manufacturers can power production lines in an eco-friendly manner. This transition to clean energy will significantly reduce the carbon footprint of the coupling's manufacturing process. Alongside green energy, facilities can implement energy-efficient practices like heat recovery systems and LED lighting to decrease their environmental impact.

Advanced material science innovations, including the development of sustainable or bio-based materials, could also influence future production. Although aluminum is already recyclable, new alloys or alternative materials could offer similar performance with even lower environmental impact. As material science progresses, manufacturers may have access to lighter, stronger, and more sustainable materials that reduce the energy required in production and extend the life of each component.

Thus the future of sustainable manufacturing for the rigid coupling will be characterized by the adoption of digitalization, circular economy principles, green energy, and innovative materials. By integrating these advancements, manufacturers can meet the demands of high-quality, cost-effective production and align with global sustainability goals. Embracing these trends will ensure that the manufacturing of rigid couplings and other components remains competitive, responsible, and environmentally friendly in the years to come.

5.6 Conclusion

Developing a manufacturing process using life cycle engineering while incorporating social, sustainable, sensing, smart, and safe dimensions offers a comprehensive approach that aligns with modern industry needs. Here are the key conclusions that can be drawn from this approach:

1. **Holistic View and Long-Term Value Creation**
   - ***Life cycle Engineering:*** By taking a holistic view of the manufacturing process, companies can create long-term value for themselves, society, and the environment. This approach fosters innovation, reduces risks, and ensures that products are designed with their entire life cycle in mind.
2. **Improved Social Responsibility**
   - ***Social Dimension:*** A focus on social responsibility ensures that manufacturing processes are developed with consideration for labor rights, community impact, and equitable access to products and services. This can lead to better brand reputation and customer loyalty.
3. **Enhanced Sustainability**
   - ***Life cycle Engineering:*** By considering the entire life cycle of a product—from raw material extraction to end-of-life disposal—companies can minimize environmental impacts, reduce waste, and enhance resource efficiency.

- ***Sustainability Dimension:*** Integrating sustainable practices at every stage ensures that the process reduces carbon footprint, conserves energy, and promotes the use of renewable resources, aligning with global sustainability goals.

4. **Real-Time Monitoring and Adaptability**
    - ***Sensing Dimension:*** Implementing sensors and data collection systems enables continuous monitoring of the manufacturing process, providing valuable insights into performance, quality, and potential issues. This adaptability leads to more responsive and agile manufacturing systems.

5. **Enhanced Decision-Making and Efficiency**
    - ***Smart Dimension:*** The integration of smart technologies, such as AI and IoT, allows for real-time monitoring, data analysis, and automated decision-making. This improves process efficiency, reduces downtime, and enables predictive maintenance, ultimately lowering operational costs.

6. **Increased Safety and Compliance**
    - ***Safe Dimension:*** Prioritizing safety in the design and operation of manufacturing processes reduces the risk of accidents, ensures regulatory compliance, and protects the well-being of workers and consumers. This leads to a safer workplace and product.

7. **Competitive Advantage and Market Differentiation**
    - Adopting this multidimensional approach gives companies a competitive edge by aligning with evolving market demands, regulatory requirements, and consumer preferences. Innovative, sustainable, and socially responsible practices enable differentiation in a crowded market.

8. **Challenges and Implementation Complexity**
    - While the benefits are significant, integrating these dimensions can be complex and resource intensive. It requires cross-functional collaboration, investment in new technologies, and a commitment to continuous improvement. However, the long-term rewards often justify the initial challenges.

Therefore, developing a manufacturing process using life cycle engineering along with social, sustainable, sensing, smart, and safe dimensions results in a more resilient, efficient, and responsible manufacturing system. It positions companies to meet future challenges, enhance their reputation, and contribute positively to society and the environment.

6

Manufacturing System: Design, Integration and Execution

This chapter explores the integration of Life Cycle Engineering (LCE) and the 5S principles—Social, Sustainable, Sensing, Smart, and Safe—in sustainable manufacturing. LCE emphasizes a comprehensive approach to designing, operating, maintaining, and decommissioning manufacturing systems, ensuring efficiency, adaptability, and sustainability throughout the entire life cycle. Manufacturers can optimize their operations, reduce waste, and enhance overall efficiency by incorporating advanced technologies such as IoT, AI, and digital twins.

A case study demonstrates the practical application of these principles, highlighting how the strategic adoption of LCE and 5S frameworks can transform manufacturing processes. The chapter discusses the challenges faced in implementing sustainable practices and the innovative solutions developed to overcome them, showcasing real-world examples of how companies can navigate the complexities of sustainable manufacturing.

Additionally, the chapter provides insights into the future of sustainable manufacturing, emphasizing the importance of continuous improvement, effective leadership, and collaboration among industry stakeholders.

Manufacturers can further enhance their sustainability efforts by focusing on renewable energy integration, improving supply chain visibility, and adopting circular economy principles. This comprehensive approach aligns with global efforts to mitigate climate change and ensures long-term economic efficiency and resilience. Through this exploration, the chapter underscores the pivotal role of sustainable practices in shaping a responsible and innovative industrial landscape.

6.1 Introduction

Sustainable manufacturing has become a cornerstone of responsible business practice in the contemporary industrial landscape. This paradigm shift aligns economic growth with environmental stewardship and social responsibility, ensuring that manufacturing processes meet current demands

DOI: 10.1201/9781003501619-6

and safeguard future resources. Sustainable manufacturing encompasses the creation of goods in an ecologically responsible and resource-efficient manner, integrating cutting-edge technologies and practices that minimize environmental impact while enhancing operational efficiency.

One of the most significant aspects of sustainable manufacturing is its environmental impact. Integrating advanced technologies such as the Internet of Things (IoT), big data analytics, and predictive maintenance allows for precise monitoring and optimization of manufacturing processes. These technologies substantially reduce emissions, waste, and energy consumption. For instance, IoT devices enable real-time data collection and analysis, facilitating efficient resource use and waste minimization. On the other hand, predictive maintenance ensures machinery operates at peak efficiency, reducing energy consumption and extending equipment lifespan.

According to the 2023 Energy Efficiency Impact Report, [3], investments in energy efficiency since 1980 have led to an $800 billion reduction in annual energy expenditures in the United States and a 16% decrease in energy consumption per household. Such data underscores the potential of sustainable manufacturing to reduce environmental footprints significantly.

Beyond environmental benefits, sustainable manufacturing offers considerable economic advantages. While the initial investment in sustainable practices and technologies can be substantial, the long-term savings and profitability often outweigh these costs. Automated quality control systems, for example, detect defects early in the production process, reducing material waste and production costs. Additionally, paperless manufacturing systems enhance operational efficiency by eliminating the need for physical documentation, further reducing costs and environmental impact.

The shift towards sustainable practices also enhances supply chain resilience. Companies can reduce dependency on volatile raw material markets by optimizing resource use and incorporating eco-friendly materials. This resilience is crucial in maintaining consistent production levels and meeting consumer demands, ultimately contributing to long-term business viability.

Sustainable manufacturing also emphasizes social responsibility, ensuring businesses operate ethically and contribute positively to society. This includes fair labor practices, safe and inclusive work environments, and transparent governance. Technologies such as AR training provide interactive and immersive training experiences, enhancing worker skills and safety without the need for physical resources. This improves workforce competence and fosters a culture of continuous improvement and innovation.

Furthermore, sustainable manufacturing is becoming a strategic priority due to its resilience across market cycles. Industry leaders view sustainability as a long-term investment that aligns with their business objectives. This perspective is reinforced by the growing trend of companies setting ambitious carbon emission targets and incorporating sustainability into their core operations, which indicates a commitment to sustainable growth and corporate responsibility [43].

The adoption of sustainable manufacturing practices is closely tied to the integration of advanced technologies. IoT and artificial intelligence (AI) are pivotal in driving the digitization of manufacturing processes. IoT devices enable precise monitoring and control of production environments, optimizing efficiency and reducing waste. AI further enhances these capabilities through predictive analytics, improving decision-making and operational performance.

Augmented reality (AR) revolutionizes manufacturing training by offering virtual simulations that enhance skill development while minimizing physical resource use. This aligns with the ethos of Industry 4.0, which supports paperless manufacturing and promotes eco-friendly practices.

Despite its benefits, sustainable manufacturing faces challenges, including high initial costs, technological integration difficulties, and regulatory compliance. Companies must balance short-term financial pressures with long-term sustainability goals. Overcoming these challenges requires strategic investments in technology, continuous innovation, and a commitment to ethical practices.

To navigate these challenges, businesses must embrace a holistic approach to sustainability, integrating environmental, social, and governance (ESG) principles into their operations. This involves adopting sustainable technologies and fostering a culture of sustainability within the organization.

6.1.1 The Significance of Sustainable Manufacturing Development in the Modern World

Global production sectors are responsible for one-fifth of carbon emissions and consume 54% of the world's energy sources. This stark reality underscores the urgent need for manufacturing companies to tackle the challenges of decarbonization [145]. Sustainable manufacturing development is paramount in the modern world due to its multifaceted benefits and the pressing need to address environmental challenges. Firstly, sustainable practices are essential for mitigating climate change, which poses the most significant ongoing challenge for the manufacturing industry. Companies can reduce their carbon footprints by adopting environmentally friendly solutions, thus contributing to the global effort to combat climate change.

Economic benefits also underscore the significance of sustainable manufacturing. Despite economic uncertainties and a lack of funding, companies increasingly recognize that sustainability can drive operational efficiency and cost savings. For instance, integrating sustainability into operations can lower maintenance and operational expenses, improving the bottom line. Moreover, sustainable practices can enhance customer loyalty and foster innovation, providing a competitive edge in the marketplace.

Furthermore, sustainable manufacturing is becoming a strategic priority due to its resilience across market cycles. Industry leaders view sustainability as a long-term investment that aligns with their business objectives. This perspective is reinforced by the growing trend of companies setting ambitious carbon emission targets and incorporating sustainability into their core operations, which indicates a commitment to sustainable growth and corporate responsibility.

Leadership and government policy are also crucial in advancing sustainable manufacturing. Companies that embed sustainability into their core values and practices are likelier to succeed in their initiatives [43]. Additionally, government policies, such as carbon taxes and incentives for renewable energy investments, can provide the necessary financial impetus for companies to adopt sustainable practices. This regulatory support can help bridge the gap between intention and implementation, ensuring sustainability becomes a standard practice in the industry. Sustainable manufacturing is critically important for several reasons:

- **Environmental Protection:** Traditional manufacturing processes often lead to significant environmental degradation, including pollution and resource depletion. Sustainable manufacturing focuses on minimizing waste, reducing emissions, and using resources more efficiently, which are crucial for protecting the natural environment.
- **Regulatory Compliance:** Governments worldwide impose stricter environmental and sustainability regulations. Sustainable manufacturing helps companies comply with these regulations, avoiding penalties and potential legal issues.
- **Economic Efficiency:** Sustainable manufacturing often reduces costs by focusing on resource efficiency and waste reduction. For example, using less material and energy can lower operating costs, and optimizing processes can reduce the need for expensive cleanups or waste disposal.
- **Consumer Demand:** Modern consumers are increasingly aware of environmental issues and often prefer to buy products from companies with strong sustainability credentials. Sustainable manufacturing can, therefore, help companies attract and retain customers, boosting their market share and profitability.
- **Innovation and Competitiveness:** Sustainable practices push companies to develop new technologies and processes. This innovation can provide a competitive edge in the global market, where efficiency and sustainability are increasingly important.

6.1.2 Overview of How Digital Technologies are Transforming Manufacturing Processes

The rapid advancement of digital technologies is profoundly transforming the manufacturing landscape, ushering in a new era of efficiency, sustainability, and innovation. This transformation is driven by the integration of cutting-edge technologies such as blockchain, the Internet of Things (IoT), artificial intelligence (AI), virtual reality (VR), and cloud computing, each contributing uniquely to modernizing manufacturing processes. Digital technologies fundamentally transform manufacturing through several key mechanisms, as illustrated by [89, 111] in TABLE 6.1.

TABLE 6.1
Examples of Digital Technologies employed during the Manufacturing Process.

Digital Technology	**Application**
Enhanced Training and Skill Development	VR provides immersive training experiences, allowing workers to learn and practice skills in a safe, controlled virtual environment.
	AI personalizes training programs, adapting them to individual learning curves and providing targeted feedback to enhance skill acquisition.
Operational Efficiency	AI algorithms optimize manufacturing processes by reducing waste and enhancing productivity. They are used for process optimization and efficiency improvements.
	VR for Virtual Prototyping and Testing reduces the need for physical models, enabling more rapid product development and iterations.
Safety and Risk Management	VR simulations train workers to handle hazardous situations without real-world risk, enhancing workplace safety.
	AI for Predictive Maintenance predicts and prevents operational failures, improving safety and reducing downtime.
Global Collaboration and Integration	VR for Global Collaboration enables virtual collaboration, allowing experts worldwide to interact remotely with the manufacturing environment.
	AI facilitates communication and coordination among global teams, streamlining international collaboration.
Flexibility and Customization	Digital Twins create detailed virtual models of manufacturing processes, allowing for real-time monitoring, simulations, and adjustments.
	AI and VR enable customization of products and processes, making it easier to meet specific customer demands and adapt to changing market conditions.
Sustainability, Circular Economy	Sustainable practices are integrated into manufacturing through digital technologies that allow for the reuse and recycling of materials, reducing waste.

Digital Technology	Application
	VR and AI in Production Planning incorporate sustainability metrics into decision-making processes, ensuring environmental considerations are part of manufacturing strategies.
Product Evaluation and Quality Control	VR environments for product evaluation offer detailed and interactive assessments, enabling thorough testing and adjustments before final production. AI for Quality Control provides real-time monitoring and feedback, ensuring products meet high-quality standards throughout manufacturing.

Blockchain technology revolutionizes data management and security in manufacturing by providing a transparent, immutable ledger for all transactions. This ensures that every step in the manufacturing process is recorded and verifiable, enhancing transparency and traceability. Blockchain's smart contracts automate and enforce compliance with predefined conditions, reducing administrative overhead and ensuring adherence to sustainability standards. This level of transparency is crucial for tracking the life cycle of products and verifying sustainable practices across the supply chain [89].

Moreover, the decentralization enabled by blockchain technology offers enhanced security and reliability. Traditional centralized systems are vulnerable to single points of failure, but blockchain's distributed ledger mitigates this risk by spreading data across multiple nodes. This decentralization ensures continuous operation and data integrity, even if part of the system is compromised. In supply chains, decentralized networks facilitate secure and efficient transactions, improving coordination and reducing delays [89].

IoT devices transform manufacturing processes by enabling real-time monitoring and control. Sensors embedded in machinery and equipment continuously collect data on performance and environmental conditions. This data is crucial for optimizing operations, reducing waste, and improving energy efficiency. One of the significant advantages of IoT is predictive maintenance, where sensors monitor equipment health and predict failures before they occur, minimizing downtime and extending machinery lifespan [89].

In addition, digital transformation encompasses adopting various digital technologies to optimize manufacturing processes. Big data analytics and AI are central to this transformation, providing the tools for process optimization, quality control, and efficiency improvements. By analyzing vast amounts of data, manufacturers can identify inefficiencies, predict market trends, and make informed decisions that enhance productivity and reduce costs. This holistic approach ensures that every manufacturing process is optimized for better performance and sustainability [89].

Thus, integrating IoT and cloud computing creates a powerful ecosystem for manufacturing. IoT devices generate vast amounts of data, which are processed and stored in the cloud. This combination facilitates seamless data sharing, collaboration, and process automation across the manufacturing floor. Cloud computing offers a scalable infrastructure that can handle the large volumes of data generated by IoT devices, providing the computational power needed for advanced analytics and real-time decision-making. This integrated system enhances operational flexibility, enabling manufacturers to adapt to changing demands and scale operations efficiently quickly.

AI is at the forefront of smart manufacturing, providing advanced analytics and automation capabilities. AI algorithms analyze data to optimize production schedules, improve quality control, and reduce waste. Intelligent systems powered by AI can make decisions in real-time, based on data inputs from various sensors and devices. This reduces human error and increases efficiency, allowing for more precise control over manufacturing processes. AI also supports the development of autonomous robots and automated systems that perform complex tasks with high precision and reliability [111].

Digital technologies are crucial in driving sustainable manufacturing practices. They enable detailed monitoring and control of energy usage, significantly improving energy efficiency and reducing carbon emissions. Advanced analytics and IoT facilitate the efficient use of raw materials, minimizing waste and promoting circular economy principles. Additionally, digital tools streamline compliance with environmental regulations by automating data collection and reporting, ensuring that manufacturing practices meet sustainability standards.

This chapter explores the development of advanced manufacturing systems using the 5S framework—Social, Sustainable, Sensing, Smart, and Safe—alongside Life Cycle Engineering (LCE) principles. The 5S framework emphasizes stakeholder engagement, resource efficiency, real-time monitoring, automation, and strict safety standards, creating a comprehensive approach to manufacturing that aligns technological advancements with ethical and environmental responsibilities. By applying these principles at different levels of automation, from manual operations to complete automation, the chapter outlines strategies to integrate sustainability and cybersecurity into the manufacturing process.

LCE principles ensure that every stage of the manufacturing system life cycle, design, operation, maintenance, and decommissioning, is optimized for sustainability, efficiency, and adaptability. Key aspects include creating modular and adaptable systems, incorporating sustainable practices in daily operations, implementing predictive and preventive maintenance, and ensuring environmentally responsible decommissioning processes. The chapter discusses the benefits of LCE, such as improved efficiency and cost savings, as well as challenges like high initial investment costs and technology integration issues.

A case study demonstrates the successful application of the 5S framework and LCE principles in a real-world manufacturing scenario, showcasing operational efficiency, cost reduction, and improvements in environmental and social impact. The chapter concludes by highlighting the future of sustainable manufacturing, emphasizing the importance of technological advancements, effective leadership, supply chain transparency, renewable energy integration, and circular economy principles. This integrated approach ensures that manufacturing processes are profitable, environmentally and socially responsible.

6.2 Manufacturing as 5S Systems

Developing a manufacturing system using the 5S framework—incorporating Social, Sustainable, Sensing, Smart, and Safe principles—creates a holistic approach that aligns advanced technology with ethical and environmental responsibilities. This strategy ensures that manufacturing operations are efficient, innovative, and socially responsible. Thus, FIGURE 6.1 shows how each of the 5S principles can be applied to material development. Here is how to integrate each of these principles into the development of a manufacturing system.

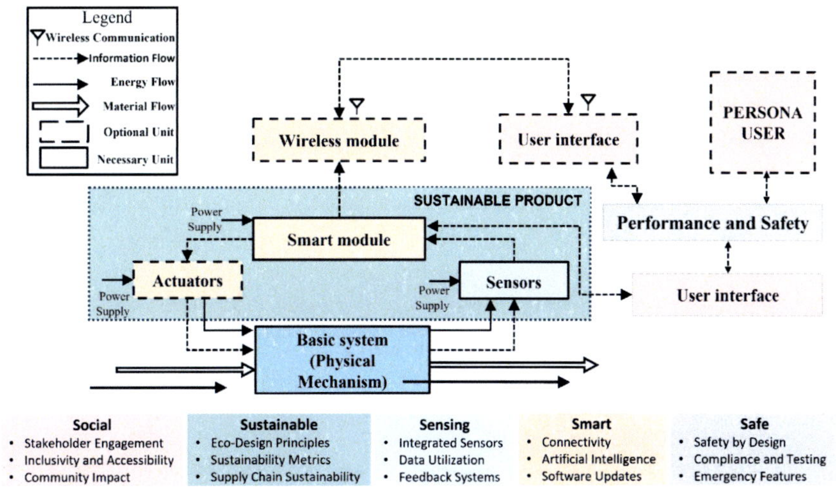

FIGURE 6.1
5S manufacturing development (social, sustainable, sensing, smart and safe).

- **Social**
  - ***Stakeholder Engagement:*** Regularly engage with all stakeholders, including employees, suppliers, customers, and local communities, to ensure their needs and concerns are addressed. This fosters a cooperative environment leading to more sustainable and accepted practices.
  - ***Community Benefits:*** Design the system to bring direct benefits to the local community, such as job creation, skill development, and community improvement initiatives, which contribute to social equity.
  - ***Inclusivity:*** Ensure the manufacturing system is accessible and offers opportunities to a diverse workforce, including those with disabilities, promoting diversity and equality within the workplace.

- **Sustainable**
  - ***Resource Efficiency:*** Focus on minimizing waste and maximizing the efficiency of resources like energy, water, and raw materials. Employ technologies and practices such as recycling, waste management systems, and efficient machinery.
  - ***Green Technologies:*** Integrate renewable energy sources, such as solar or wind power, into the manufacturing process to reduce carbon footprints and dependence on non-renewable energy.
  - ***Sustainable Supply Chain:*** Develop a supply chain that prioritizes suppliers who also adhere to sustainable practices, ensuring that sustainability is considered from the extraction of raw materials to the delivery of finished products.

- **Sensing**
  - ***Real-Time Monitoring:*** Implement sensors throughout the manufacturing system to monitor various parameters such as production efficiency, energy usage, and environmental conditions, providing data for continuous improvement.
  - ***Predictive Maintenance:*** Use sensor data to anticipate maintenance needs, reducing downtime and extending the lifespan of equipment.
  - ***Feedback Loops:*** Establish feedback loops that use sensor data to dynamically adjust processes, improving responsiveness to changing conditions or detecting faults before they cause larger issues.

- **Smart**
  - ***Automation and Robotics:*** Employ automation and robotics to increase precision, reduce labor costs, and improve safety. This also allows human workers to focus on more complex or creative tasks.
  - ***AI:*** Integrate AI to analyze data collected from sensors to optimize manufacturing operations, improve decision-making, and tailor production processes to specific requirements dynamically.

  - ***Connectivity and Integration:*** Ensure all systems and processes are interconnected through an IoT platform, facilitating seamless communication and data sharing across the manufacturing system.

- Safe
  - ***Health and Safety Standards:*** Design the manufacturing system to comply with the highest safety standards. This includes safe machinery layouts, emergency protocols, and the use of protective gear.
  - Regularly conduct risk assessments to identify potential hazards and implement corrective measures to mitigate risks.
  - ***Safety Training:*** Provide continuous safety training to all employees, ensuring they are aware of how to operate safely within the new system and how to respond in case of an emergency.

By applying the 5S principles to developing a manufacturing system, companies can ensure that their operations are not only technologically advanced but also responsible and sustainable. This approach leads to a robust system that meets high ethical standards, is environmentally friendly, and ensures all stakeholders' safety and well-being while being economically efficient.

6.2.1 5S Taxonomy Levels

The 5S features provide invaluable support toward more intelligent, resilient, and environmentally conscious manufacturing processes. By wholeheartedly adopting the 5S framework, companies can adapt and flourish within the ever-changing landscape of Industry 4.0. They can make innovation, sustainability, and efficiency their guiding principles, ensuring long-term success. On the other hand, the process towards Industry 4.0, characterized by digitalization, connectivity, and automation, offers manufacturing companies unprecedented opportunities for growth, efficiency, and sustainability. However, the path to achieving these goals is not a one-size-fits-all approach. To effectively navigate the complex landscape of Industry 4.0, organizations must adopt a structured framework that aligns with their automation levels and goals. This comprehensive strategy guide explores implementing the 5S framework across different automation levels, from manual operations to full automation [16, 17]. The 5S framework represents a holistic approach integrating Sensing, Smart, Sustainable, Social, and Safety dimensions to optimize manufacturing processes. By tailoring strategies for each automation level, companies can ensure a seamless transition towards Industry 4.0 while addressing environmental, social, and safety aspects. Below are the strategies for each level, considering their unique challenges and opportunities.

- **Level 0, Manual Operations:** The first step is introducing sensing technologies at this basic level of manual operations. Install IoT sensors to monitor critical parameters such as temperature, pressure, and machine status. Begin collecting data to gain insights into equipment performance.
- **Level 1, Machine Assistance:** In Level 1, focus on enhancing smart capabilities. Leverage the data collected from sensors to develop predictive maintenance models. Implement automation for routine tasks, such as data collection and reporting, using AI and machine learning.
- **Level 2, Partial Automation:** In Level 2, work towards a more integrated approach. Expand on sensing and smart capabilities. Install additional sensors to cover a wider range of parameters. Utilize AI algorithms to optimize production scheduling, resource allocation, and energy usage.
- **Level 3, Conditional Automation:** Sustainability practices can be integrated effectively at this level. Implement renewable energy sources like solar panels to reduce carbon footprint. Begin waste reduction initiatives, including recycling and reusing materials. Focus on continuous improvement in sustainability measures.
- **Level 4, High Automation:** With a high level of automation, enhance the social dimension. Invest in employee training and development programs. Create a culture of sustainability and innovation among the workforce. Collaborate with local suppliers to ensure ethical sourcing practices and support community engagement initiatives.
- **Level 5, Full Automation:** Achieving full automation implies high technological maturity. At this stage, prioritize safety. Strengthen cybersecurity measures to protect critical data and systems. Conduct regular security audits, enforce strict access controls, and provide comprehensive employee training in cybersecurity best practices.

FIGURE 6.2 depicts the five levels of automation that guide the process of fully automated manufacturing processes. A comprehensive long-term strategy plan is necessary to transition from Level 1 to Level 5. Below, an example of a long-term plan is presented:

- First, a thorough evaluation of current operations is conducted in the initial assessment and planning phase spanning Years 1–2. This entails identifying areas for improvement and potential bottlenecks. A detailed roadmap is then developed, outlining the steps to progress through the levels. Clear objectives and key performance indicators (KPIs) are set to measure progress throughout the transition process.
- Next, in the investment in technology phase, which spans Years 2–3, resources are allocated to invest in advanced technologies and infrastructure. This includes upgrading sensor networks and data collection systems to enhance sensing capabilities and data accuracy. Advanced AI

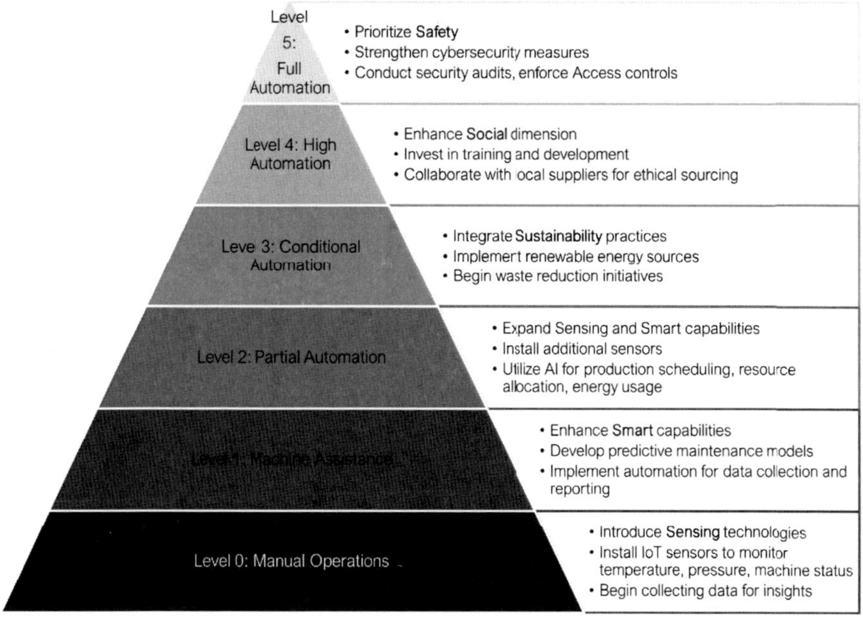

FIGURE 6.2
Five levels of automation using the 5S framework.

and machine learning algorithms are implemented to optimize operations, improve predictive maintenance models, and automate decision-making processes.

- Following this, in the process optimization phase spanning Years 4–5, processes and workflows are streamlined to leverage the full potential of automation and advanced technologies. Systems are integrated across departments and functions to enable seamless data flow and communication. Performance metrics are continuously monitored and analyzed to identify areas for further optimization and refinement.
- Sustainability practices are integrated into operations in the sustainability integration phase, which extends from Years 6–7. This includes investing in renewable energy sources and energy-efficient technologies to reduce carbon footprint and operational costs. Waste management strategies, including recycling and reuse initiatives, are implemented to minimize environmental impact and enhance resource efficiency.
- Subsequently, in the human capital development phase covering Years 7–8, investments are made in employee training and development programs. The goal is to enhance digital literacy and technological skills among the workforce. A culture of innovation, collaboration, and continuous learning is fostered to adapt to evolving technologies and market dynamics.

- In the cybersecurity enhancement phase, spanning Years 9–10, cybersecurity measures are prioritized to protect critical data and systems from potential threats and vulnerabilities. Network security infrastructure is strengthened, robust access controls are implemented, and regular security audits and assessments are conducted. Ongoing cybersecurity training and awareness programs ensure that employees can handle cybersecurity risks effectively.

Throughout this transition process, ongoing monitoring and adaptation are crucial. Progress is continuously monitored against established KPIs and strategies are adjusted to address emerging challenges and opportunities. Keeping abreast of technological advances and market trends ensures that organizations remain competitive and agile in a rapidly evolving landscape. A culture of continuous improvement and innovation drives sustained growth and success over the long term.

6.2.2 The Theory of Integrated Manufacturing Systems

In manufacturing, the 5S framework is integrated by embedding sensors across machinery for real-time data gathering, which, when analyzed by artificial intelligence, enhances predictive maintenance and operational efficiency. Sustainable practices are embedded to minimize environmental impact through waste and energy reduction. The framework also focuses on improving the work environment and stakeholder engagement for social benefits. Additionally, robust cybersecurity measures protect sensitive data and intellectual property, ensuring a secure manufacturing ecosystem. Implementing the 5S framework in manufacturing companies offers numerous advantages, bolstering their overall success and competitiveness in the Industry 4.0 era. Some of the main advantages of integrating 5S in manufacturing are the following:

- **Operational Efficiency:** By integrating Sensing and Smart technologies, real-time monitoring, predictive maintenance, and automation become achievable. This leads to enhanced production efficiency, reduced downtime, and optimized resource allocation, resulting in cost savings and heightened productivity.
- **Innovation:** The 5S framework cultivates an environment conducive to innovation, leveraging data analytics, AI, and other cutting-edge technologies. This fosters a culture of continuous improvement and the development of novel products and processes, enabling manufacturers to remain competitive and adaptive in swiftly evolving markets.
- **Sustainability:** Sustainability practices advocated by the 5S framework, including energy conservation and waste reduction, yield environmental benefits while also delivering cost savings. Manufacturers can reduce their environmental footprint, meet regulatory mandates, and appeal to eco-conscious consumers.

- **Employee Satisfaction:** The social dimension of 5S centers on improving the work environment and fostering stakeholder engagement. Investments in employee training, health and safety increase job satisfaction, morale, and retention rates. In turn, a motivated workforce contributes to increased productivity and product quality.
- **Supply Chain Resilience:** Embracing the 5S framework encourages end-to-end supply chain integration and supplier collaboration. This improves the resilience of the supply chain, improves risk management, and increases the response to market changes and disruptions.
- **Data Security:** With the Safety pillar, companies safeguard their critical data and systems from cyber threats. Robust cybersecurity measures ensure data integrity, maintain customer trust, and prevent costly data breaches or operational disruptions.
- **Competitive Advantage:** Manufacturers wholeheartedly adopting the 5S framework position themselves to outperform competitors. They distinguish themselves through superior operational efficiency, sustainability practices, and innovation, ultimately gaining a competitive edge in the market.
- **Long-Term Sustainability:** Through adopting sustainability practices, waste reduction, and community engagement, companies cultivate a positive reputation that can lead to long-term sustainability and sustained growth. Ethical and environmentally responsible manufacturing practices resonate with consumers and investors alike.
- **Cost Reduction:** The 5S framework facilitates cost reduction through energy conservation, waste minimization, and optimized resource utilization, ultimately enhancing a company's profitability.
- **Adaptability to Industry Changes:** Equipped with the tools provided by the 5S framework, companies can effectively respond to the rapidly evolving landscape of Industry 4.0. They can navigate technological advancements, adapt to market shifts, and comply with regulatory requirements, ensuring continued relevance and competitiveness.

Adopting the 5S framework in manufacturing companies brings many benefits, spanning operational efficiency, innovation, sustainability, employee satisfaction, and competitive advantage. It positions companies to thrive within the dynamic Industry 4.0 landscape while positively contributing to the environment and society. The importance of implementing the 5S framework in manufacturing and industry lies in its ability to address the complex challenges and opportunities of the Industry 4.0 era, leading to enhanced operational efficiency, innovation, sustainability, workforce satisfaction, supply chain resilience, data security, competitive advantage, long-term viability, cost reduction, and adaptability. It provides a structured approach to navigate the evolving business landscape and achieve success. Implementing the 5S methodology in manufacturing companies comes with several significant challenges that need to be addressed:

1. **Initial Investment Costs:** Integrating Sensing, Smart, Sustainable, Social, and Safety technologies and practices typically requires a substantial upfront investment, including acquiring new technologies, employee training, and process reconfiguration.
2. **Technology Integration:** Integrating various technologies seamlessly can be complex, as different solutions may not always be compatible, necessitating customization or integration solutions.
3. **Data Management and Privacy:** Managing large volumes of data from sensors and other sources can be overwhelming. Ensuring data accuracy, privacy, and compliance with data protection regulations is essential but challenging.
4. **Workforce Training:** Transitioning to advanced technologies and practices requires a skilled workforce. Training employees to operate, maintain, and troubleshoot these technologies effectively is crucial but can face resistance.
5. **Cybersecurity Risks:** As manufacturing becomes more connected, the risk of cyberattacks increases, making robust cybersecurity measures essential to protect sensitive data and systems.
6. **Regulatory Compliance:** Adhering to industry regulations and standards, which can vary by location and product, can be complex and time-consuming.
7. **Supply Chain Integration:** Achieving end-to-end supply chain integration can be challenging, especially with multiple suppliers and partners, requiring negotiation and coordination efforts.
8. **Cultural Change:** Shifting the organization's culture to embrace the Social dimension of 5S may cause resistance or reluctance to adopt new practices.
9. **Environmental Regulations:** Meeting sustainability goals amid evolving environmental regulations and standards requires constant monitoring and adaptation.
10. **ROI Uncertainty:** Calculating the return on investment (ROI) can be challenging, as quantifying tangible benefits, particularly in the early stages, may be difficult.
11. **Technological Obsolescence:** Rapid technological advancements mean that technologies adopted today may become outdated quickly, necessitating planning for future upgrades and staying agile in adopting innovations.

FIGURE 6.3 illustrates a flow diagram of the challenges faced in the manufacturing sector when 5S is deployed. Overcoming these challenges demands strategic planning, ongoing commitment, and collaboration with technology providers and industry partners.

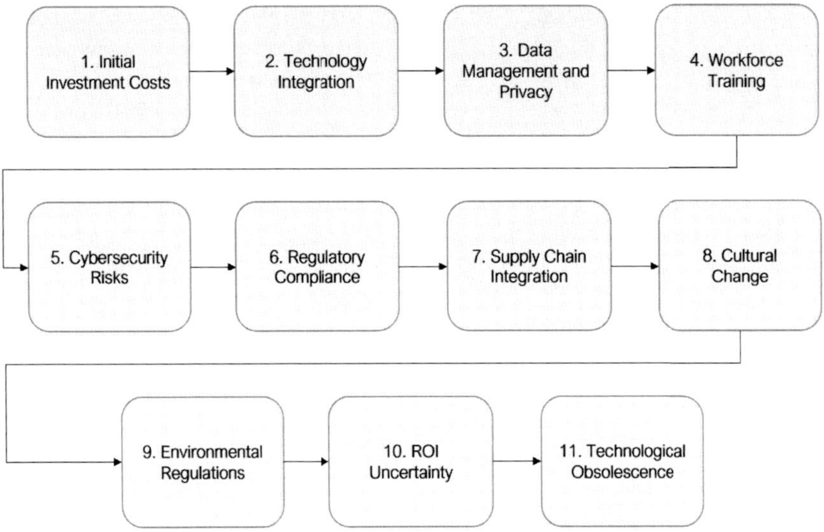

FIGURE 6.3
Flow diagram challenges in integrating 5S in manufacturing.

Manufacturers must carefully assess their specific needs, resources, and goals when implementing the 5S methodology and remain adaptable to changes in technology and industry landscapes. A strategic approach is paramount to elevating the 5S framework by integrating diverse technologies, focusing on AI. This approach must navigate the dual landscape of leveraging inherent benefits while mitigating the challenges of such technological fusion. The strategic initiative aims to conduct a rigorous evaluation and enhancement of existing systems and workflows, focusing on Sensing, Smart technologies, Sustainability, Social impact, and Safety, collectively called the 5S framework. The objective is to execute an exhaustive analysis to identify inefficiencies, gaps, and potential improvements through AI-powered analytics, leveraging data from current infrastructures, stakeholder feedback, and industry benchmarks.

6.2.3 5S Systems to Enable Industry 5.0

A strategic blueprint will be crafted to create a foundation for integrating AI and other advanced technologies into the 5S framework. This plan outlines clear objectives, timelines, and resource allocations, starting with pilot schemes in key sectors for a phased rollout. The goal is to establish measurable targets for each aspect of the 5S framework, ensuring strategic alignment and execution.

Technological fusion is another critical component, focusing on seamlessly integrating scalable and flexible technologies like IoT for advanced sensing and AI for intelligent operations. By implementing IoT sensors to collect real-time operational data and applying AI and machine learning, the initiative seeks to optimize processes and decision-making. Advanced data management will capitalize on the extensive datasets from IoT and other sources to enhance operational efficiency, sustainability, and strategic decision-making. Sophisticated data analytics, including Fuzzy TOPSIS and genetic algorithms, could integrate expert judgments and improve optimization resilience and adaptability. The comprehensive involvement of stakeholders is essential for fostering a culture of social responsibility and innovation. This includes extensive training on new technologies and methodologies for the workforce, clients, and the broader community, alongside initiating community projects to highlight the social contributions of the 5S framework. Safety is paramount, with the initiative committed to forging and maintaining stringent measures to protect sensitive information and infrastructures. This involves the integration of advanced cybersecurity protocols, consistent security evaluations, and adherence to current standards and regulations. The philosophy of perpetual enhancement will drive continuous improvement. Regular monitoring, assessment, and refinement of performance indicators powered by AI-generated insights will establish iterative feedback mechanisms for ongoing process optimization. This commitment to relentless improvement underpins the entire strategic initiative, ensuring its adaptability and longevity in achieving its goals within the 5S framework. Adhering to this refined strategy enables organizations to implement the 5S framework adeptly, harnessing AI and other technologies to optimize electric drive systems and beyond. This strategy promises a comprehensive adoption that augments performance and environmental stewardship and reinforces safety while addressing the inherent challenges of technological integration. FIGURE 6.4 depicts the general strategy flow diagram for deploying 5S.

Integrating the 5S framework with Industry 4.0 and 5.0 principles marks a pivotal shift in manufacturing, heralding a future where efficiency, innovation, and societal well-being are at the forefront. This strategic initiative leverages the core components of Sensing, Smart technologies, Sustainability, Social impact, and Safety to create a manufacturing ecosystem that is efficient, productive, sustainable, socially responsible, and secure. At the heart of this transformative approach is the seamless fusion of IoT and AI technologies, which provide real-time data collection and analysis. This capability is crucial for enhancing operational efficiency by reducing downtime, streamlining production lines, and facilitating predictive maintenance. The outcome is a significant increase in throughput and a reduction in operational costs, setting a new standard for manufacturing excellence. Sustainability and environmental stewardship are also central to this integration. By prioritizing eco-friendly practices, manufacturers can dramatically lower their environmental impact. Advanced data analytics is key here, optimizing resource use and minimizing waste. The 5S framework

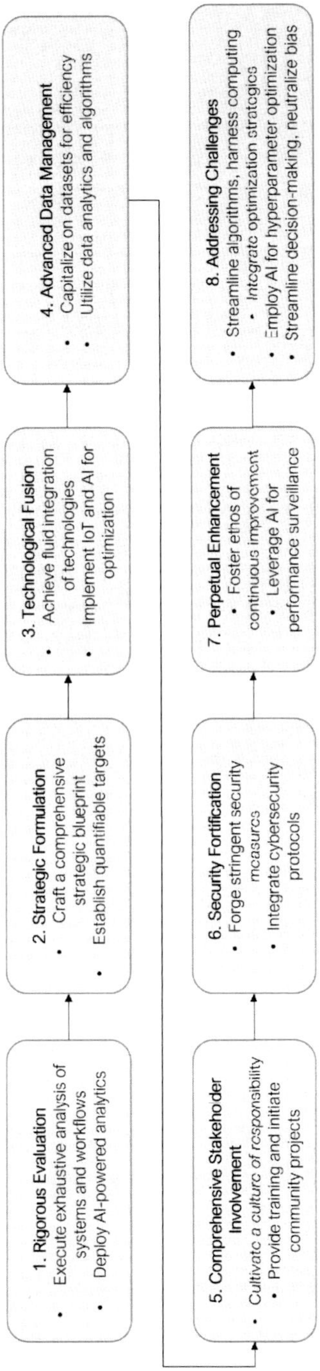

FIGURE 6.4
Flow diagram strategies in integrating 5S in manufacturing.

also underscores the importance of social impact and community engagement. Manufacturers can forge stronger connections with their workforce and the wider community by initiating training programs and community projects. This prepares a skilled workforce for future manufacturing challenges and showcases a commitment to corporate social responsibility.

In today's digital age, where cyber threats loom large, the safety aspect of the 5S framework ensures robust protection for sensitive data and manufacturing systems. Adopting cutting-edge cybersecurity measures safeguards the integrity of manufacturing operations, instilling trust among stakeholders and customers alike. Moreover, continuous improvement and innovation principles are deeply ingrained in the 5S framework. AI-driven insights fuel a culture of perpetual enhancement, allowing manufacturers to refine their processes, products, and services continually. This culture of innovation aligns perfectly with the objectives of Industry 4.0 and 5.0, which advocate leveraging digital technologies and human creativity to advance manufacturing. Agility and adaptability are further enhanced through the strategic integration of smart technologies and data management. This adaptability is essential for navigating market fluctuations, tailoring products to specific customer preferences, and exploring innovative business models. The confluence of the 5S framework with Industry 4.0 and 5.0 principles equips manufacturers with a holistic approach that transcends traditional manufacturing paradigms. This integrated approach not only boosts the efficiency and productivity of manufacturing processes but also ensures they are conducted in a sustainable, socially responsible, and secure manner. Manufacturers with this forward-thinking approach are better positioned to anticipate and mitigate operational challenges, innovate in response to market demands, and adapt to societal expectations.

Ultimately, adopting the 5S framework, in conjunction with Industry 4.0 and 5.0, paves the way for a new manufacturing era where technology and human ingenuity converge to create smarter, more resilient, and sustainable production ecosystems that benefit not just the industry but society.

6.3 Life Cycle Engineering Description

LCE focuses on the comprehensive and systematic approach to designing, operating, maintaining, and decommissioning manufacturing systems. It emphasizes sustainability, efficiency, and adaptability across the entire lifecycle of the manufacturing system. The LCE framework ensures that each stage of the system's life—from ideation and conceptualization to design, creation, and operation—is meticulously planned and executed to meet the dynamic needs of modern manufacturing. Key aspects of LCE in manufacturing system development include:

1. **System Design:** Emphasizes creating manufacturing systems that are efficient, modular, and adaptable. This ensures that the systems can quickly respond to changes in production requirements and market demands.
2. **Operation:** Incorporates sustainable practices into daily operations, focusing on proactive energy and waste management to minimize environmental impact and improve overall efficiency.
3. **Maintenance:** Utilizes predictive and preventive maintenance strategies to enhance system longevity and performance, reducing downtime and maintenance costs.
4. **Decommissioning:** Ensures that the decommissioning processes are conducted sustainably, minimizing environmental impact and effectively managing the disposal or repurposing of system components.

Digital technology innovations play a pivotal role in LCE by enabling rapid development and optimization of manufacturing systems. Technologies provide powerful tools for system design, monitoring, and continuous improvement.

6.3.1 Benefits and Limitations of Applying Life Cycle Engineering Principles in this Entity

LCE principles offer a comprehensive approach to manufacturing system development, emphasizing efficiency, sustainability, and adaptability throughout the entire life cycle of manufacturing systems. The following sections outline the detailed benefits and limitations of implementing LCE principles in manufacturing system development.

Implementing LCE principles in manufacturing system development can lead to numerous advantages that enhance both operational efficiency and sustainability. These benefits are critical for organizations seeking to remain competitive and responsive in today's dynamic manufacturing landscape.

- **Enhanced Efficiency**
  - ***Optimized Resource Use:*** LCE promotes the efficient use of materials, energy, and other resources throughout the manufacturing process, leading to reduced waste and lower costs.
  - ***Improved Workflow:*** Systematic planning and optimization of workflows reduce production bottlenecks and enhance overall productivity, ensuring smoother operations and higher output.
- **Modularity and Adaptability**
  - ***Scalable Systems:*** Modular design allows for easy scalability and reconfiguration of manufacturing systems to meet changing production demands, providing flexibility in manufacturing operations.

  - ***Rapid Response to Market Changes:*** Adaptable systems can quickly adjust to new product designs, market trends, and technological advancements, enabling manufacturers to stay ahead of the competition.

- **Sustainability**
  - ***Reduced Environmental Impact:*** Emphasizing sustainable practices minimizes waste, lowers emissions, and reduces the overall ecological footprint of manufacturing operations, contributing to environmental conservation.
  - ***Resource Conservation:*** Efficient resource management and recycling practices help in conserving resources, ensuring long-term sustainability and reducing costs associated with raw material procurement.

- **Predictive Maintenance**
  - Utilizing IIoT and AI technologies enables predictive maintenance, reducing unexpected downtime and extending the lifespan of equipment by predicting failures before they occur.
  - ***Preventive Strategies:*** Preventive maintenance schedules ensure regular upkeep, minimizing the risk of equipment failure and maintaining optimal performance, thus reducing maintenance costs over time.

- **Cost Savings**
  - ***Reduced Operating Costs:*** Efficient resource use and optimized workflows lead to significant cost savings in daily operations by minimizing waste and improving productivity.
  - ***Lower Maintenance Costs:*** Proactive maintenance strategies, such as predictive and preventive maintenance, reduce the frequency and cost of repairs, resulting in long-term savings.

- **Improved Quality**
  - ***Consistent Product Quality:*** Detailed planning and quality control measures ensure consistent product quality, reducing defects and rework, and enhancing customer satisfaction.
  - ***Enhanced Customer Satisfaction:*** High-quality products and timely delivery improve customer satisfaction and loyalty, leading to repeat business and a strong market reputation.

- **Digital Integration**
  - ***Advanced Analytics:*** Utilizing big data and advanced analytics provides valuable insights for continuous improvement and decision-making, enabling data-driven strategies for optimization.

- ***Real-Time Monitoring:*** Digital twins and IIoT enable real-time monitoring and adjustments, ensuring optimal system performance and quick response to any issues.

While applying LCE principles offers significant benefits, it has challenges and limitations. These must be carefully considered to ensure successful implementation and maximize LCE's benefits.

- **Resource Allocation:** Allocating resources for the initial setup can strain budgets, particularly for small and medium-sized enterprises (SMEs).
- **Integration Challenges:** Integrating various digital technologies and ensuring seamless data flow can be complex and time-consuming, requiring specialized knowledge and skills.
- **Organizational Resistance:** Employees and management may resist changes to established processes and systems, hindering the adoption of LCE principles.

Understanding these benefits and limitations can help organizations make informed decisions about adopting Life Cycle Engineering principles in their manufacturing system development. This balanced approach maximizes the advantages while effectively managing the challenges.

6.3.2 How Life Cycle Engineering Contributes to Sustainable Manufacturing in this Entity

LCE contributes to sustainable manufacturing in product transfer, technology transfer, and facility design through several key strategies [43, 89, 111, 145]:

- **Product Transfer:**
  - ***Enhanced Resource Efficiency:*** LCE emphasizes optimizing resource use, reducing material waste, and lowering costs by integrating efficient workflows and resource management practices. This includes using sustainable materials and ensuring that products are designed for durability and recyclability.
  - ***Modular and Scalable Designs:*** Creating modular products that can be easily updated or modified to meet changing demands helps extend the product life cycle, reducing the need for complete replacements and thus minimizing waste.
- **Technology Transfer:**
  - ***Integration of Digital Technologies:*** Digital tools like IoT, AI, and VR are employed to enhance the design, monitoring, and optimization of manufacturing processes. These technologies enable predictive maintenance, real-time monitoring, and efficient production planning, which reduce downtime and energy consumption.

- ***Sustainable Innovation:*** By fostering continuous innovation and adaptation of new technologies, LCE supports the development of more efficient manufacturing processes and systems, promoting long-term sustainability.

- **Facility Design:**
  - ***Energy Efficiency:*** LCE incorporates sustainable practices in facility design, such as optimizing energy use, employing renewable energy sources, and enhancing overall facility efficiency. This reduces the environmental footprint of manufacturing operations.
  - ***Circular Economy Principles:*** Designing facilities that support recycling, refurbishing, and reusing materials helps create a circular economy in manufacturing. This approach minimizes waste and promotes the sustainable use of resources throughout the facility's life cycle

Overall, LCE supports sustainable manufacturing by embedding sustainability principles into the entire life cycle of manufacturing systems, from design and production to maintenance and decommissioning. This comprehensive approach ensures that manufacturing processes are efficient and adaptable, environmentally responsible, and economically viable in the long term.

6.4 Life Cycle Diagram Flow

The LCE Diagram Flow depicted in FIGURE 6.5 provides a comprehensive framework for developing and implementing a manufacturing system. This framework is divided into three main elements: Product Transfer, Technology Transfer, and Facility Design. Each element encompasses specific stages: Ideation/Imagination, Basic Development/Conceptualization, Advanced Development/Design, and Launch/Creation/Operation. The detailed descriptions of each element and their respective stages are outlined below.

6.4.1 Entity Description: Manufacturing System

The manufacturing entity focuses on systematically developing and deploying a manufacturing system, integrating three crucial elements: Product Transfer, Technology Transfer, and Facility Design. Each element contributes to the overall success of the manufacturing system by addressing specific aspects of development and implementation.

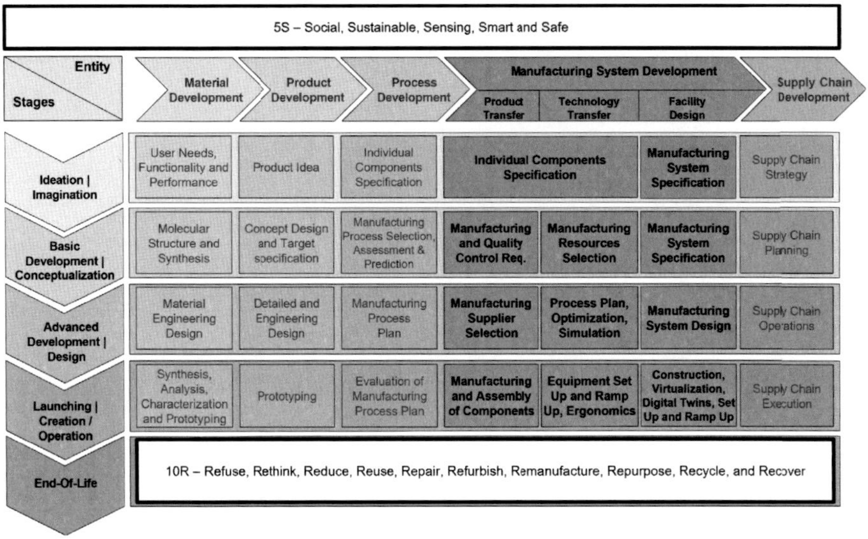

FIGURE 6.5
Life Cycle diagram flow: Manufacturing system development and its stages.

- **Product Transfer:** This element involves the detailed specification, development, and manufacturing of individual components or products. It encompasses the ideation and imagination of product requirements, the conceptualization of manufacturing and quality control needs, the selection of suppliers, and the actual manufacturing and assembly of components.
- **Technology Transfer:** This element addresses the specification, selection, and implementation of manufacturing resources and technologies. It includes the ideation and imagination of component specifications, the conceptualization of manufacturing resources, the planning and optimization of manufacturing processes, and the setup and ramp-up of production equipment.
- **Facility Design:** This element focuses on the design and construction of the manufacturing facility itself. It covers the ideation and imagination of product and process specifications, the conceptualization of the manufacturing system, the design and simulation of manufacturing strategies, and the construction and ramp-up of the facility.

6.4.2 Stages

The life cycle of a manufacturing system is divided into distinct stages, each with specific activities and outputs that contribute to the overall development process. These stages include Idea/Imagination, Basic Development/Conceptualization, Advanced Development/Design, and Launching/Creation/Operation. Each stage encompasses different aspects of

Product Transfer, Technology Transfer, and Facility Design. Below, a detailed overview of each stage and the associated activities, inputs, and outputs is provided.

6.4.2.1 Idea—Imagination

In the Idea—Imagination stage, the initial concepts and specifications for the product, technology, and facility are developed. This stage is crucial as it sets the groundwork for all subsequent stages by gathering essential information and defining preliminary requirements.

- **Product Transfer:** Individual Component Specification: This stage involves the collection of product information and the creation of a Bill of Materials (BOM). The BOM identifies materials, standard components, quality requirements, and delivery times necessary for the product.
  - ***Activity Inputs:*** Market requirements, customer needs, preliminary design concepts, existing product data.
  - ***Activity Outputs:*** Detailed BOM, material specifications, component standards, documentation of quality requirements, estimated delivery timelines.
- **Technology Transfer:** Individual Component Specification: Information about the component or family of components is captured, focusing on three aspects: geometry (drawings), materials (specifications), and production rates (batch size).
  - ***Activity Inputs:*** Design drawings, material specifications, production rate requirements.
  - ***Activity Outputs:*** Detailed component specifications, geometric data (drawings), material data sheets, production batch size documentation.
- **Facility Design:** Product and Manufacturing Process Specification: Technical information about the product and manufacturing processes is used to initiate the design of a new facility.
  - ***Activity Inputs:*** Product design specifications, process flow diagrams, production volume forecasts.
  - ***Activity Outputs:*** Initial facility layout plans, manufacturing process specifications, technical requirement documentation.

6.4.2.2 Basic Development—Conceptualization

In the Basic Development—Conceptualization stage, more detailed specifications and requirements are defined. This stage focuses on refining the initial ideas and preparing for the detailed design and development phases.

- **Manufacturing and Quality Control Requirements:** During this stage, manufacturing and quality control specifications are defined in terms of volume, cost, and time. Product attributes such as material, dimensions, tolerances, and surface finish are specified. Control variables for acceptance or rejection are also established.
  - ***Activity Inputs:*** BOM, product attributes (material, dimensions, tolerances, surface finish), production volume and cost estimates.
  - ***Activity Outputs:*** Detailed manufacturing specifications, quality control parameters and acceptance criteria, control variable definitions.
- **Manufacturing Resources Specification and Selection:** Available technology, including machines, material handling equipment, tooling, and jigs/fixtures, is evaluated to select the best equipment for manufacturing the component.
  - ***Activity Inputs:*** Component specifications, available technology data (machines, Material Handling Equipment, tooling, jigs/fixtures).
  - ***Activity Outputs:*** Technology and equipment selection reports, equipment specification sheets, preliminary process flowcharts.
- **Manufacturing System Specification:** Available technology, including machines, material handling equipment, tooling, and jigs/fixtures, is evaluated to select the best equipment for the manufacturing system.
  - ***Activity Inputs:*** Facility layout plans, manufacturing process specifications, available technology data.
  - **A*ctivity Outputs:*** Detailed manufacturing system specifications, equipment and technology selection reports, updated facility layout plans.

6.4.2.3 Advanced Development—Design

The Advanced Development—Design stage involves detailed planning and designing of the manufacturing processes, systems, and partnerships. This stage ensures that all components and processes are well-integrated and ready for production.

- **Manufacturing Supplier Selection:** This stage involves evaluating and selecting partners based on their capacities and capabilities. The goal is to integrate the competencies of different companies to develop the project effectively.
  - ***Activity Inputs:*** Manufacturing specifications, supplier capability assessments, supplier qualification criteria.
  - ***Activity Outputs:*** List of selected suppliers, supplier agreements and contracts, supplier capability and capacity reports.

- **Process Plan:** A detailed process chart defines raw materials, tools, fixtures, gauges, and other devices necessary for manufacturing the component. Documents for process quality control and standard operation procedures are also developed.
  - ***Activity Inputs:*** Component specifications, selected technology and equipment, quality control documentation.
  - ***Activity Outputs:*** Detailed process charts, list of required raw materials, tools, fixtures, and gauges, standard operation procedures (SOPs), quality control process documentation.
- **Manufacturing System Design:** Manufacturing strategies are defined, and the configuration of stations, cells, shop-floor, and the factory is planned. Capacity plans are evaluated using simulation tools to ensure efficient production.
  - ***Activity Inputs:*** Manufacturing system specifications, capacity plans, simulation data.
  - ***Activity Outputs:*** Detailed manufacturing system designs, layout configurations for stations, cells, and shop-floor, capacity planning reports, simulation results and analysis.

6.4.2.4 Launching—Creation/Operation

The Launching—Creation/Operation stage focuses on implementing and operationalizing the manufacturing system. This stage ensures that all components are manufactured and assembled and that the facility is ready for full-scale production.

- **Manufacturing and Assembly of Components:** In this final stage, components are manufactured, assembled, and delivered to the facility. Quality controls are checked and documented to ensure the components meet the specified requirements.
  - ***Activity Inputs:*** Detailed manufacturing and assembly instructions, quality control specifications, supplier delivery schedules.
  - ***Activity Outputs:*** Assembled components, quality control reports, delivery and inventory documentation.
- **Equipment Set-up and Ramp-up of Production:** Equipment is checked, tested, and set up. Ramp-up and production start-up processes are initiated to begin full-scale manufacturing.
  - ***Activity Inputs:*** Equipment setup instructions, test protocols, production schedules.
  - ***Activity Outputs:*** Operational equipment, test results and validation reports, initial production output.

- **Facility Construction, Set-up, and Ramp-up of Production:** The manufacturing system is configured, constructed, and installed. Production ramp-up begins, and the facility's performance is evaluated and adjusted as necessary to meet production goals.
  - ***Activity Inputs:*** Facility construction plans, system installation instructions, ramp-up schedules.
  - ***Activity Outputs:*** Constructed and operational facility, performance evaluation reports, adjusted production processes as needed.

6.4.2.5 End of Life: 5S—Social, Sustainable, Sensing, Smart, Safe

Incorporating the 5S principles ensures that the manufacturing system is decommissioned ethically, sustainably, technologically advanced, and safely. Each principle has specific activity inputs and outputs that contribute to a responsible end-of-life management.

- **Social:** Social considerations at the end of a manufacturing system's life cycle involve ensuring that decommissioning processes are ethical and socially responsible. Engage communities in decommissioning and promote the reuse and repurposing of equipment and facilities.
  - ***Activity Inputs:*** Ethical decommissioning guidelines, community engagement strategies.
  - ***Activity Outputs:*** Ethically decommissioned facilities, community engagement reports, repurposed equipment and infrastructure.
- **Sustainable:** Sustainable end-of-life management includes minimizing waste and maximizing the reuse and recycling of materials and equipment. Implement strategies to recover valuable resources and reduce environmental impact during decommissioning.
  - ***Activity Inputs:*** Environmental impact assessments, resource recovery plans.
  - ***Activity Outputs:*** Recovered materials and equipment, reduced waste, life cycle assessment reports.
- **Sensing:** Utilize sensing technologies to monitor the decommissioning process and optimize the reuse and recycling of manufacturing components. Real-time data can enhance efficiency and ensure the effective handling of materials.
  - ***Activity Inputs:*** Advanced sensing and monitoring technologies, real-time data analytics.
  - ***Activity Outputs:*** Enhanced monitoring systems, real-time data reports, optimized decommissioning processes.

- **Smart:** Smart end-of-life management involves using advanced technologies to disassemble and repurpose manufacturing systems efficiently. Develop systems that can be easily reconfigured or upgraded for new applications.
  - ***Activity Inputs:*** Smart technology integration plans, reconfiguration protocols.
  - ***Activity Outputs:*** Reconfigurable manufacturing systems, smart disassembly procedures, upgraded components.
- **Safe:** Safe end-of-life practices ensure manufacturing systems' safe decommissioning and disposal. Implement rigorous safety protocols to prevent environmental contamination and protect human health.
  - ***Activity Inputs:*** Safety standards and testing protocols, hazardous material handling plans.
  - ***Activity Outputs:*** Safety compliance certifications, non-toxic disposal methods, safe decommissioning reports.

By applying the 5S principles to the end-of-life stage of a manufacturing system, companies can ensure that their operations conclude responsibly and sustainably, maintaining high ethical standards and minimizing environmental impact.

6.4.3 Techniques and Methods Required During the Development

Techniques like virtual design and simulation, along with process planning, significantly reduce the time and cost of development by allowing issues to be addressed before physical production begins. By incorporating rigorous quality control systems and thorough supplier management, manufacturing systems can consistently produce high-quality products, reducing waste and the need for reworks. Besides, effective facility design and technology selection provide a manufacturing system with the flexibility to adapt to product design or demand changes, enhancing the system's scalability.

Companies can minimize production costs through strategic supplier selection and technology evaluation while maximizing output efficiency. Comprehensive planning and simulation help identify potential risks in the manufacturing process, allowing for preemptive measures to mitigate them.

Here are some techniques and methods that are integral to the development of manufacturing systems, focusing on advanced technologies, management strategies, and operational efficiencies:

- **Advanced Technologies**
  - ***Additive Manufacturing (3D Printing):*** Allows for rapid prototyping and production of complex geometries, reducing time and waste associated with traditional manufacturing methods.

  - ***IoT:*** Embedding sensors and smart devices in manufacturing equipment to collect and analyze data, enhancing predictive maintenance and real-time production line monitoring.
  - ***Artificial Intelligence and Machine Learning:*** Utilized for predictive analytics to forecast maintenance needs, optimize production processes, and enhance quality control.
  - ***Digital Twin Technology:*** Creating virtual replicas of physical systems to simulate, analyze, and optimize manufacturing processes and product performance in a virtual environment.

- **Management Strategies**
  - ***Lean Manufacturing:*** Focuses on minimizing waste within manufacturing systems while maximizing productivity.
  - ***Six Sigma:*** A data-driven approach to eliminate defects in any process, aiming to achieve near-perfect quality.
  - ***Agile Manufacturing:*** Adapting quickly to market changes with flexible production processes that can adjust to new conditions without significant time or cost penalties.
  - ***Total Quality Management (TQM):*** A company-wide approach to ensure quality standards are met across all manufacturing process operations.
  - ***Performance Monitoring with KPIs:*** Implementing KPIs to measure and enhance manufacturing efficiency, quality, and performance. Key indicators might include Overall Equipment Effectiveness (OEE), cycle times, yield rates, and customer satisfaction metrics.

- **Operational Techniques**
  - ***Kaizen (Continuous Improvement):*** A strategy where employees at all company levels work together proactively to achieve regular, incremental improvements in the manufacturing process.
  - ***Just-in-Time Manufacturing:*** Reduces inventory costs by scheduling material deliveries exactly when needed in the production process.
  - ***Supply Chain Management:*** Efficient supply chain management to ensure timely delivery of materials, which helps reduce production delays and costs.
  - ***Capacity Planning:*** Using statistical and forecasting tools to plan the production capacity needed to meet changing demands effectively.

- **Sustainability and Compliance**
  - ***Sustainable Manufacturing:*** Implementing eco-friendly practices that minimize environmental impact and conserve natural resources.

  - ***Regulatory Compliance:*** Ensuring all manufacturing processes and products meet local, national, and international regulatory standards and safety requirements.

- **Collaboration and Communication**
  - ***Cross-Functional Teams:*** Facilitating collaboration between different departments (e.g., engineering, procurement, operations) to enhance product designs and manufacturing efficiency.
  - ***Stakeholder Engagement:*** Regular interaction with all stakeholders, including suppliers, customers, and employees, to gather feedback and ensure the manufacturing process aligns with broader business objectives.

Integrating these techniques and methods into manufacturing development optimizes the production process, ensures adaptability, enhances efficiency, reduces costs, and improves product quality. These strategies also support creating a sustainable and responsive manufacturing environment capable of meeting the demands of rapidly changing markets.

6.4.4 The Ten Core Technologies Applied to the Entity

Integrating the ten core technologies into this entity is applied strategically, as depicted in FIGURE 6.6.

1. **Artificial Intelligence and Machine Learning (AI/ML)**
   (a) ***Ideation:*** Applied in Product Transfer and Facility Design to automate early-stage material and process selection, analyze data trends, and predict the feasibility of components.
   (b) ***Advanced Development:*** Used in Technology Transfer to optimize workflows, enhance predictive maintenance, and refine process plans dynamically.
   (c) ***Launching:*** AI-driven insights improve operational efficiency and quality control by enabling real-time decision-making.

2. **Simulation Tools**
   (a) ***Ideation:*** Supports virtual testing of facility layouts, workflows, and initial product design in Facility Design.
   (b) ***Basic Development:*** Simulates material flows and evaluates manufacturing process feasibility for Technology Transfer.
   (c) ***Advanced Development:*** Models manufacturing system behavior and optimizes layouts during facility planning.

3. **Industrial IoT (IIoT)**
   (a) ***Basic Development:*** IIoT sensors enable real-time monitoring of resources, machinery, and quality in Technology Transfer.

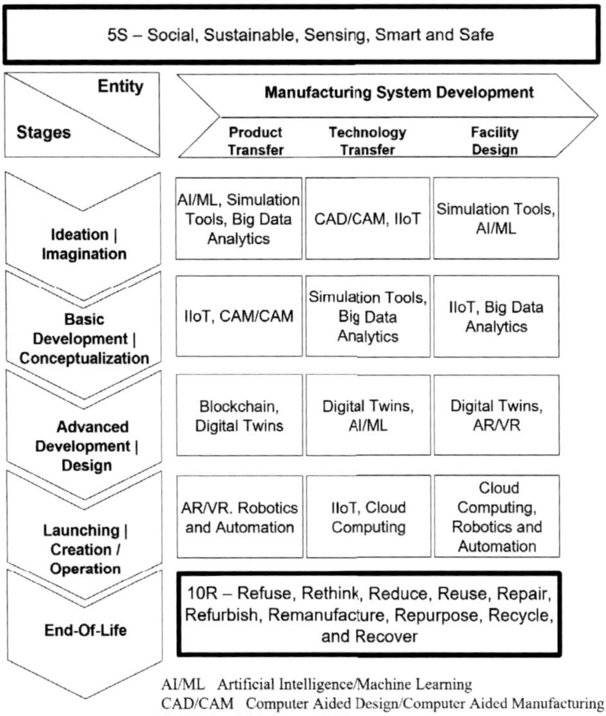

AI/ML Artificial Intelligence/Machine Learning
CAD/CAM Computer Aided Design/Computer Aided Manufacturing
IIoT Industrial Internet of Things
AR/VR Augmented Reality/Virtual Reality

FIGURE 6.6
Core technologies applied to this entity.

(b) ***Launching:*** IIoT ensures continuous feedback during equipment ramp-up, optimizing performance in Product Transfer.

4. **Digital Twins**
   (a) ***Advanced Development:*** Creates virtual replicas of production systems to optimize performance in Technology Transfer and Facility Design.
   (b) ***Launching:*** Allows real-time monitoring and adjustment of manufacturing processes, ensuring operational efficiency.

5. **Additive Manufacturing (3D Printing)**
   (a) ***Ideation:*** Accelerates prototyping of individual components during Product Transfer.
   (b) ***Advanced Development:*** Enables the creation of custom fixtures and tooling for facility layouts.

6. **Computer-Aided Tools**
   (a) ***Ideation:*** CAD/CAM supports component design and BOM generation in Technology Transfer.

(b) ***Basic Development:*** Facilitates process planning and ensures manufacturability through CAPP integration in Product Transfer.

(c) ***Advanced Development:*** Enhances the design of complex manufacturing systems.

7. **Big Data Analytics**
   (a) ***Basic Development:*** Provides data-driven insights for manufacturing resource allocation and optimization in Technology Transfer.
   (b) ***Launching:*** Analyzes performance metrics to identify inefficiencies and enable continuous improvement.

8. **Cloud Computing**
   (a) ***Launching:*** Facilitates collaboration across teams during facility construction and operation, enabling seamless process data integration in Facility Design.
   (b) ***Advanced Development:*** Supports scalable infrastructure for process simulation and optimization.

9. **Augmented and Virtual Reality (AR/VR)**
   (a) ***Advanced Development:*** Assists in virtual walkthroughs of facility layouts, identifying potential bottlenecks in Facility Design.
   (b) ***Launching:*** Enhances operator training and improves assembly accuracy through immersive instructions in Product Transfer.

10. **Robotics and Automation**
    (a) ***Launching:*** Applied in Product Transfer for automated assembly, ensuring precision and reducing manual labor.
    (b) ***Facility Design:*** Supports the deployment of scalable automation systems to handle diverse manufacturing tasks efficiently.

Integrating these technologies not only propels the manufacturing process towards greater efficiency and reduced waste but also significantly bolsters the sustainability of operations, aligning with global efforts to promote environmentally responsible manufacturing practices.

6.5 Case Study Applying the Life Cycle Engineering of this Entity

Manufacturing Co. is a vanguard industry in a manufacturing landscape increasingly defined by Mexico's need for innovation and sustainability. This entity embarked on a transformative journey to integrate Industry 4.0 technologies with a deep-seated commitment to sustainability. Challenged by the imperative to reduce operational costs and comply with stringent environmental standards, Manufacturing Co. recognized that a shift towards more sustainable practices underpinned by advanced technologies was essential for its continued competitiveness and environmental stewardship. At a critical moment, with escalating competition and growing demands for environmentally responsible, high-quality products, Manufacturing Co. confronted the limitations of traditional manufacturing processes. These processes, characterized by significant resource and energy consumption, were increasingly at odds with an industry trend favoring efficiency and sustainability.

The necessity for a strategic overhaul was clear: maintaining its industry standing while empowering new standards in sustainable manufacturing. The transformation was initiated by adopting IoT sensors across Manufacturing Co.'s operations, marking a pivotal move towards enhanced operational insight. This technological leap enabled real-time monitoring of machinery performance and energy usage, uncovering inefficiencies and preventing potential failures. The resulting decrease in downtime and maintenance costs illustrated the immediate benefits of a data-driven approach to manufacturing, setting the stage for further efficiency gains. Harnessing the power of the data collected, Manufacturing Co. adopted AI and advanced analytics to refine its operations further.

Predictive maintenance models and AI-optimized production schedules led to a marked reduction in waste and energy usage, aligning operational efficiency with environmental sustainability. Elevating sustainability to a core operational pillar was central to Manufacturing Co.'s transformation. Initiatives ranging from installing energy-efficient lighting to adopting solar energy solutions underscored the company's commitment to reducing its environmental impact.

Waste reduction efforts complemented these initiatives, showcasing a sustainable operational model's practical and economic benefits. The social dimension of the 5S Framework brought into focus the crucial role of Manufacturing Co.'s workforce and broader community engagement. Investment in employee training and development fostered a culture of innovation and continuous improvement. Ethical sourcing practices and active participation in community initiatives underscored the company's broader commitment to social responsibility, enhancing its reputation and cementing its role as a community pillar. In the digital area, Manufacturing Co. approached cybersecurity with the seriousness it demanded. Through robust security protocols, regular audits, and comprehensive employee training, the company safeguarded its operations against digital threats, ensuring the integrity of its critical data and systems. Manufacturing Co.'s strategic embrace of the 5S Framework has redefined its operational approach and established it as a benchmark in the industry for integrating sustainability with technological innovation. This case study illustrates the potential for Industry 4.0 technologies to revolutionize manufacturing practices, emphasizing the importance of sustainability and efficiency. Manufacturing Co.'s journey offers a compelling blueprint for the industry, showcasing the strategic value and transformative potential of adopting a holistic approach to sustainable manufacturing. Through this endeavor, Manufacturing Co. has secured its competitive edge and championed a future where industrial progress and environmental impact go hand in hand. FIGURE 6.7 shows how the industry is empowered in the manufacturing industry.

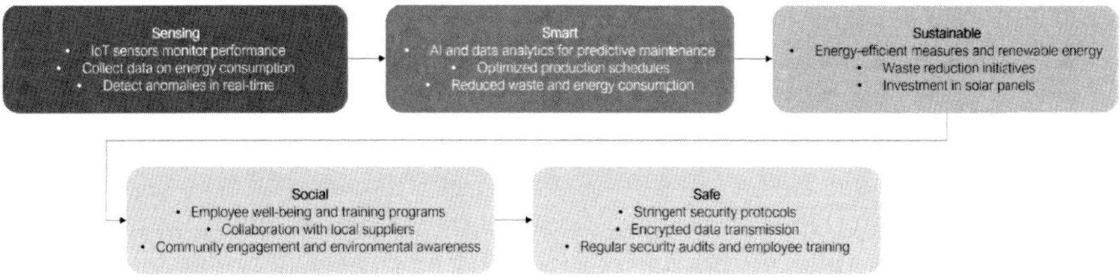

FIGURE 6.7
Empowering industry by 5S.

Implementing predictive maintenance and production optimization technologies at Manufacturing Co. could lead to a reduction in downtime and a boost in overall production efficiency. IoT sensors and AI analytics allow for the early detection of potential equipment failures, enabling repairs before breakdowns occur, thus minimizing unplanned downtime. Similarly, AI-driven optimization of production schedules ensures that machinery and resources are utilized more effectively, streamlining operations and increasing output. These improvements enhance the company's ability to meet customer demand more reliably and reduce the costs associated with idle machinery and inefficiencies, contributing directly to the bottom line. Manufacturing Co.'s strategic focus on sustainability has yielded a reduction in energy consumption, thanks to the implementation of energy-efficient technologies and renewable energy sources. This reduction in energy use directly contributes to a lower carbon footprint, aligning with global efforts to combat climate change.

Furthermore, waste reduction initiatives have decreased waste generation, emphasizing the company's commitment to environmental stewardship.

These initiatives preserve natural resources and position Manufacturing Co. as a leader in sustainable manufacturing practices, appealing to increasingly environmentally conscious consumers and stakeholders. The company's investment in ongoing training and development programs has profoundly impacted employee satisfaction, skill levels, and morale. This commitment to employee growth has decreased employee turnover and increased productivity. By providing employees with opportunities for professional development and recognizing their contributions, Manufacturing Co. has fostered a more engaged and motivated workforce. Higher employee satisfaction and reduced turnover translate into consistent operational knowledge and skills within the company, driving productivity and innovation. Manufacturing Co.'s collaboration with local suppliers and active participation in community initiatives have significantly enhanced its reputation and strengthened its ties with the local community. Engaging with local suppliers supports the local economy and promotes sustainable supply chains, while participation in community initiatives demonstrates the company's commitment to social responsibility. This engagement contributes to the community's social fabric and establishes Manufacturing Co. as a valued community partner, enhancing its social license to operate and contributing to long-term business sustainability.

Manufacturing Co.'s robust cybersecurity measures have been instrumental in protecting critical data and systems in an increasingly prevalent era of cyber threats. These measures maintain customer trust by safeguarding sensitive information and preventing potentially costly data breaches. Effective cybersecurity protocols prevent financial losses associated with data breaches and protect the company's reputation, ensuring it remains a trusted partner for customers, suppliers, and stakeholders. This trust is essential for maintaining competitive advantage and securing long-term business relationships in the digital age. Manufacturing Co.'s strategic initiatives in operational efficiency, sustainability, employee satisfaction, community engagement, and cybersecurity have yielded impressive results and positioned the company for long-term success. These initiatives demonstrate a comprehensive approach to business management that balances profitability with environmental stewardship and social responsibility, setting a benchmark for sustainable industry practices. Manufacturing Co.'s implementation of the 5S framework within the context of Industry 4.0 has led to a sustainable transformation that improved operational efficiency and reduced costs and positioned the company as an environmental and social steward in its community. This case study demonstrates how integrating advanced technologies and sustainability practices can lead to a thriving and responsible Industry 4.0 journey. By integrating the 5S framework into their operations, Manufacturing experiences improved operational efficiency, reduces energy costs, minimizes waste, and contributes positively to their community and the environment.

They have embraced Industry 4.0, using technology and innovation to remain competitive and sustainable in the long run.

6.5.1 Walkthrough of the Entire Life Cycle Process: From Idea to Launching

This case study exemplifies the utilization of Life Cycle Engineering (LCE) in evaluating the sustainability impacts of 3D printing technologies, specifically in the production process of eight turbine blades at Honeywell Aerospace's plant in Mexico. The LCE approach ensures that every stage of the material's life cycle is optimized for sustainability, from the initial idea to the final launch.

- **Idea—Imagination:**
  - ***Product Transfer:*** Manufacturing Co. began with a comprehensive understanding of market demands and customer needs. This stage involved brainstorming and initial concept development. Essential product information was gathered, creating a detailed BOM, outlining the necessary materials, standard components, quality requirements, and delivery timelines.
  - ***Technology Transfer:*** The initial component designs were developed, focusing on geometry, material specifications, and production rates. These details were crucial for the subsequent stages of technology and facility design.
  - ***Facility Design:*** Technical information about the product and manufacturing processes was utilized to draft the initial facility layout plans. These plans were essential for visualizing the new manufacturing facility and aligning with the forecasted production volumes.

- **Basic Development—Conceptualization:**
  - ***Manufacturing and Quality Control Requirements:*** Manufacturing Co. refined its initial ideas in this stage, focusing on detailed specifications and requirements. Manufacturing and quality control specifications were defined, ensuring that the product attributes such as material, dimensions, tolerances, and surface finish met the necessary standards.
  - ***Manufacturing Resources Specification and Selection:*** Available technologies, including machines, material handling equipment, tooling, and jigs/fixtures, were evaluated to select the best equipment for manufacturing the components. This selection process ensured that the chosen technologies aligned with the company's goals for efficiency and sustainability.
  - ***Manufacturing System Specification:*** Detailed specifications for the manufacturing system were developed, incorporating the selected

technologies and equipment. The facility layout plans were updated to reflect these specifications, ensuring a cohesive and efficient manufacturing system design.

- **Advanced Development—Design:**
  - ***Manufacturing Supplier Selection:*** Manufacturing Co. evaluated and selected suppliers based on their capacities and capabilities. This selection process was critical for integrating the competencies of various partners, ensuring effective project development.
  - ***Process Plan:*** A detailed process plan was developed, outlining the raw materials, tools, fixtures, gauges, and other necessary devices for manufacturing the components. This plan included standard operation procedures (SOPs) and quality control documentation to ensure consistent and high-quality production.
  - ***Manufacturing System Design:*** Manufacturing strategies were defined, and the configuration of stations, cells, shop-floor, and the factory was planned. Capacity plans were evaluated using simulation tools to ensure efficient production and optimize the overall system design.
- **Launching—Creation/Operation:**
  - ***Manufacturing and Assembly of Components:*** In this stage, the components were manufactured, assembled, and delivered to the facility. Quality controls were checked and documented, ensuring that the components met the specified requirements.
  - ***Equipment Set-up and Ramp-up of Production:*** Equipment was checked, tested, and set up, followed by the ramp-up and production start-up processes. This stage ensured that all equipment was operational and ready for full-scale manufacturing.
  - ***Facility Construction, Set-up, and Ramp-up of Production:*** The manufacturing system was configured, constructed, and installed. Production ramp-up began, and the facility's performance was evaluated and adjusted as necessary to meet production goals.
- **End of Life: 5S—Social, Sustainable, Sensing, Smart, Safe**
  - ***Social:*** At the end of the manufacturing system's life cycle, social considerations were prioritized. Ethical decommissioning processes were implemented, engaging the community and promoting the reuse and repurposing of equipment and facilities.
  - ***Sustainable:*** Sustainable end-of-life management minimized waste and maximized the reuse and recycling of materials and equipment. Strategies were implemented to recover valuable resources and reduce environmental impact.

- ***Sensing:*** Sensing technologies monitored the decommissioning process, optimizing the reuse and recycling of manufacturing components. Real-time data enhanced efficiency and ensured effective material handling.
- ***Smart:*** Smart end-of-life management utilized advanced technologies to disassemble and repurpose manufacturing systems efficiently. Systems were developed to be easily reconfigured or upgraded for new applications.
- ***Safe:*** Safe end-of-life practices ensured manufacturing systems' safe decommissioning and disposal. Rigorous safety protocols prevented environmental contamination and protected human health.

By following this detailed life cycle process, Manufacturing Co. was able to transform its operations, aligning with Industry 4.0 technologies and sustainability practices. This comprehensive approach ensured that the company not only improved its operational efficiency and reduced costs but also contributed positively to its community and the environment, setting a benchmark for sustainable industry practices.

6.6 The Future of Sustainable Manufacturing

A blend of technological advancements, practical approaches, and collaborative efforts among industry leaders is shaping the future of sustainable manufacturing. Technological innovations at the heart of this transformation drive operational efficiency, thereby supporting sustainability goals by reducing waste and resource consumption. For instance, advanced sensors, controls, and automation systems optimize industrial processes and minimize energy waste through real-time monitoring and data analysis. The increasing use of AI to monitor industrial machines and detect inefficiencies further underscores the integration of digital technologies in achieving sustainability [43].

Adopting a holistic approach that embeds sustainability into core operations rather than treating it as a separate initiative is crucial. This involves factoring environmental impacts into cost-benefit analyses, reducing material usage, and designing products with their end-of-life in mind to ensure recyclability. These strategies enhance sustainability and improve the bottom line by increasing efficiency and reducing long-term operational costs.

Effective leadership is essential for advancing sustainability initiatives. Besides, policies like the Inflation Reduction Act in the U.S. provide necessary financial incentives to support these efforts. Companies are more likely to invest in sustainability when they see it aligned with their business growth and profitability.

Improving supply chain visibility is another critical aspect. Companies are striving for closer relationships with a smaller network of suppliers who adhere to sustainable practices. This includes incorporating sustainability criteria in supplier contracts and continuously requesting impact data to emphasize its importance.

Investment in renewable energy sources is growing. This focus is on enhancing energy security by diversifying energy sources and utilizing a mix of solar, wind, and hydropower. This transition reduces reliance on finite resources and protects against regulatory and geopolitical shocks.

Embracing a circular economy is also pivotal. In this economy, products are designed for durability, reuse, and recyclability, minimizing waste and resource extraction. The adoption of digital twins and advanced simulation technologies supports this by optimizing process efficiency and product life cycle management. Additionally, decentralized production models, facilitated by innovations such as 3D printing, can reduce the carbon footprint associated with logistics and enable localized manufacturing tailored to specific community needs.

These strategies, complemented by regenerative practices, reduce environmental impact and actively enhance the ecosystems they interact with. Engaging stakeholders through education and aligning with global sustainability standards ensure that these practices are widely adopted and integrated into business strategies, securing a sustainable future for the manufacturing sector.

Thus, the future of sustainable manufacturing relies on integrating technology, fostering effective leadership, enhancing supply chain transparency, investing in renewable energy, and adopting a long-term, collaborative approach. These efforts collectively contribute to a more sustainable and profitable manufacturing industry.

6.7 Conclusion

Sustainable manufacturing has become essential for responsible business practices, balancing economic growth, environmental stewardship, and social responsibility. Integrating advanced technologies such as IoT, AI, and AR training is revolutionizing manufacturing processes by optimizing resource use, reducing waste, and enhancing operational efficiency. These innovations improve the bottom line and contribute significantly to environmental protection and supply chain resilience. Sustainable manufacturing fosters fair labor practices and safe working environments, aligning business operations with ethical standards and societal well-being.

The shift towards sustainable practices offers substantial economic benefits, from long-term cost savings to enhanced customer loyalty.

Companies that embrace sustainability are better positioned to navigate market cycles and meet consumer demands, ensuring long-term viability. The digitization of manufacturing through IoT and AI enables precise monitoring, predictive maintenance, and real-time decision-making, further driving efficiency and sustainability. Despite challenges such as high initial costs and technological integration difficulties, the long-term advantages of sustainable manufacturing make it a strategic priority for forward-thinking businesses.

Adopting the 5S framework—Social, Sustainable, Sensing, Smart, and Safe—provides a holistic approach to developing efficient, innovative, and socially responsible manufacturing systems. This framework ensures that manufacturing processes are optimized for performance and environmental and social impacts. Companies can overcome challenges and achieve sustainability goals by focusing on continuous improvement and integrating ESG principles.

Looking ahead, the future of sustainable manufacturing will be shaped by ongoing technological advancements, effective leadership, and collaborative efforts among industry stakeholders. Embracing renewable energy sources, improving supply chain visibility, and adopting circular economy principles will be crucial. Using digital twins, AI, and decentralized production models will further enhance sustainability efforts, reducing the environmental footprint and fostering innovation.

Ultimately, sustainable manufacturing represents a strategic investment that aligns with global efforts to mitigate climate change and promote economic efficiency. Companies that integrate sustainable practices into their core operations will thrive in an increasingly competitive market and contribute positively to the environment and society. This holistic approach ensures that the benefits of sustainable manufacturing extend beyond the factory floor, fostering a resilient and responsible industrial landscape.

7

Supply Chain Development

This chapter explores the critical role of sustainable supply chain development in modern manufacturing, emphasizing a strategic, integrated LCE (life cycle Engineering) approach from ideation to launching and considering end-of-life management based on the 5S principles (Social, Sustainable, Sensing, Smart, Safe). Supply chains can enhance efficiency, transparency, and real-time decision-making by adopting advanced technologies such as IoT, AI, blockchain, and digital twins. The framework begins with setting ambitious sustainability goals and scenario planning in the ideation phase, followed by detailed planning that involves accurate demand forecasting and collaborative supplier engagement. During the operational phase, sustainable procurement, production, and logistics practices are implemented to minimize the carbon footprint. The end-of-life stage maximizes environmental and social benefits through community engagement, material recovery, and smart disassembly.

A case study of Schneider Electric illustrates the practical application of this framework. Schneider Electric's approach to supply chain decarbonization, particularly targeting Scope 3 emissions, is detailed through strategic ideation, supplier collaboration, digital tool implementation, and operational execution. The case study highlights how Schneider Electric integrates the 5S principles into its supply chain to achieve significant sustainability milestones.

The chapter concludes that the future of sustainable manufacturing relies on commercial innovation, deeply embedded sustainable operations, real-time decision execution, and human-centric work design. By integrating these strategies, supply chain leaders can create a competitive advantage, ensuring resilience and sustainability in an evolving market landscape. This comprehensive approach meets current consumer and regulatory demands and positions companies to thrive in the future.

7.1 Introduction

Supply chain development is integral to sustainable manufacturing, as it encompasses the strategic coordination of business functions across the globe

DOI: 10.1201/9781003501619-7

to reduce environmental impact and promote social welfare. Effective supply chain strategies can significantly reduce carbon emissions, waste, and energy use. Adopting green supply chain practices, such as sourcing renewable materials and using energy-efficient transportation, aligns with the growing consumer and regulatory demand for sustainability in manufacturing [55, 121].

Digital technologies are revolutionizing supply chain development by increasing transparency, efficiency, and responsiveness. Technologies like the Internet of Things (IoT) offer real-time tracking of goods and assets, enhancing visibility and control. Artificial Intelligence (AI) and machine learning can analyze huge amounts of data/information to improve demand forecasting and decision-making, while blockchain technology secures and streamlines supply chain transactions [99]. These innovations enable more sustainable operations by optimizing resource use and reducing waste [110, 149].

Gartner, Inc. [50] is a well-known research and advisory firm offering insights, advice, and tools across various domains, including supply chain management. Famous for its Magic Quadrants, which rank vendors in technology and service markets, Gartner's extensive supply chain expertise covers planning, procurement, logistics, and manufacturing. Their research and analysis help businesses optimize operations, enhance efficiency, and increase resilience. Additionally, Gartner hosts an annual Supply Chain Executive Conference, attracting leaders worldwide to discuss industry trends, innovations, and challenges, making it a pivotal resource for organizations aiming to refine their supply chain strategies and embrace new technologies.

In [48], they offer several insights into the evolving nature of supply chain management. Moreover, Figure 7.1 illustrates the current landscape of partner engagement among supply chain executives, with Tier 1 suppliers, third-party logistics providers (3PLs), and technology vendors being the primary collaborators. It also forecasts the expected partnerships by 2025, where the focus is projected to shift towards Tier 2 and beyond suppliers, end-users, and contract manufacturers as the top three engagement cohorts.

Besides, key insights include:

- **Ecosystem Approach:** A move towards creating a supply chain ecosystem where all participants are connected through a common data platform, providing near-real-time access to critical information and supporting informed decision-making across the network.
- **Digitalization:** Over the next two years, more than 50% of chief supply chain officers (CSCOs) plan to invest in digitalization, including data processes and platforms, to accelerate the creation of and participation in supply chain ecosystems.
- **Engagement Shift:** A significant shift towards engaging with Tier 2 and beyond suppliers exists. The number of supply chain leaders engaging with

this group is expected to double over the next three years. This is coupled with a notable shift away from engaging with Tier 1 suppliers and 3PLs (Third-Party Logistics), where engagement is anticipated to be less than half by 2025 compared to today.

- **Broader Trading Network:** Supply chain leaders are increasingly engaging with a broader trading network, which includes startups, government agencies, and even competitors. This contrasts today's focus, which is more on Tier 1 suppliers, 3PLs, and technology vendors.
- **Expected Benefits:** The engagement with a broader network is expected to result in a more cost-effective end-to-end supply chain that supports new business model creation. Other benefits include greater agility, more effective risk management, enhanced customer experience, and business growth.
- **Business Growth:** Engaging in a digital supply chain ecosystem is seen as a way to enable faster responses to unexpected changes or risks, increase customer experience, reduce costs, support new business model creation, shorten delivery time, access the latest product technology innovations, co-develop new products or services, help achieve a broader purpose, and access to new markets.

7.1.1 Accelerating Capabilities to Enhance the Supply Chain

A resilient and robust supply chain is increasingly recognized as a business strategic differentiator. Leading organizations harness their network relationships, advanced supply chain technologies, and innovative digital transformation and service delivery approaches to enhance customer retention and revenue expansion.

These top-performing supply chains drive growth and advancement in risk management. They exhibit a heightened capacity for swiftly rebounding from ongoing disruptions and are adept at forecasting and sidestepping potential threats before they materialize.

For any industry, the success of future supply chains will be contingent on current investments in developing supply chain capabilities that will make a company more robust and financially successful than its rivals. In times of volatility, this antifragile approach enables leading supply chain entities to seize growth opportunities and reinforce the supply chain's significance within the broader business context while maintaining agility and resilience.

On the other hand, attaining sustainability objectives necessitates collaboration, data sharing, and united efforts. Supply chain executives are now held responsible for activities that stretch across extensive value chains; without collective action, achieving these significant goals becomes a formidable task.

Top supply chain entities that utilize many-to-many ecosystems instead of traditional one-to-many networks are successfully scaling their sustainability efforts. Areas where this collaborative dynamic is particularly evident include:

- **Net Zero Emission:** By incorporating training and incentives for renewable energy use into their supplier relationship and performance management strategies, supply chains are making strides toward net zero emissions.
- **Living Wages:** Through partnerships with NGOs and governments, supply chains are working to improve living standards, ensuring that all contributors within the value chain receive a living wage.
- **Regenerative Agriculture:** In sectors like food & beverage and apparel, companies are coming together to establish regenerative agriculture standards and investing in the necessary training and infrastructure to meet these benchmarks, often with the help of third-party organizations.

As sustainability becomes an ever-greater imperative, the demand for many-to-many ecosystem approaches is set to rise, necessitating a strategic transformation in supply chain operations. The most successful supply chain organizations will develop robust partnering and data-sharing frameworks, establish effective governance, and foster mindsets geared toward inclusive collaboration.

In addition, in a globalized marketplace, Global Chief Supply Chain Officers (GCSCOs) recognize the imperative to evolve their teams' capabilities, relationships, and structures to foster high performance and engagement. This is particularly evident in supply chain transformation strategies, especially regarding using technology to enhance interactions with customers and staff.

Leading consumer product companies have clearly understood retail on-shelf availability by leveraging technologies such as inbound traceability, machine vision, and predictive analytics. Similarly, in B2B settings, companies track inventory replenishment and usage in real-time with their customers, allowing for more seamless daily operations and strategic planning. These firms also place teams within customer organizations to spur joint innovation and address challenges collaboratively.

Technological solutions are revolutionizing employees' roles, maintaining situational awareness, and augmenting productivity. These tools foster innovation and cooperation, integrating machinery as a complementary force within the workforce. To enhance efficiency and product quality, substantial investments are being channeled into automating and digitizing factory and warehouse operations, with analytics playing a significant role in these advancements.

7.2 The Supply Chain as a 5S Systems

The challenges faced by the design of a next generation of supply chain as described before requires a new system of system approach such as the concept of 5S systems. Defining a supply chain as a 5S system incorporates principles of being Social, Sustainable, Sensing, Smart, and Safe into every stage of the supply chain, from sourcing to end-user delivery. Here's how each principle can be effectively integrated:

- **Social:**
  - ***Fair Labor Practices:*** Ensure that all entities within the supply chain adhere to ethical labor practices, such as fair wages, safe working conditions, and reasonable working hours.
  - ***Community Engagement:*** Actively engage with communities affected by supply chain operations, offering support for local development projects and ensuring the supply chain operations do not adversely affect the local populace.
  - ***Inclusivity and Diversity:*** Promote diversity and inclusivity within the supply chain, supporting businesses owned by underrepresented groups and ensuring products are accessible to diverse markets.

- **Sustainable:**
  - ***Eco-friendly Materials and Practices:*** Include suppliers who use sustainable materials and environmentally friendly practices. Optimize logistics and adopt greener technologies to minimize waste and carbon footprint.
  - ***Resource Conservation:*** Implement practices that reduce water and energy usage, and manage waste through recycling and reuse initiatives.
  - ***Sustainability Metrics and Goals:*** Set clear sustainability goals and track them using reliable metrics to measure progress and make necessary adjustments.

- **Sensing:**
  - ***Data-driven Optimization:*** Install sensors and implement IoT solutions to gather real-time data on inventory levels, product conditions, and transportation environments to enhance decision-making and efficiency.
  - ***Supply Chain Visibility:*** Use data to create transparent and traceable supply chain processes, allowing for better monitoring of goods from source to store.

  - ***Predictive Analytics:*** Leverage data for predictive analytics to foresee supply chain disruptions, demand fluctuations, and inventory needs, facilitating preemptive actions to mitigate risks.

- **Smart:**
  - ***Automation and AI Integration:*** Integrate automation in warehousing, sorting, and transportation. Use AI to optimize routes, manage stocks, and predict trends, reducing human errors and increasing efficiency.
  - ***Smart Contracts and Blockchain:*** Employ blockchain technology for smart contracts to ensure transparency and trust among parties, automating payments and compliance with contractual obligations.
  - ***Adaptive Supply Chain Management:*** Develop systems that can adapt dynamically to changing market conditions and consumer demands, ensuring flexibility and resilience.

- **Safe:**
  - ***Safety Standards and Compliance:*** Adhere to the highest safety standards in the handling, storage, and transportation of goods, ensuring all products are safe for consumption or use.
  - ***Risk Management Strategies:*** Implement robust risk management strategies to address potential threats, from cybersecurity breaches to physical disruptions.
  - ***Emergency Preparedness and Response:*** Establish clear protocols for emergency situations, ensuring quick and effective responses to preserve the integrity of the supply chain and safeguard stakeholders.

Integrating these 5S principles transforms a traditional supply chain into a dynamic, responsive, and responsible system. It not only improves operational efficiency and customer satisfaction but also boosts the supply chain's resilience to disruptions and enhances its contribution to economic, social, and environmental goals. This approach ensures that the supply chain becomes a cornerstone of sustainable and ethical business practices, capable of adapting to future challenges and innovations. FIGURE 7.1 represents the concepts and principles of a 5S supply chain where all the interaction of core process of a company (Create, Sale, Make, Service) interacts with customers and suppliers to achieve a sustainable supply chain (zero footprint, zero emissions), sensing (use of sensors across the supply chain, data-driven optimization, supply chain visibility, predictive analytics), smart (automation and AI integration, demand forecasting, inventory management, intelligent transport, supply chain management, control and automation, smart contracts and blockchain, adaptive supply chain management), and ensures safety delivery of products and services (safety standards, and compliance, risk management strategies, emergency preparedness and response).

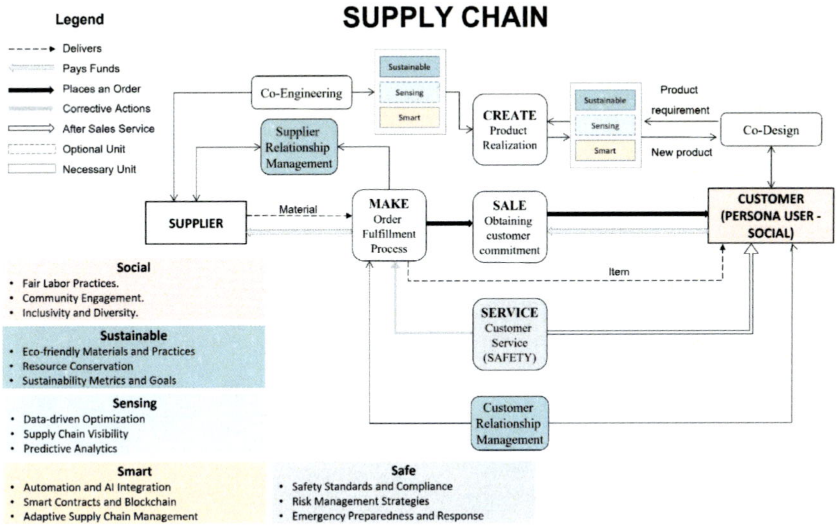

FIGURE 7.1
5S supply chain (social, sustainable, sensing, smart and safe).

7.3 Life Cycle Engineering Description

The framework provided in FIGURE 7.2 outlines a strategic approach to Life Cycle engineering within Supply Chain Management (SCM) to design and create 5S supply chains. This interdisciplinary approach integrates competitive strategy, value chain management, production/service strategy, and core process improvement to enhance overall supply chain performance.

- **Competitive Strategy:** Determines how a company can gain competitive advantage by innovating, achieving operational excellence, or focusing on customer intimacy. Competitive strategies must be measured using KPIs like new product sales, operational improvements, and customer loyalty.
- **Value Chain Strategy:** Defines how a company will organize its value chain, including relations with suppliers and customers, to compete effectively in the selected industry. Strategies include vertical integration, horizontal integration, and collaboration, each aiming to optimize resource utilization and enhance flexibility and communication within the value chain.
- **Production/Service Strategy:** Focuses on organizing core processes and aligning competencies to support competitive and value chain strategies. This involves decisions about product characteristics, customer and supplier characterizations, and process definition. Production strategies

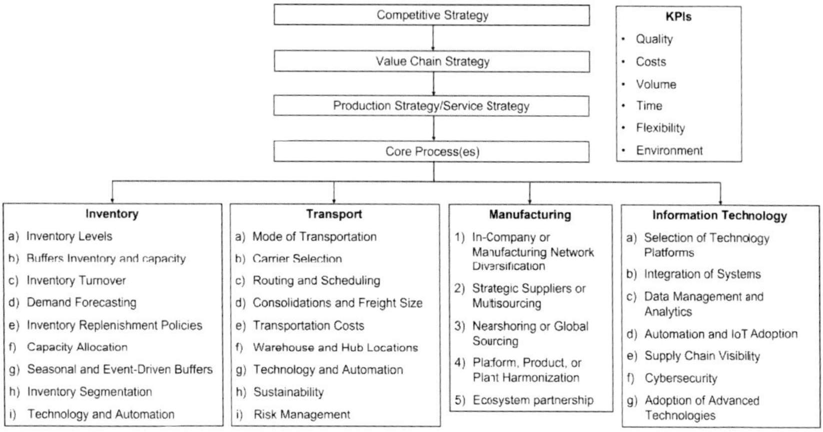

FIGURE 7.2
Drivers-strategies for supply chain resilience.

consider Make to Stock (MTS), Make to Order (MTO), Assemble to Order (ATO), Configure to Order (CTO), Build to Order (BTO), and Engineering to Order (ETO), depending on the market's demands and the product's life cycle stage.

- **Core Process Selection/Definition:** Evaluates and selects core processes that are essential to the company's operations, considering the product/service life cycle, core processes, and core competencies. Core processes must be aligned with the chosen competitive strategy and can include customer-driven design, co-engineering, supplier relationship management, customer service, order fulfillment, obtaining customer commitment and product realization.
- **KPIs (Key Performance Indicators):** Metrics used to assess the effectiveness of the core processes. These indicators are often tied to quality, costs, volume, time, flexibility and environment, providing a quantitative measure of the supply chain's performance.
- **Drivers:** Inventory, transport, manufacturing, and information technology are critical operational components that underpin core processes. Effective management in these areas is crucial for meeting KPIs and achieving strategic objectives.

This framework emphasizes the interconnectivity of strategy, process improvement, and performance measurement. Tools and models (such as System Dynamics, Discrete Event Simulation, Program Logic Models, etc.) support strategic planning and decision-making within this framework, ensuring that organizations can effectively analyze, simulate, and optimize their supply chains. TABLE 7.1 summarizes each supply chain strategy by its characteristics, key focus, and performance measures.

TABLE 7.1
Strategy Formulation Process

Characteristics	Key Focus	Performance Measures
Competitive Strategy-Supply Chain Strategy		
• Examination of industry environment • Intra-industry analysis • Evaluation of competitive advantage	• product innovation • Operational excellence • Customer focus	• % Sales of new products • ROI, operational costs • Customer loyalty
Value Chain Strategy—Supply Chain Planning		
• Analysis of value chain • Assessment of organizational structure • Appraisal of management systems	• Vertical integration • Horizontal integration • Strategic business units • Collaboration	• Volume • Operational costs • Quality, customer loyalty • Time, flexibility, environment
Production/Service Strategy-Supply Chain Operations		
• Product/Service description • Customer and Supplier characterization • Process definition	• Production Strategy: MTS (Make to Stock), MTO (Make to Order), ATO (Assemble to Order), CTO (Configure to Order), Build to Order (BTO), ETO (Engineering to Order) • Service Strategy: Services on Catalogues (SoC), Configuration of Services (CoS) and Design of Services (DoS)	• Quality • Volume • Time • Costs • Flexibility • Environment
Drivers of Supply Chain—Supply Chain Operations		
• Geopolitical shifts • Trade tensions • Resilience	• Inventory and capacity buffers • Manufacturing network diversification • Multisourcing • Nearshoring • Platform, product, or plant harmonization Ecosystem partnerships	• Order fulfillment rate • Time reduction • Operational efficiency • Risk mitigation index • Standardization rate • Geographic proximity index • Customer satisfaction

7.3.1 Competitive Strategy—Supply Chain Strategy

The competitive strategy is deeply rooted in examining the industry environment, intra-industry analysis, and competitive advantage evaluation.

This strategy is subdivided into three key focuses that a company may adopt:

1. **Product Innovation:** This strategy aims to keep the company at the forefront of the market by continuously introducing groundbreaking products or services. An increase in the percentage of sales from new products reflects the company's commitment to innovation and its market receptivity.
2. **Operational Excellence:** Revolves around streamlining operations to achieve the highest efficiency and production costs. Improvements in operational costs and ROI measure it. A company pursuing operational excellence may leverage advancements such as automation, process mining, and smart supply chain solutions to enhance productivity.
3. **Customer Focus:** This strategy focuses on cultivating deep relationships with customers and personalizing offerings to meet their specific needs. Its effectiveness is reflected in customer loyalty metrics and the ability to command a premium for tailored services or products.

The competitive strategy is closely linked with the Value Chain Strategy, which involves decisions on vertical integration, horizontal integration, strategic business units, and collaboration networks. This aspect of strategy deals with organizing the company's value chain from suppliers to end customers, ensuring that each link in the chain adds value and operates efficiently.

Furthermore, the Production/Service Strategy ties into how the company organizes its core processes to execute the competitive and value chain strategies. This includes determining whether to adopt a Make to Stock (MTS), Make to Order (MTO), Assemble to Order (ATO), or any other production approach based on product life cycle and customer demand. For service industry (Service based on Catalogues—SoC, Configuration of Services—CoS and Design of Services—DoS).

All these strategic elements are geared towards achieving key Performance Measures. These measures, such as volume, quality, time, costs, flexibility, and environmental impact, provide quantitative data to assess how well the competitive strategy and its sub-components contribute to the company's goals. In today's context, competitive strategy also encompasses:

- **Digital Integration:** Leveraging digital tools and platforms to enhance product development, optimize operations, and create a seamless customer experience.
- **Agile Methodologies:** Adapting quickly to market changes with flexible production systems and supply chain management practices.

- **Sustainability:** Incorporating eco-friendly practices and sustainable resources into product development and operational processes can serve as a differentiator and drive down long-term costs.
- **Data-Driven Decision Making:** Utilizing big data and analytics to inform strategic decisions, tailor customer experiences, and optimize the value chain.

The success of this complex strategy is then measured against performance indicators that reflect not only financial success but also customer satisfaction and sustainable practices.

7.3.2 Value Chain Strategy—Supply Chain Planning

In today's fast-paced and interconnected market, a Value Chain Strategy is about crafting an agile and responsive organizational model that strategically integrates the company within the industry's value chain. This strategy aims to optimize internal resources and capabilities and collaborate externally to form a robust value creation network. The strategy has three main directions:

- **Vertical Integration:** Companies might pursue vertical integration to gain better control over their supply chains, ensuring more resilient and responsive operations. This includes control over supply sources, manufacturing processes, and distribution channels.
  - ***Cost-Related Synergy:*** Backward integration can lead to cost reductions, while forward integration can enhance customer experience and increase the ability to customize offerings.
- **Horizontal Integration:** Extends beyond mere acquisition of competitors. It also involves strategic alliances to broaden market reach and incorporate innovative processes and technologies that can complement existing competencies.
  - ***Access to New Markets and Technologies:*** It allows for rapid scaling, diversification, and enhancement of product portfolios.
- **Collaboration:** In the modern context, digital platforms often facilitate communication and coordination. It involves strategic partnerships, joint ventures, and networked ecosystems that promote innovation and value creation.
  - ***Virtual Enterprises and Industry Networks:*** As businesses move towards models like the virtual enterprise, collaboration becomes key to accessing new markets, sharing risks, and driving innovation. An emerging concept, especially in dynamic markets, is the formation of virtual enterprises—temporary alliances of businesses that come together to exploit fast-changing opportunities, often facilitated by technology.

Types of Networks:

- **Vertical Networks:** These involve collaboration among companies at different stages of the supply chain, aiming for a symbiotic relationship that enhances the capabilities and competitive advantage of all participants. In today's context, vertical networks can be enhanced by technologies such as IoT and blockchain to increase transparency and efficiency.
- **Horizontal Networks:** Such networks consist of peer companies that may collaborate on specific projects, share resources, or jointly market their products. Modern horizontal networks often leverage collective intelligence and shared platforms to achieve economies of scale and stronger bargaining power in the market.

The Value Chain Strategy also integrates digital transformation, sustainability considerations, and customer-centric approaches to create a competitive advantage. Businesses now also focus on transparency and ethical sourcing within their value chains and on building relationships that foster trust and long-term partnerships.

The ultimate goal of a Value Chain Strategy is to find a suitable organizational model that can effectively navigate competition challenges. It involves strategic planning to ensure that every link in the value chain contributes to delivering high-quality products and services, improving operational efficiency, and enhancing customer satisfaction. The strategy should be flexible enough to adapt to market changes, incorporate new technologies, and meet evolving consumer demands.

The performance measures provide quantifiable targets to assess and refine the value chain's effectiveness continuously, ensuring alignment with the company's broader strategic objectives. Performance Measures:

- **Quality:** Reflects the adherence to customer expectations and standards, influencing customer satisfaction and loyalty.
- **Operational Costs:** Directly related to profitability, operational costs are key efficiency indicators within the value chain.
- Volume measures the amount of output or sales, indicating market share and the effectiveness of the company's value delivery.
- **Time:** Encompasses lead times, production times, and time to market, which are crucial for responding to customer demands and market changes.
- **Flexibility:** The ability of a company to adapt to changes in demand, supply, and market conditions, reflecting the responsiveness of the value chain.
- **Environment:** Measures the sustainability aspect of the value chain, including the use of eco-friendly materials, waste reduction, and energy efficiency.

The Value Chain Strategy bridges the company's chosen path for competitive advantage (competitive strategy) and the execution of delivering products and services (production/service strategy). The Value Chain Strategy directly supports the competitive strategy by ensuring that the organizational model enhances the company's ability to innovate, operate efficiently, or focus on customer needs. For instance, if the competitive strategy emphasizes cost leadership, the value chain strategy would focus on reducing operational costs and improving volume through efficient processes. In addition, decisions made within the value chain strategy influence the production/service strategy, such as choosing between Make to Stock (MTS) or Make to Order (MTO), and impact the time, flexibility, and quality of outputs. A well-aligned value chain strategy ensures that the production/service approach is optimal for market demand and internal capabilities.

7.3.3 Production/Service Strategy—Supply Chain Operations

In the current business climate, a Production/Service Strategy encapsulates a company's approach and methodology to bring its products or services to the market. This strategy is tailored to meet customer demands while maintaining efficiency and managing costs. It considers several critical factors:

1. **Product/Service Description:** Involves a detailed characterization of the products or services, including their features, benefits, and life cycle stage. This description helps determine the appropriate production or service strategy based on the product's complexity, customization level, and volume.
2. **Customers and Suppliers Characterization:** This requires a deep understanding of customer expectations and needs. It also involves setting the standards and requirements for suppliers and managing relationships and collaborations with suppliers to ensure they are aligned with the company's production goals and quality standards.
3. **Process Definition:** Specifies the activities and sequences necessary for producing a product or delivering a service. This includes identifying key performance measures such as quality, efficiency, and timeliness that will drive process improvement.

Production strategies are plans and methods a company employs to manage the manufacturing of its products. They dictate how products are designed, created, and delivered to customers and are selected based on various factors like demand predictability, product customization, and lead times. Key production strategies include [94]:

- **Make to Stock (MTS):** Employed when customer demand can be accurately forecasted and the product has few variations. This strategy is often associated with mature products at the peak of their production volume.
- **Make to Order (MTO):** Used when customer demand is unpredictable but allows the production process to start after receiving the order. This strategy is common in customized housing construction and some computer industry segments.
- **Assemble to Order (ATO):** Appropriate when customer lead time is shorter than manufacturing lead time and there is a predictable option mix. The automotive and electronics industries frequently employ this strategy.
- **Configure to Order (CTO):** Designed for customers who require a mix of predefined options to tailor a product to their specific needs. This strategy offers a balance between customization and lead time, as products are assembled from modular components selected by the customer. Commonly used in industries like automotive and consumer electronics, where customers can choose from configurable options like hardware specifications or aesthetic features.
- **Build to Order (BTO):** Employed for customers seeking highly personalized products manufactured only after an order is placed. This strategy minimizes inventory costs and ensures that production meets exact customer requirements. It is often used in industries like aerospace and high-tech electronics, where customers demand unique specifications and high-quality standards.
- **Engineering to Order (ETO):** Suitable for customized, complex products early in their life cycle or those requiring significant customer-specific engineering design. Industries like aerospace, shipbuilding, and large-scale construction often use this strategy.

Each strategy requires different levels of inventory, lead time, and flexibility, and the choice of strategy significantly impacts a company's supply chain management and ability to meet customer demands efficiently. Production strategies are crucial operational approaches that align with how a company manufactures and stocks its products, based on customer demand and market dynamics. They are measured and refined through several performance metrics:

- **Quality:** Ensures that the final product meets predefined standards and customer expectations, which is critical for maintaining brand reputation and customer satisfaction.
- **Volume:** Refers to the quantity of products produced, which should align with market demand to optimize inventory levels and avoid shortages or surpluses.

- **Time:** Involves the duration from order placement to product delivery, including production lead times, which is essential for meeting customer expectations and improving turnaround.
- **Costs:** Encompasses the expenses associated with the production process, including materials, labor, and overhead. Effective strategies aim to optimize costs while maintaining product quality.
- **Flexibility:** The ability of a production system to adapt to changes, such as varying customer demands or shifts in market conditions, without significant performance loss or cost increase.
- **Environment:** The sustainability aspect, considering the environmental impact of the production process and the product itself, from resource usage to emissions and waste.

These performance measures are integral to selecting and evaluating the appropriate production strategy, whether it is ETO, MTO, ATO, or MTS. They ensure that the strategy not only supports the company's competitive edge but also meets broader operational and sustainable goals.

The choice of production strategy is a direct response to a company's competitive strategy. For example, if a company competes on quality and customization (differentiation), it may lean towards an ETO or MTO approach. If the competitive advantage is based on cost leadership, an MTS strategy might be more appropriate due to economies of scale. Furthermore, production strategies are operationalized within the framework of the value chain strategy. They dictate how each value chain segment—from sourcing raw materials (backward integration) to delivering the final product to the consumer (forward integration)—is managed. A company might choose ATO to align with a value chain strategy emphasizing rapid response to market demand while maintaining efficiency.

7.3.4 Drivers of Supply Chain—Supply Chain Operations

In light of challenges such as COVID-19, geopolitical shifts, and trade tensions, six strategies (see FIGURE 7.2) have been proposed for balancing efficiency with resilience [46]:

1. **Inventory:** This strategy involves maintaining additional production capacity and inventory beyond safety stock requirements to enhance resilience. While costly, leading companies implement buffers as surge capacity for new product launches or use contract manufacturers for strategic surge needs. Decisions to be made regarding inventory are related to: (a) Inventory Levels (b) Buffers Inventory and capacity (c) Inventory Turnover (d) Demand Forecasting; (e) Inventory Replenishment Policies (f) Capacity Allocation (g) Seasonal and Event-Driven Buffers (h) Inventory

Segmentation (i) Technology and Automation (Blockchain, Sensors, Tracking, Warehouse Automation, IoT).

2. **Transportation:** In supply chain management, transportation decisions are critical as they influence cost, efficiency, customer satisfaction, and overall supply chain performance. Key decisions related to transportation include: (a) Mode of Transportation (b) Carrier Selection (c) Routing and Scheduling (d) Consolidations and Freight Size (e) Transportation Costs (f) Warehouse and Hub Locations (g) Technology and Automation (h) Sustainability and (i) Risk Management. Regarding digital technologies, use of GPS, RFID, TMS (Transportation Management Systems), and IoT devices [24].
3. **Manufacturing:** Decisions related to manufacturing are:
    (a) ***In-Company or Network Diversification:*** Companies have started diversifying their sourcing and manufacturing bases in response to disruptions like the U.S.-China trade war. This could mean switching to suppliers outside China or diversifying across multiple locations, viewed as a necessary cost of doing business rather than inefficiency.
    (b) ***Strategic Suppliers or Multisourcing:*** After major natural disasters revealed the risks of single sourcing, multisourcing emerged to mitigate these risks. It involves knowing supplier networks in detail and diversifying supply sources by engaging additional suppliers or utilizing existing ones capable of producing from various locations.
    (c) ***Nearshoring or Global Sourcing:*** Some companies aim to reduce geographic dependence by nearshoring or moving production closer to the end consumer. Though potentially more expensive due to added complexity, it allows for greater inventory control and shorter cycle times.
    (d) ***Platform, Product, or Plant Harmonization:*** Harmonizing technology across plants and standardizing components across products enable seamless movement across a regionalized network. This strategy simplifies sourcing and enhances resilience by allowing higher volume placements among multiple suppliers.
    (e) ***Ecosystem Partnerships:*** The pandemic underscored the importance of diversified sourcing and collaborative relationships with key suppliers and external service partners. Companies leverage these partnerships to diversify production and distribution, engaging with contract manufacturers and logistics partners to support product moves and enhance future preparedness and resilience.

4. **Information Technology:** In supply chain management, decisions regarding information technology (IT) are crucial as they impact data flow, operational efficiency, and overall decision-making. Here are the key IT-related decisions:
   (a) ***Selection of Technology Platforms:*** Cloud-based vs. on-premises solutions and choosing appropriate IT systems (e.g., ERP, WMS, TMS, SCM software). The factors to be considered: Scalability, cost, integration capability, and user-friendliness.
   (b) ***Integration of Systems:*** integrating IT systems across all supply chain functions (e.g., procurement, production, transportation, and distribution). Ensuring compatibility with suppliers' and partners' systems. Choosing the right application solutions: Programming Interfaces (APIs), middleware, and EDI (Electronic Data Interchange).
   (c) ***Data Management and Analytics:*** Implementing data collection, storage, and processing systems. Choosing between structured (e.g., databases) and unstructured data systems (e.g., big data). These technologies enables data-driven decision-making through advanced analytics and business intelligence.
   (d) ***Automation and IoT Adoption:*** Applying IoT devices for real-time tracking and monitoring (e.g., RFID, GPS, and sensors). Automating repetitive tasks using robotics or AI. All these solutions enhances real-time visibility, reduces errors, and improves efficiency.
   (e) ***Supply Chain Visibility:*** Using tools to track inventory, shipments, and performance metrics across the supply chain. Choosing platforms with real-time visibility and predictive analytics capabilities. The objective is to improve responsiveness to disruptions and enhance collaboration.
   (f) ***Cybersecurity:*** Protecting sensitive supply chain data with robust security measures (e.g., encryption, firewalls, and secure access protocols). Implementing disaster recovery and business continuity plans. The goal is to comply to industry standards (e.g., GDPR, CCPA).
   (g) ***Adoption of Advanced Technologies:*** Investing in emerging technologies like Artificial Intelligence (AI), Blockchain, and Machine Learning (ML). Use of Blockchain for traceability, AI for demand forecasting, and ML for predictive maintenance. This IT will drive innovation and improve supply chain agility.

These strategies aim to fortify supply chains against future disruptions by enhancing visibility, agility, and the ability to adapt sourcing, manufacturing, and distribution practices rapidly.

7.3.5 Benefits and Limitations of Applying Life Cycle Engineering Principles in this Entity

Integrating LCE principles into the Supply Chain Development entity can enhance environmental sustainability. By considering the entire life cycle of products, from raw material extraction to end-of-life disposal, LCE encourages the creation of supply chains that minimize ecological impact through reduced resource consumption and waste generation. This holistic approach aligns with the increasing global emphasis on sustainability and can improve a company's image and market competitiveness.

However, applying LCE principles comes with its challenges. It requires a comprehensive understanding of the environmental impacts at each stage of the product life cycle, which can be complex and resource-intensive to assess. Moreover, the upfront costs of developing sustainable supply chain processes can be significant. Companies may need to invest in new technologies, reconfigure manufacturing processes, or source more sustainable but potentially expensive materials.

Furthermore, implementing LCE in supply chain development can sometimes encounter resistance due to the difficulty of measuring and valuing long-term environmental benefits against immediate financial costs. This can be especially challenging when attempting to secure stakeholder buy-in. The dynamic nature of supply chains also means that continuous adaptation and reevaluation are necessary, which can complicate the consistent application of LCE principles. Despite these limitations, the growing imperative for sustainable practices will likely drive further adoption and innovation in LCE across supply chains.

Thus, LCE significantly contributes to sustainable manufacturing within the supply chain by advocating for the design and development of products and processes with consideration for their entire life cycle. By analyzing and optimizing each stage—from material extraction through production, use, and disposal—LCE aims to minimize environmental impacts, reduce waste, and conserve resources. This approach aligns with sustainable manufacturing goals by promoting efficiency, reducing the carbon footprint, and ensuring products are designed for durability, maintainability, and recyclability, ultimately leading to a more sustainable supply chain network.

LCE contributes by encouraging manufacturers to design products with their entire life cycle in mind. This includes selecting sustainable materials, designing products for ease of recycling, and planning for efficient logistics. By doing so, the entire supply chain becomes more eco-friendly, as products are manufactured to reduce their ecological footprint throughout their life cycle.

Moreover, LCE supports creating closed-loop systems where end-of-life products are reintegrated into the production cycle. This approach conserves resources and reduces the need to extract new raw materials, decreasing the overall environmental impact. LCE's contribution to sustainable manufacturing is critical as it aligns with global efforts to mitigate climate change and supports the transition to a circular economy within supply chain development.

7.4 Life Cycle Diagram Flow

According to Molina et al. [92], the Life Cycle Engineering (LCE) applied to the Supply Chain Development entity emphasizes a strategic, integrated approach across all stages of product life, from conception to delivery (See FIGURE 7.3) In the Ideation/Imagination stage, the focus is on formulating a supply chain strategy that ensures an efficient flow of materials and adapts to market shifts, which is critical for early-stage responsiveness. During Basic Development/Conceptualization, supply chain planning involves detailed forecasting and inventory management to meet customer demands. The Advanced Development/Design stage addresses supply chain operations, streamlining processes to ensure seamless production to distribution. Finally, the Launching/Creation/Operation stage executes the supply chain, leveraging technology for real-time tracking and supplier collaboration.

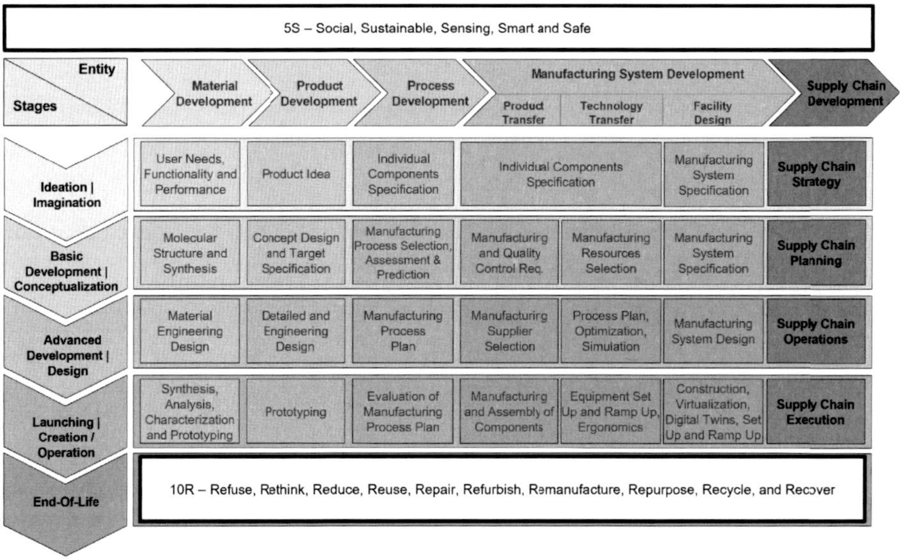

FIGURE 7.3
Life cycle diagram flow: Supply chain development and its stages.

7.4.1 Entity Description: Supply Chain Development

Supply Chain Development (SCD) encapsulates the strategic, tactical, and operational elements of managing and optimizing the flow of goods, services, and information from origin to consumer. Effective SCD requires a deep understanding of each component of the supply chain, an ability to foresee and react to market changes, and an integration of technology to streamline operations [27].

In the strategic phase, called Ideation or Imagination, companies must develop a supply chain strategy that aligns with their business goals and can adapt to new technologies and shifting market demands. The strategy must be collaborative, agile, and resilient, ensuring all stakeholders are involved, risks are mitigated, and disruptions can be managed effectively. It should also leverage digital tools like ERP systems, IoT, AI, and machine learning to increase visibility and improve decision-making processes.

Supply chain planning, also known as basic development or conceptualization, is essential at the tactical level. It involves forecasting, inventory management, production planning, and distribution coordination. The planning must be detailed enough to meet customer demand while maintaining efficiency and managing costs.

The operational phase, Advanced Development or Design, is concerned with the daily management of supply chain activities. This includes procurement, production, distribution, and logistics. Supply chain operations must be streamlined to ensure smooth production flows and timely product delivery.

Finally, the execution phase, Launching or Creation/Operation, involves carrying out the supply chain activities according to the strategic plan. This phase is critical for maintaining customer satisfaction and competitive advantage. Real-time tracking and collaboration tools ensure that operations run efficiently and customer needs are met promptly.

Applying supply chain strategies has shown significant benefits for large organizations. For instance, Boeing utilized the ASCM Supply Chain Operations Reference (SCOR) model to improve its supply chain strategy, which led to a notable decrease in critical shortages within their fighter planes program [17, 109].

7.4.2 Stages

The stages of the supply chain development begin with Ideation, where strategic planning focuses on efficiency and adaptability. Basic Development follows this, involving detailed planning to align with customer demand. The Advanced Development stage manages daily operations, optimizing procurement and logistics. Finally, the Launching stage executes the supply chain, ensuring efficient order fulfillment and customer satisfaction, leveraging technology for real-time tracking and collaboration.

7.4.2.1 Idea—Imagination

- Strategy formulation for efficient material and information flow, focusing on timely delivery, risk mitigation, and technological adaptability. Provides the foundation for supply chain resilience and responsiveness to market conditions.
  - ***Activity Inputs:*** Market analysis, risk assessment, technology trends, business objectives.
  - ***Activity Outputs:*** Strategic plan, risk management strategy, technology integration plan.

7.4.2.2 Basic Development—Conceptualization

- Integration of forecasting, inventory management, production planning, and distribution coordination. Balances supply and demand, optimizes inventory, and ensures cost efficiency.
- ***Activity Inputs:*** Sales forecasts, inventory levels, production capacity, distribution logistics.
- ***Activity Outputs:*** Production and procurement schedules, inventory strategy, distribution plan.

7.4.2.3 Advanced Development—Design

- Daily procurement, production, and distribution management to streamline the supply chain. Maintains product flow and meets delivery commitments through operational efficiency.
  - ***Activity Inputs:*** Supplier agreements, production data, logistics plans, customer orders.
  - ***Activity Outputs:*** Procured materials, production reports, delivery schedules, operational analytics.

7.4.2.4 Launching—Creation/Operation

- Execution of the supply chain plan, including order fulfillment and logistics management. Transforms plans into customer value, focusing on coordination and real-time operations management.
  - ***Activity Inputs:*** Order confirmations, fulfillment strategies, logistics arrangements, supplier collaborations.
  - ***Activity Outputs:*** Delivered goods, customer satisfaction data, performance metrics, supply chain insights.

7.4.2.5 End of Life: 5S—Social, Sustainable, Sensing, Smart, Safe

- Social considerations at the end of a product's life cycle involve implementing community-focused programs for recycling, refurbishing, and job creation in the remanufacturing sector. The goal is to minimize the negative social impact and enhance the positive effects on communities.

 - ***Activity Inputs:*** Community engagement plans, educational materials on recycling and reuse, infrastructure for recycling and refurbishing.
 - ***Activity Outputs:*** Increased community participation in recycling programs, job creation in the recycling and refurbishing sectors, positive social impact reports.

- Sustainable end-of-life management includes recycling, upcycling, and the recovery of materials. It focuses on designing products for easy disassembly and repurposing to minimize waste and reduce the ecological footprint.
  - ***Activity Inputs:*** Sustainable materials, recycling technologies, waste management systems.
  - ***Activity Outputs:*** Reduced landfill waste, recovered materials for reuse, environmental impact assessments.
- Sensing involves integrating sensors and data collection capabilities to monitor product performance and usage throughout its life cycle, including end-of-life. This data helps in optimizing recycling processes and improving future product designs.
  - ***Activity Inputs:*** Sensors and IoT devices, data collection and analysis tools, monitoring systems.
  - ***Activity Outputs:*** Performance data reports, insights for product design improvements, enhanced recycling process efficiency.
- Smart end-of-life management incorporates intelligent systems for automation and interconnectivity. This includes smart sorting systems for recyclables and automated disassembly lines.
  - ***Activity Inputs:*** Automated sorting systems, smart disassembly lines, integrated software for process management.
  - ***Activity Outputs:*** Increased automation in recycling processes, higher efficiency and lower costs, better tracking and management of materials.
- Safe end-of-life practices ensure that all processes involved in recycling and disposal are safe for workers and do not harm the environment. This includes strict adherence to safety standards and regulations.
  - ***Activity Inputs:*** Safety protocols and standards, protective equipment for workers, compliance monitoring tools.
  - ***Activity Outputs:*** Safe working environment, compliance with health and safety regulations, reports on safety performance.

7.4.3 Techniques and Methods Required During the Development

Several best practices are recommended in developing a supply chain entity. These include recruiting and developing supply chain professionals to address the skill gaps in the industry, aligning the supply chain team across various functions to ensure efficiency, and establishing strong alliances with suppliers for improved reliability and shared success. Purchasing supplies in volume can reduce costs by taking advantage of economies of scale while diversifying supplier relationships can help avoid delays by mitigating the risk of supply chain disruptions. Communication with suppliers is crucial for the proactive management of potential delays.

In supply chain development, demand forecasting is a critical technique that uses statistical tools and data analysis to predict future customer demand. This process allows businesses to align their production and inventory levels closely with market needs, minimizing the risk of overproduction or shortages. By accurately predicting demand, companies can optimize their operations, reduce costs, and improve customer satisfaction by ensuring the right products are available at the right time.

Inventory management, another essential technique, involves strategies like Just-In-Time (JIT) and Economic Order Quantity (EOQ) to maintain optimal inventory levels. JIT aims to reduce inventory costs by receiving goods only as they are needed in the production process, while EOQ determines the optimal order quantity to minimize total inventory costs, including holding, ordering, and shortage costs. Effective inventory management helps reduce waste, lower storage costs, and increase operational efficiency.

Supplier relationship management (SRM) is focused on developing strategic partnerships with suppliers. It emphasizes mutual trust, open communication, and long-term engagements rather than short-term transactions. SRM enables companies to secure reliable sources of materials and services, negotiate better terms, and foster innovation through collaborative efforts. Strong supplier relationships contribute to a resilient supply chain capable of responding to market changes and challenges.

When integrated into supply chain development, these techniques play a significant role in optimizing operations. Demand forecasting ensures that production aligns with market demand, inventory management strategies minimize costs associated with holding and ordering stock, and effective SRM strengthens the supply chain's foundation. Together, they contribute to a more efficient, responsive, and sustainable supply chain, supporting the organization's strategic goals and adapting to the dynamic business environment.

Some of the most common techniques and methods in supply chain development include:

1. **Demand Forecasting:** Utilizes historical data and analytics to predict future product demand, optimizing inventory and production planning.
2. **Inventory Management:** Implements strategies like Just-In-Time (JIT) and Economic Order Quantity (EOQ) to maintain optimal inventory levels and minimize costs.
3. **Supplier Relationship Management (SRM):** Focuses on building strong partnerships with suppliers for reliable sourcing and innovation.
4. **Lean Manufacturing:** Aims to eliminate waste within manufacturing processes, enhancing efficiency and reducing costs.
5. **Six Sigma:** Uses data-driven techniques to improve quality and reduce defects in processes.
6. **Sustainable Sourcing:** Involves selecting suppliers and materials that minimize environmental impact and promote ethical practices.

7.4.4 The Ten Core Technologies Applied to the Entity

Integrating the ten core technologies into this entity is applied strategically, as depicted in FIGURE 7.4.

1. **Ideation-Imagination:** Formulating a supply chain strategy that ensures the efficient flow of materials and information, aligns with market shifts, and integrates advanced technologies.
   (a) ***AI/ML:*** Analyzes market trends and customer data to define strategic supply chain priorities. Provides predictive analytics for risk assessment and demand forecasting.
   (b) ***Cloud Computing:*** Facilitates collaborative strategy development among stakeholders by enabling data sharing.
   (c) ***Simulation Tools:*** Models supply chain scenarios to test adaptability and risk mitigation plans.
2. **Basic Development—Conceptualization:** Detailed planning to integrate forecasting, inventory management, production, and distribution. Balances supply and demand while optimizing costs and securing resources.
   (a) ***Big Data Analytics and Blockchain:*** Analyzes historical sales data for accurate forecasting. Ensures transparency and traceability in supplier transactions.
   (b) ***IoT:*** Monitors real-time inventory levels and transportation conditions for planning.
   (c) ***Simulation Tools:*** Tests distribution and inventory strategies to predict performance under different scenarios.

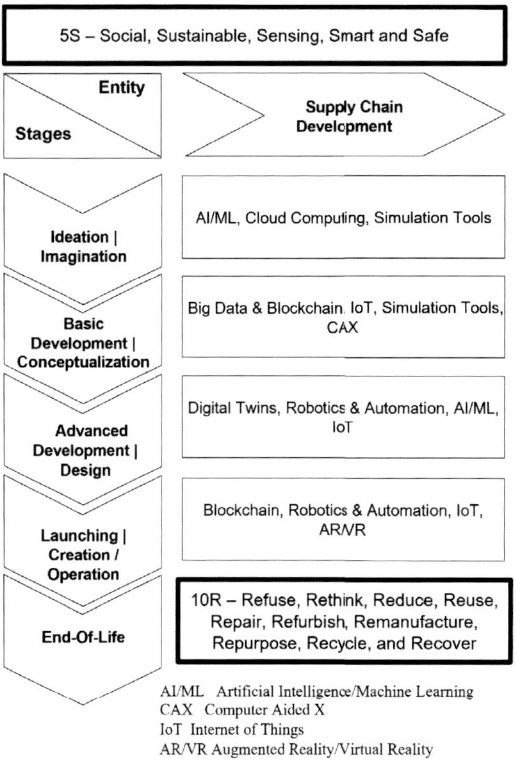

FIGURE 7.4
Core technologies applied to this entity.

   (d) ***CAX:*** Supports process design for inventory and distribution plans.

3. **Advanced Development—Design:** Managing daily supply chain operations, including procurement, production, and logistics, to streamline processes and meet delivery commitments.
   (a) ***Digital Twins:*** Simulates and optimizes real-time supply chain operations dynamically. Allows for virtual testing of procurement and distribution strategies.
   (b) ***Robotics and Automation:*** Automates warehousing operations, including sorting and packing.
   (c) ***AI/ML:*** Optimizes transportation routes and enhances supplier performance monitoring.
   (d) ***IoT:*** Provides live tracking of goods and equipment to ensure operational efficiency.

4. **Launching—Creation/Operation:** Executing the supply chain plan, including order fulfillment, logistics management, and supplier

performance monitoring. Ensures efficient operations through real-time tracking.

(a) ***Blockchain:*** Guarantees secure and immutable transaction records for supplier collaboration and trust-building.
(b) ***Robotics and Automation:*** Supports automated order fulfillment and quality assurance.
(c) ***IoT:*** Real-time tracking of shipments and inventory levels to prevent disruptions.
(d) ***AR/VR:*** Enhances training for supply chain operators and visualizes logistical workflows.

These technologies enable precise demand forecasting and real-time inventory management, reducing overproduction and excess inventory, thus minimizing waste. Furthermore, they support sustainability by ensuring the optimal use of resources and promoting transparency in sourcing materials. Companies can favor suppliers who adhere to eco-friendly practices, contributing to a lower environmental footprint across the supply chain.

Embracing these cutting-edge technologies improves operational efficiency, reduces costs, fosters sustainability, and builds trust among stakeholders. This holistic approach to technology integration is key to developing a future-ready, resilient, and responsible supply chain.

7.4.5 Digital Supply Chain Roadmap Implementation

In the context of the Supply Chain Development entity, the strategic alignment of business and supply chain strategies plays a crucial role in the successful digital transformation of the supply chain [49]. This transformation, pivotal for mitigating risks and optimizing costs, hinges on harmonizing advanced technical skills, such as analytics, with core business competencies, including cross-functional collaboration and data-driven decision-making. Additionally, the capacity for adaptability and a willingness to embrace risk are essential traits in this rapidly evolving landscape.

Chief supply chain officers (CSCOs) face the challenge of steering their supply chains through digitalization, often contending with the difficulty of converting their digital ambitions into a coherent, actionable, and business-aligned transformation roadmap. This multiyear strategic plan roadmap should articulate the necessary technology investments in the supply chain to bolster business growth and performance. It must comprehensively address the capabilities, talent development, and process enhancements necessitated by digital technologies across the entirety of the supply chain.

Organizations must craft and execute an integrated digital transformation roadmap to capitalize on the benefits of digital technologies for the supply chain. This roadmap should not only focus on immediate operational enhancements but also on sustaining a strategic vision for the long term, ensuring the Supply Chain Development entity remains a vital driver of business performance in an increasingly digital world.

Thus, FIGURE 7.5 depicts the roadmap for creating a digital supply chain from its present state (A) to a desired future state (B) with improved digital capabilities. This linear flow diagram considers the four stages of the Supply Chain Development Entity that a supply chain must go through for digital transformation:

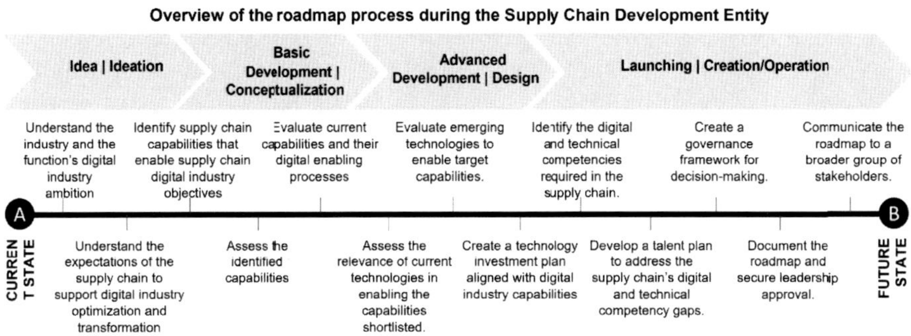

FIGURE 7.5

Digital supply chain roadmap adapted from [49].

- **Idea—Ideation:** This is the initial stage where the supply chain's objectives for supporting digital optimization and transformation are identified. It involves understanding the enterprise's digital business ambition or function and setting the supply chain expectations to support this digital transformation.
- **Basic Development—Conceptualization:** The supply chain's capabilities that enable digital business objectives are determined here. This includes evaluating existing capabilities and the health of these capabilities, as well as assessing the relevance of current technologies in enabling digital objectives.
- **Advanced Development—Design:** This stage involves prioritizing investments in technology. It evaluates emerging technologies enabling targeted supply chain capabilities and identifies the digital and technical competencies required in the supply chain.
- **Launching—Creation/Operation:** The final stage is creating a plan to address the supply chain's digital talent gaps and determining a governance framework. This includes developing a talent plan, creating a governance framework for decision-making, and communicating the roadmap to a broader group of stakeholders. It concludes with documenting the roadmap and securing leadership approval.

Delineating roles and responsibilities, provided in TABLE 7.2, within a project team focused on supply chain management is crucial for several reasons [49]:

- Primarily, it ensures clear role definition, which is fundamental for avoiding overlaps and gaps in responsibilities, thereby fostering efficient collaboration and accountability among team members. Such clarity is indispensable for successfully executing complex supply chain operations and initiatives.
- Moreover, the roles promote strategic alignment across the supply chain, ensuring that activities directly contribute to the organization's broader strategic objectives. This alignment is vital for the coherent and unified execution of business strategies, particularly in supply chain management's dynamic and often volatile domain. By explicitly outlining the responsibilities, the table facilitates effective team communication and external stakeholder communication, streamlining decision-making processes and enhancing operational efficiency.
- The emphasis on specific roles regarding technology integration and financial management highlights the importance of digital transformation and cost-effective operation within the supply chain. In the context of rapidly evolving market demands and technological advancements, this focus is essential for maintaining competitiveness and ensuring that supply chain strategies are sustainable and aligned with the company's digital business objectives.
- Lastly, this table underscores the role of the supply chain in creating enterprise value. Detailing how various functions contribute to the overall business strategy and objectives illustrates the integral role of supply chain management in driving organizational growth, profitability, and shareholder value. This perspective is crucial for recognizing the supply chain as a strategic asset rather than a mere operational necessity.

7.5 Case Study Applying the Life Cycle Engineering of this Entity

The Gartner Supply Chain Top 25 for 2023 [47] is a prestigious annual ranking showcasing the world's leading supply chains. It highlights companies that combine efficiency with responsiveness to customers' needs, leveraging innovative practices beyond traditional supply chain operations. These companies are recognized for superior performance across several dimensions, including financial achievements, corporate social responsibility, and the ability to foster innovation within the supply chain.

The ranking is based on quantitative and qualitative measures, including financial performance indicators, opinions from the supply chain community,

TABLE 7.2
Roles and Responsibilities of the Project Team Suggested by [49]

Role	Responsibility
CSCO	Present an overview of the industry's and supply chain's mission, operational processes, and performance goals.
Head of Supply Chain Strategy Realization	Manage to collect and examine information on needs and alternatives, determine the roadmap's contents, and secure stakeholder consent for the necessary investments and resources.
Supply Chain Technology Leadership	Guarantee the effectiveness of digital supply chain initiatives by managing process and technology harmonization.
Enterprise Architect	Explain how business capabilities will contribute to creating enterprise value.
Business Unit Manager(s)	Confirm that the business capabilities accurately represent the activities of their respective groups.
Supply Chain Finance Leadership	Manage the financial operations, encompassing cost-benefit evaluations, ROI calculation, and the distribution of budgetary resources to advance digital business goals.

and an evaluation of corporate social responsibility efforts. Gartner's assessment aims to identify, celebrate, and profile companies demonstrating excellence in supply chain management. Organizations across industries look to these leaders for peer benchmarks to build a business case for supply chain transformation and adopt best practices to accelerate their supply chain performance.

The top companies in this ranking are celebrated for their commitment to sustainable practices, customer value investment, demand management, and innovation against environmental, social, and governance (ESG) criteria. They leverage relationships, technology, and new service models to retain and grow revenue, develop robust supply chain risk management strategies, and invest in antifragile strategies that make their supply chains more resilient and capable of thriving under competitive pressures. Through fostering many-to-many ecosystems, these leaders accelerate progress on ESG initiatives, highlighting a collective move towards sustainability, innovation, and enhanced collaboration within the supply chain community.

The Top 10 companies for 2023 that excelled in their supply chain management for several key reasons, as highlighted by their specific strategies and initiatives:

1. **Schneider Electric:** leads with its Zero Carbon Project, reducing carbon emissions through collaboration with suppliers, promoting sustainable practices, and driving transformational change with its STRIVE program focused on sustainability, resilience, and efficiency.
2. **Cisco Systems:** addresses the semiconductor shortage by developing capabilities that close operational gaps, using scenario planning, and creating playbooks to mitigate risks, ensuring business continuity in volatile environments.
3. **Colgate-Palmolive:** emphasizes environmental sustainability and community well-being, applying end-to-end supply chain segmentation, modular designs, and new technologies to enhance agility and innovation, including a unique partnership with NASA for product testing in space.
4. **Johnson and Johnson:** showcases leadership in evolving products and solutions across its diversified portfolio, employing digital control towers for real-time decision-making and establishing a supply chain resilience dashboard for continuous product tracking.
5. **PepsiCo:** utilizes digital twins and AI in manufacturing to reduce downtime, develops digital tools for enhanced retail collaboration, and fosters regenerative agriculture principles for a sustainable food supply.
6. **Pfizer:** leverages lessons from the COVID-19 pandemic to improve go-to-market strategies and cost-to-serve solutions, demonstrating industry innovation with initiatives like Accord for a Healthier World aimed at closing the health equity gap.
7. **Microsoft:** focuses on supply chain innovation and transformation, generating actionable insights for fast decision-making and developing a sentient supply chain for end-to-end visibility, prioritizing ESG progress and investment.
8. **Lenovo:** balances lean operations with agility, organizing for a better customer experience, showcasing manufacturing capabilities for revenue growth, and driving innovation in partnership with design, supported by AI and analytics.
9. **Walmart:** drives industry-wide initiatives, such as implementing RFID chips across the supply chain and investing in network automation, aiming for impact on the wider ecosystem and enhancing e-commerce fulfillment.
10. **L'Oréal:** transforms its fulfillment and factory strategies with automated, demand-driven distribution centers, investing in analytics for real-time decision-making and engaging suppliers for sustainable sourcing.

These companies exemplify excellence in supply chain management through innovative use of technology, commitment to sustainability, strategic risk management, and a focus on customer and supplier collaboration, setting industry benchmarks for others to follow.

Besides, The Supply Chain Masters are companies that attained top-five composite scores in the Gartner Supply Chain Top 25 for at least seven out of the last 10 years (2013 – 2023). The five-composite scores are [46]: (Peer Opinion × 27.5%) + (Gartner Opinion × 27.5%) + (Bond Rating × 15%) + (Quality of Patient Care × 20%) + (ESG × 10%). Hence, the Supply Chain Masters companies were:

1. Amazon strongly focuses on Environmental, Social, and Governance (ESG) initiatives, aiming to operate the world's first zero-carbon operation certified by the International Living Future Institute (ILFI) with a new same-day site in California. It continues to grow its third-party marketplace, diversifying its sales strategy and reducing organizational redundancies by consolidating technology.
2. Apple remains focused on diversifying and enhancing the resilience of its supply chain. It has significantly improved its ESG performance by increasing clean energy usage by 30% in one year and getting over 200 suppliers to reduce 1 million metric tons of carbon by 2025. Additionally, Apple mitigates risks through organizational restructuring.
3. Procter & Gamble (P&G) demonstrates a strong commitment to digital transformation, operational excellence, and productivity through deliberate segmentation and rigorous governance. It fostered internal innovation by formally recognizing individual and team contributions towards ESG goals and was recognized for its equity and inclusion initiative at its Goa plant as the people breakthrough of 2023 by the Gartner Power of the Profession™ Supply Chain Awards.
4. For sustainable growth, Unilever has redesigned its supply chain organization by prioritizing specific products and transitioning them to new, expanding categories. It has restructured into smaller divisions to accelerate decision-making speed and quality, continuing to demonstrate leadership in sustainability. Notably, Unilever achieved a 13% progress toward its 2030 GHG emissions reduction targets in 2022.

These Supply Chain Masters lead the way due to their comprehensive strategies that blend operational excellence, technological innovation, and a strong commitment to sustainability and social responsibilities. They serve as benchmarks for leveraging supply chains as a competitive advantage while positively impacting the environment and society. Moreover, the key trends among 2023 supply chain leaders emphasize the strategic integration of relationships, technology, and innovative service models to enhance customer revenue and manage risks effectively. These trends include:

- **Seizing New Opportunities:** The leaders leverage their relationships, advanced technology, and new service models to both retain and grow customer revenue. They are also proactive in developing robust supply chain risk management strategies to mitigate disruptions, showcasing a dynamic approach to navigating the complexities of modern supply chains.
- **Investing in Antifragile Strategies:** These organizations focus on strengthening their supply chains in a way that not only withstands disruptions but also thrives because of them. This antifragile approach elevates the role of the supply chain within the business, ensuring it contributes significantly to resilience and competitive advantage.
- **Driving Collective Progress on ESG Initiatives:** By adopting a many-to-many ecosystem approach rather than traditional one-to-many networks, these leaders can fast-track progress on critical Environmental, Social, and Governance (ESG) initiatives. This includes efforts towards achieving net-zero emissions, developing regenerative supply chains, and ensuring living wages, among others, through collaborative partnerships across the supply chain with NGOs, governments, and even competitors.
- **Accelerating Cultural Transformation and Innovation:** At the heart of these trends is the creative integration of people and technology, accelerating cultural transformation and innovation within these organizations. This involves transforming the technology used for customer interactions, enhancing daily operations and planning, and fostering an environment where employees can innovate and collaborate more effectively.

These trends highlight the strategic priorities of supply chain leaders in navigating the challenges and opportunities of the current business landscape, emphasizing sustainability, resilience, technological innovation, and collaborative ecosystems as key drivers of success.

7.5.1 Walkthrough of the Entire Life Cycle Process: From Idea to Launching. Schneider Electric as a Case Study

The Greenhouse Gas (GHG) Protocol Corporate Standard, co-developed by the World Resources Institute (WRI) and the World Business Council on Sustainable Development (WBCSD), serves as the global framework for measuring and managing GHG emissions from corporate activities. It categorizes all emissions contributing to a company's carbon footprint into three distinct 'scopes' based on the source [36].

This system delineates emissions from corporate activities into direct and indirect categories:

- Direct emissions arise from sources that are owned or controlled by the company, such as a corporate facility.
- Indirect emissions originate from company activities but occur at sources over which the company does not have direct control, like those associated with purchased utilities or business travel.

The GHG Protocol further divides these emissions into:

- **Scope 1:** Direct emissions from owned or controlled sources. It includes all direct emissions from company-owned and controlled resources. Actions to manage these emissions include upgrading to energy-efficient equipment, retrofitting facilities with LED lighting, and using advanced data management software to enhance operational efficiency. Switching to lower-carbon fuels, such as replacing coal with natural gas or biofuels, offers another method to reduce Scope 1 emissions. For unavoidable emissions, companies may invest in carbon offsets, which balance out emissions by supporting projects that reduce, avoid, or remove greenhouse gases from the atmosphere.
- **Scope 2:** Indirect emissions from the generation of purchased electricity, heat, or steam. This Scope accounts for indirect emissions from purchased electricity, heat, steam, and cooling consumption. Though these emissions occur at the facility generating the energy, they are considered the responsibility of the company consuming the energy. Companies can reduce their Scope 2 footprint by purchasing energy attribute certificates, investing in onsite renewable energy, and entering into power purchase agreements.
- **Scope 3:** All other indirect emissions not covered under Scope 2. Scope 3 emissions are the most complex, encompassing all other indirect emissions from activities not owned or controlled by the company. These can include emissions related to the supply chain, employee commuting, business travel, and waste disposal. Reducing Scope 3 emissions might involve using virtual meetings to replace business travel, improving material efficiency, and engaging with suppliers to encourage them also to reduce their carbon footprint. Like Scope 1, companies can use carbon offsets to counterbalance unavoidable Scope 3 emissions.

In that sense, the '*Supply Chain Decarbonization: An Essential Step Towards Net-Zero*' report, prepared by Schneider Electric [37], emphasizes the critical role of supply chain management in achieving climate neutrality. It particularly focuses on addressing Scope 3 emissions, which are emissions from activities not owned or directly controlled by the reporting organization but linked to their value chain. These emissions are often the most significant but the hardest to manage due to their indirect nature.

Hence, to integrate the insights from the report into the LCE framework for a structured approach to decarbonization, Schneider Electric can map its strategies according to the different stages of the product and supply chain life cycle (see FIGURE 7.6):

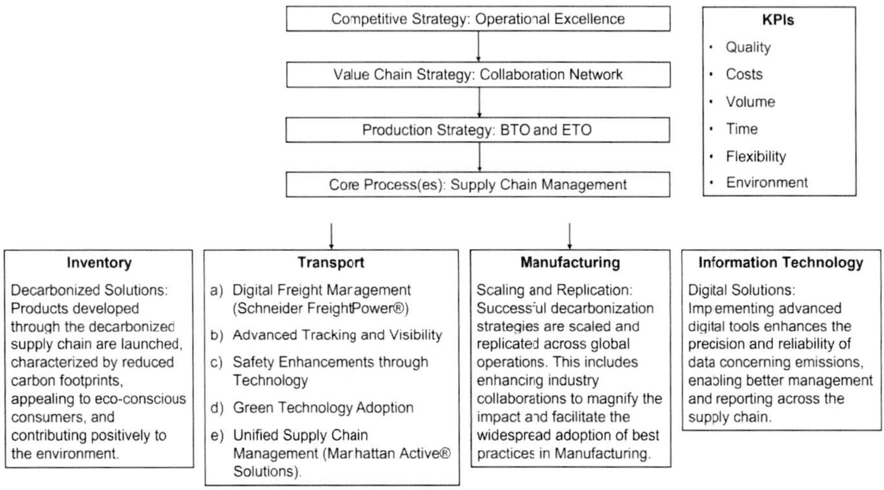

FIGURE 7.6
Drivers—strategies for supply chain resilience: Schneider electric exemplification.

- Idea—Imagination, Strategic Ideation and Goal Setting: **Operational Excellence**. The decarbonization journey at Schneider Electric starts with the Ideation stage, where the focus is placed on setting actionable and ambitious goals, particularly targeting Scope 3 emissions, which involve indirect emissions from activities not owned or controlled by the company but over which it can exert influence.
  - ***Objective Setting:*** The company defines its decarbonization goals with an understanding of the intricate dynamics between upstream and downstream supply chain activities, aiming to address the most significant emissions sources comprehensively.
  - ***Scenario Planning:*** Potential future scenarios are considered, allowing Schneider Electric to plan for various contingencies, ensuring flexibility and resilience in its strategies.

- Basic Development—Conceptualization - Supplier Engagement and Collaboration: **Collaboration Network**. During the Conceptualization stage, Schneider Electric develops its approach to incorporating ESG criteria into supplier relationships, emphasizing the need for collaborative efforts to meet sustainability targets.
  - ***Environmental, Social, and Governance (ESG) Integration:*** Criteria for environmental, social, and governance factors are integrated into supplier selection processes, ensuring that all partners align with Schneider Electric's sustainability ethos.

  - ***Support and Collaboration:*** The company sets clear emissions reduction targets for suppliers, offering technical and financial support to help them achieve these objectives, thereby ensuring collective progress towards broader environmental goals.

- Advanced Development—Design - Digital Tools Implementation and Regulatory Adaptation: **Engineer to Order (ETO) and Build to Order (BTO)**. In the Advanced Development stage, Schneider Electric focuses on deploying digital tools to improve the accuracy of emissions tracking and compliance with evolving regulations.
  - ***Regulatory Readiness:*** The company stays abreast of regulatory changes and adapts its operations accordingly to maintain compliance across different regions, safeguarding the supply chain against potential disruptions due to legislative changes.
- **Launching—Creation/Operation - Operational Execution and Scaling:** The Launching stage sees Schneider Electric moving from strategy to action, implementing the plans across its operations, and scaling successful practices. **Core process focus: Supply Chain Management.**
  - ***Inventory—Decarbonized Solutions:*** Products and solutions developed through the decarbonized supply chain are launched, characterized by reduced carbon footprints, appealing to eco-conscious consumers, and contributing positively to the environment.
  - ***Transport—Technology and Automation:*** Schneider has strategically integrated technology and automation into its transportation operations to enhance efficiency, visibility, and customer satisfaction. Key initiatives include: (a) Digital Freight Management (Schneider FreightPower®), (b) Advanced Tracking and Visibility, (c) Safety Enhancements through Technology, (d) Green Technology Adoption and (e) Unified Supply Chain Management (Manhattan Active® Solutions).
  - ***Manufacturing—Scaling and Replication:*** Successful decarbonization strategies are scaled and replicated across global operations. This includes enhancing industry collaborations to magnify the impact and facilitate the widespread adoption of best practices in Manufacturing.
  - ***Information Technology—Digital Solutions:*** Implementing advanced digital tools enhances the precision and reliability of data concerning emissions, enabling better management and reporting across the supply chain.

Thus, throughout the LCE framework, regular monitoring and reporting mechanisms are established to maintain transparency with stakeholders and facilitate ongoing optimization of processes based on real-time data and feedback. Furthermore, the company seeks and employs financial strategies and instruments that support its long-term decarbonization goals, aligning financial investments with sustainable development objectives.

Additionally, the 5S principles at the end of the product life cycle can maximize sustainability and minimize environmental impact:

- **Social:** Schneider could engage with local communities to facilitate the recycling of electronic and electrical components. This includes ***educational programs*** about the importance of recycling and providing resources or incentives for proper disposal. Moreover, it can implement programs focusing on refurbishing and repurposing old equipment, ***creating jobs*** in local communities, especially where their products are heavily used.
- **Sustainable:** Ensure their products are designed for easy disassembly. Promoting the ***recycling and upcycling*** of parts can help reduce landfill waste and support circular economy initiatives. Focus on recovering valuable materials (*material recovery*) like metals and plastics from old equipment that can be reused in new products, reducing the need for virgin materials and decreasing the overall environmental footprint.
- **Sensing:** Utilize sensors and IoT technology to track products' health and usage patterns throughout their life cycle (***product life cycle tracking***). This data can inform when a product is nearing the end of its life and can be optimally recycled or upgraded. Implement systems to monitor the end-of-life processing of products (***performance monitoring***) to ensure they are disposed of or recycled in the most efficient and environmentally friendly way.
- **Smart:** Develop smart tools and systems that facilitate the disassembly of products at the end of their life cycle (***smart disassembly***). This could include robotic systems that automatically sort and disassemble products to extract reusable components and recyclable materials. Develop programs where end-of-life products are integrated into ***smart grids*** as storage units or other functional components until fully depleted.
- **Safe:** Ensure that all end-of-life processes adhere to the highest safety standards to prevent any harm to workers involved in the disposal and recycling processes (***safe disposal practices***). Design products with non-toxic materials that are safe to recycle and will not leach harmful chemicals into the environment during recycling.

This comprehensive approach ensures that Schneider Electric achieves its internal sustainability goals and drives the industry towards more eco-friendly practices, setting a benchmark in corporate responsibility and environmental stewardship. In addition, by applying the 5S principles to the end-of-life phase of their products, Schneider Electric can significantly enhance the sustainability of their operations. This approach helps reduce the environmental impact and strengthens the company's commitment to social responsibility and technological innovation.

7.6 The Future of Sustainable Manufacturing for this Entity

Envisioning the future of sustainable manufacturing involves a deep dive into each stage of the supply chain entity, integrating advanced technologies and sustainability principles at every step.

During the Idea—Ideation stage, the focus is on developing a supply chain strategy that ensures the efficient flow of materials and information and incorporates sustainability from the outset. This involves planning for using renewable resources, minimizing energy consumption, and designing for product recyclability. Technology integration is crucial in enhancing adaptability to market shifts and demand fluctuations, ensuring supply chains can quickly respond to changes without compromising on sustainability goals. Furthermore, this stage is about envisioning a supply chain that is resilient and sustainable. It is where strategic decisions are made to adopt circular economy principles, aiming to keep materials in use for as long as possible and recover and regenerate products at the end of their life cycle. Advanced digital twins and simulation models can be used here to predict the outcomes of different supply chain designs, incorporating environmental impact assessments to choose the most sustainable options.

In the Basic Development—Conceptualization stage, supply chain planning becomes more detailed, emphasizing using data analytics for accurate forecasting and inventory management. This stage is critical for balancing supply with demand in an eco-friendly manner, reducing overproduction, and minimizing waste. Strategies to secure sustainable raw materials and components are prioritized to ensure the entire production process contributes to environmental conservation. Therefore, at this juncture, detailed planning involves sophisticated data analytics for more accurate demand forecasting and inventory optimization, reducing excess production and waste. Incorporating life cycle assessment tools during this stage helps identify the environmental footprint of products, guiding decisions towards more sustainable materials and methods. Collaborative platforms enable closer supplier integration, focusing on sustainable sourcing and transparency.

The Advanced Development—Design stage sees the operationalization of the supply chain, where the focus shifts to executing sustainable practices in procurement, production, distribution, and logistics. Key considerations include streamlining these processes to reduce carbon footprint, optimizing routes to decrease fuel consumption, and implementing energy-efficient production technologies. The design phase is where theoretical sustainability strategies are implemented, making it a critical juncture for impacting the supply chain's overall environmental performance. Here, the operational design of the supply chain is refined with a focus on sustainability. This includes selecting manufacturing processes that reduce emissions and waste, implementing energy-efficient technologies, and planning logistics to minimize transportation impacts. Sustainable packaging design is also considered, using biodegradable materials or easier to recycle. The role of AI in optimizing routing and load planning comes to the forefront, significantly reducing the carbon footprint of transportation.

Finally, the Launching—Creation/Operation stage involves executing the supply chain, leveraging real-time tracking and collaboration technologies to ensure efficient and sustainable operations. This stage demonstrates the culmination of the planning and design efforts, showcasing a supply chain that meets customer satisfaction and does so by adhering to the highest sustainability standards. For instance, the execution phase employs real-time IoT sensors and AI-driven analytics to monitor and adjust operations for maximum efficiency and minimum environmental impact. Supply chain visibility tools track products from source to consumer, ensuring that sustainability standards are maintained. Blockchain technology underpins the traceability of goods, offering an immutable record of sustainable practices and materials used, enhancing consumer trust.

Although the perception of the CSCO as an equal strategic partner has reverted to pre-pandemic levels [45], with 47% in 2022 versus 56% in 2020. However, 72% of CSCOs are still considered as strategic business partners among the highest-performing supply chain organizations. This indicates a differentiation in how top-performing companies continue to value and leverage supply chain leadership as a strategic component of their business, even as overall industry perceptions may have shifted back towards pre-pandemic views. Furthermore, it is envisioned that there are four keys to creating a competitive advantage in supply chain management in the future:

- **Commercial Innovation:** Transitioning focus from operational excellence to enabling top-line growth through services and solutions that allow customers to achieve their goals, supported by supply chain capabilities "as a service", a consolidated service menu, and operational flexibility.
- **Sustainable Operations:** Embedding sustainability across the supply chain to drive revenue growth and cost savings, not just as an isolated effort but as an integrated part of business operations focusing on developing a sustainability roadmap and building partnerships for a circular economy.
- **Real-Time Decision Execution:** Leveraging digital technology, data, and talent to make and execute decisions in real-time, ensuring the supply chain can rapidly respond to changing conditions with the help of advanced analytics, AI, and machine learning.

- **Human-Centric Work Design:** Prioritizing the employee experience and value proposition to attract and retain talent by creating a work environment that offers flexibility, personal development, and choice in work design, using technologies like AR/VR, AI, and robotics to enhance job flexibility.

The future of sustainable manufacturing in supply chain development lies in this holistic integration of eco-friendly practices, from ideation to operation, ensuring that each step of the process contributes positively to the environment. These technologies and methodologies will optimize the supply chain for efficiency and cost-effectiveness and ensure that operations are aligned with global sustainability goals. Integrating renewable energy sources, biodegradable materials, and closed-loop systems will become standard practices driven by consumer demand, regulatory requirements, and a collective commitment to environmental stewardship.

7.7 Conclusion

Developing a sustainable supply chain is critical for the future of manufacturing. This process involves a strategic, integrated approach from ideation to launching stages, and considering the end of life based on the 5S principles is essential for the future of manufacturing. This process involves a strategic, integrated approach that incorporates advanced technologies and sustainability principles at every stage, ensuring that each phase of the supply chain contributes positively to environmental goals, operational efficiency, and global sustainability objectives.

Beginning with ideation, establishing ambitious and actionable sustainability goals is crucial. Scenario planning and advanced digital tools enable the development of a flexible and resilient supply chain strategy. Accurate demand forecasting and inventory management, supported by data analytics during the planning phase, help reduce waste and optimize resources. Engaging suppliers through ESG criteria and collaborative platforms ensures a collective effort toward sustainability, enhancing overall supply chain efficiency.

Operational execution involves implementing sustainable practices in procurement, production, and logistics to minimize the carbon footprint. Leveraging IoT, AI, blockchain, and digital twins enhances efficiency, transparency, and real-time decision-making, ensuring that the supply chain adapts to changing market conditions without compromising sustainability goals.

Sustainable end-of-life management, guided by the 5S principles (Social, Sustainable, Sensing, Smart, Safe), maximizes the environmental and social benefits of products at the end of their life cycle. This includes community engagement, material recovery, performance monitoring, smart disassembly, and safe disposal practices.

The future of sustainable manufacturing hinges on several key factors. Commercial innovation will drive top-line growth through flexible supply chain capabilities and services. Sustainable operations, deeply embedded across the supply chain, will contribute to environmental goals while driving revenue growth and cost savings. Real-time decision execution, powered by digital technologies and advanced analytics, will enable swift and informed decision-making. Human-centric work design, prioritizing employee experience, and leveraging technologies to enhance job flexibility and satisfaction will be essential for attracting and retaining talent.

By embracing these strategies, supply chain leaders can create a competitive advantage, ensuring resilience and sustainability in an ever-evolving market landscape. The commitment to continuous improvement, technological integration, and collaborative efforts will drive the transformation toward more sustainable, efficient, and responsible supply chains. This approach meets the demands of today's consumers and regulatory environments and positions companies to thrive in the future.

8

Competitive Intelligence in Industry

Competitive Intelligence (CI) is a systematic and ethical approach to collecting, analyzing, and applying information about the business environment to support strategic decision-making. It encompasses understanding competitors, market trends, technologies, regulatory frameworks, and customer needs to help companies maintain or gain a competitive edge. CI involves a detailed and strategic process to understand competitors' actions, resources, and decision-making, supporting informed strategic decisions. Key aspects include analyzing historical behaviors, strategic goals, resources, capabilities, and scenario planning to anticipate competitors' reactions. CI also integrates with Industry 4.0 technologies, enhancing its impact through real-time data collection, predictive analytics, and improved operational efficiency. The relationship between CI and Industry 4.0 is significant, offering insights into technological advancements, operational efficiencies, and strategic opportunities. Implementing CI within the Life cycle Engineering (LCE) framework further enhances material development, product innovation, manufacturing processes, and supply chain management, ensuring companies remain competitive. Ethical and legal considerations are crucial in CI practices to maintain credibility and compliance. Future trends in CI emphasize AI, machine learning, big data, ethical considerations, and globalization, shaping its evolution towards a more data-driven, real-time, and collaborative approach. Integrating CI across various business functions is essential for navigating market complexities, driving growth, and achieving long-term success.

8.1 Introduction

Competitive Intelligence (CI) in the industry is the systematic and ethical collection, analysis, and application of information about the business environment. This includes understanding competitors, market trends, technologies, regulatory frameworks, and customer needs. The primary goal of CI is to support strategic decision-making, helping companies maintain or gain a competitive edge.

CI involves a detailed and strategic process to *get inside your competitor's head*, understanding their actions, resources, and decision-making processes.

DOI: 10.1201/9781003501619-8

This comprehensive approach helps stay ahead and make informed strategic decisions. Competitors can be differentiated into several categories based on their strategies, industry positions, and market approaches [61]:

- **Traditional Competitors:** Well-established entities competing on similar products and services.
- **Adjacent Invaders:** Companies from adjacent industries that leverage their capabilities to enter new markets.
- **Fast-Moving Innovators:** New entrants or existing players typically employ disruptive innovations.
- **Entrepreneurs and Start-ups:** Bring fresh ideas and are often more agile and risk-taking.
- **Public vs. Private Competitors:** Public companies disclose more due to regulatory requirements, while private companies keep their strategies more confidential.
- **Disruptive Innovators:** Create significant market changes through groundbreaking innovations.

Moreover, predicting how competitors will react involves understanding their motivations, resources, and typical behaviors:

- **Analyze Historical Behavior:** Examine past actions in similar situations to identify response patterns. Look at how they have made decisions in the past—whether they are quick to react or take a more cautious approach.
- **Understand Their Strategic Goals:** Identify their short-term and long-term goals. Review their public communications to understand their strategic priorities.
- **Evaluate Their Resources and Capabilities:** Assess financial strength and operational flexibility. Companies with strong financial resources can afford to react more aggressively.
- **Consider the Human Factor:** The personality and history of key decision-makers influence how the company might react. Organizational culture also plays a role—companies with a culture of innovation might respond with creative solutions.
- **Scenario Planning:** Develop best-case and worst-case scenarios based on possible competitor actions. Simulate competitive responses in a controlled environment to test different strategies.
- **Monitor Real-Time Data:** Continuously gather data on competitors' actions and market conditions. Monitor competitors' social media activity and news mentions for immediate insights.
- **Leverage Competitive Insight Tools:** Use tools designed to track and analyze competitors' digital footprints, marketing activities, and consumer feedback. Benchmark key performance metrics with competitors to understand where they might have an edge or be vulnerable.

- **Identify Triggers for Reaction:** Identify market changes that could trigger a competitive reaction. Monitor changes in customer preferences that can prompt competitors to react.

CI involves collecting data from various sources, analyzing it to uncover actionable intelligence, and using it to anticipate market shifts, identify emerging opportunities, and mitigate potential threats. Industry 4.0 is characterized by the integration of digital technologies into manufacturing processes. This includes using Internet of Things (IoT) devices, artificial intelligence (AI), machine learning, big data analytics, and cyber-physical systems to create smart factories. The relationship between CI and Industry 4.0 is significant and multifaceted.

Over the last two decades, there have been diverse changes in the direction of foreign and direct investments (FDIs) due to political, economic, social, and environmental drivers [120].

Thus, FIGURE 8.1 conceptualized a framework for understanding various investment localization strategies businesses may employ. The central theme is how companies position their operations concerning their home country, ranging from local (reshoring) to regional (nearshoring) to international (offshoring):

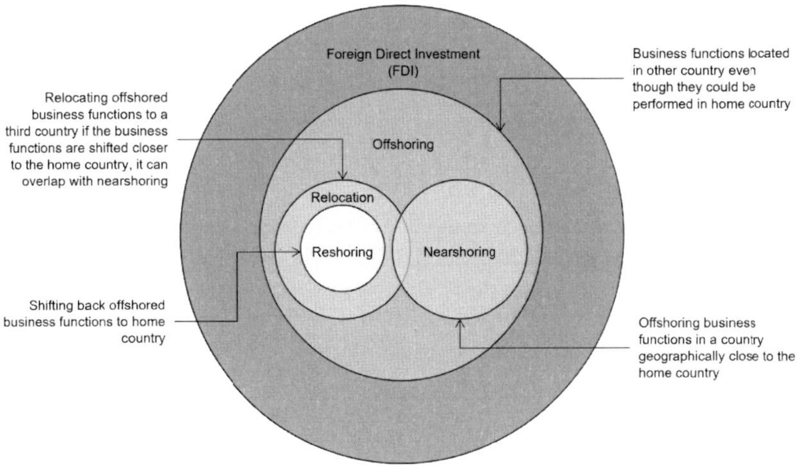

FIGURE 8.1
Conceptual framework of investment localization strategies based on [120].

- **Foreign Direct Investment (FDI):** This is represented by the outermost circle and involves company investment into business interests in another country. Typically, this includes establishing business operations or acquiring business assets in a foreign country.
- **Offshoring:** Indicated within the FDI circle is relocating business processes from a company's home country to a foreign country, usually to leverage cost

advantages. It does not necessarily mean the target country is geographically close to the home country; however, it could be even in the home country.

- **Nearshoring:** A subset of offshoring, represented as an inner circle. This is when a company relocates its business functions to a country that is geographically closer to the home country. This strategy may be chosen to reduce travel times, improve cultural alignment, or take advantage of similar time zones.
- **Reshoring:** Refers to shift back to the home country. Companies that had previously offshored their operations may bring them back for various reasons, such as to tighten control, enhance quality, or respond to consumer preferences for local products.
- **Relocation:** The relocation process, where a company might move operations either back to the home country (reshoring) or to a third country (which could be nearshoring if the third country is closer to the home country).
- **Overlap Areas:** The figure shows areas of overlap between these strategies For instance, if a business moves operations from a far-off country to one closer but not to the home country, it could be considered both nearshoring and relocation.

In addition, the steps required for a comprehensive competitive intelligence analysis are:

1. **Define and Establish CI Operations:**
   (a) ***Define Intelligence Needs:*** Identify and define the intelligence cycle's specific intelligence needs, questions to answer, and objectives.
   (b) ***Establish a Competitive Intelligence Function:*** Form a dedicated team responsible for the CI cycle, with clear goals such as understanding competitor strategies and identifying market trends.
   (c) ***Develop a CI Framework:*** Develop standardized processes for collecting, analyzing, and reporting competitive intelligence. Choose appropriate CI tools and technologies that fit your organization's needs.
2. **Gather and Analyze Intelligence:**
   (a) ***Intellectual Property Search:*** Review patents, trademarks, and copyrights using databases such as Goldfire, Carrot2, etc.
   (b) ***State of the Art and Open Source Search:*** Evaluate current technology, processes, and research from academic databases, industry publications, and other open sources.
   (c) ***Competitive Analysis:*** Research direct and indirect competitors, market positioning, SWOT analysis, market share, and competitive landscape.

   (d) ***Customer Insights and Trends:*** Use surveys, social media, and market research to collect information on customer needs, preferences, and emerging trends.
   (e) ***Regulatory Environment Review:*** Investigate legal and regulatory factors influencing product development and market strategies.

3. **Synthesis and Strategic Integration:**
   (a) ***Collect and Analyze Data:*** Collect data from diverse sources such as market research reports, competitor websites, social media, customer feedback, and industry news. Analytical tools are used to process this data to identify trends and generate actionable insights.
   (b) ***Analyze and Benchmark:*** Compare business processes, performance metrics, product offerings, and best practices within the industry using sources such as Gartner, Deloitte, and McKinsey.
   (c) ***Synthesize Findings:*** Integrate the gathered data to identify patterns, opportunities, and insights. This should clearly show the company's potential moves in the market.
   (d) ***Disseminate Competitive Insights:*** Create regular reports and implement dashboards that summarize key findings tailored to the needs of different stakeholders.

4. **Strategic Development and Implementation:**
   (a) ***Develop Strategic Options:*** Based on intelligence and analysis, identify possible products, innovations, or solutions that align with the company's capabilities and market needs.
   (b) ***Embed CI into Strategic Planning:*** Regularly include competitive insights in strategic planning meetings. Use CI for scenario planning and war games to anticipate competitor moves and develop strategic responses.
   (c) ***Create Implementation Roadmap:*** Develop a detailed action plan for bringing the identified solutions to market, including timelines, resource allocation, and milestones.

5. **Culture and Continuous Improvement:**
   (a) ***Foster a CI Culture:*** Provide training on the importance of competitive intelligence and how to gather and use it effectively. Encourage a collaborative environment where departments share insights and collaborate on CI efforts.
   (b) ***Feedback and Iteration:*** Establish a feedback mechanism to refine the intelligence based on outcomes, new information, and changing market conditions. Ensure continual improvement and relevancy of the intelligence process.

(c) ***Monitor and Adapt:*** Establish a continuous monitoring system to keep track of competitor actions and market changes. Regularly review and adjust the effectiveness of CI processes.

Thus, industries increasingly turn to competitive intelligence to safeguard their future and optimize their operations in an era of rapid technological evolution and intensifying global competition. This chapter delves into a collection of pivotal research studies that collectively provide a roadmap for leveraging competitive intelligence through technological foresight, strategic innovation, and robust predictive modeling.

The concept of competitive intelligence transcends mere data collection, encompassing a broader strategic framework that empowers businesses to predict changes, anticipate challenges, and seize opportunities. As demonstrated in the 2015 study on developing technological front routes, businesses are encouraged to adopt structured methodologies for scenario planning to navigate the uncertainties of technological progress and market dynamics [54]. This approach facilitates better preparedness and enhances the ability to pivot and adapt strategies effectively.

The advent of additive manufacturing, explored in the 2017 paper, exemplifies a disruptive technology with the potential to redefine manufacturing paradigms [9]. The scenarios presented in this study offer insights into how industries can integrate such innovations to maintain a competitive edge, illustrating the critical role of competitive intelligence in fostering technological adaptability.

Further compounding the importance of methodological advancements, the 2019 studies provide significant enhancements to system dynamics models, crucial for simulating the economic impacts of quality management in manufacturing [38]. Additionally, the integration of critical design and technology into educational frameworks highlights the imperative to cultivate a future workforce capable of driving and sustaining innovation [32].

The exploration into RFID technology in 2020 further enriches our understanding, demonstrating how Industry 4.0 technologies can significantly boost operational productivity when strategically implemented [39]. This study underscores the importance of adopting a nuanced approach to new technologies, ensuring they align with broader strategic objectives to enhance competitive intelligence.

By 2023, the focus shifts towards decentralized manufacturing systems, advocating for a model that emphasizes agility and innovation, which are vital in a rapidly changing industrial landscape [120]. This progression towards decentralization not only reflects an adaptive industry response to emerging market demands but also a strategic realignment towards more resilient manufacturing practices.

This chapter aims to equip industry leaders, policymakers, and academics with the insights needed to forge a path towards a more innovative and competitive future.

- **Data-Driven Insights:** The vast amounts of data generated by Industry 4.0 technologies provide a rich source for CI. Analyzing data from IoT devices and digital ecosystems enables companies to uncover insights about competitor activities, market trends, and technological advancements, as highlighted by the transformation in the traditional fashion industry towards more personalized and efficient production methods [64].
- **Real-Time Monitoring:** The real-time data collection and analysis capabilities inherent in Industry 4.0 technologies enable businesses to conduct CI more dynamically and timely. This allows for quicker responses to competitive threats and market changes.
- **Predictive Analytics:** AI and machine learning, key components of Industry 4.0, enhance CI by enabling predictive analytics. Companies can use these technologies to forecast future market trends, competitor moves, and potential disruptions, allowing for proactive strategy development.
- **Customization and Personalization:** Supported by CI, Industry 4.0's emphasis on customization and personalization empowers companies to understand better and predict customer needs and preferences. This leads to developing tailored products and services, offering a competitive edge by meeting specific consumer demands more accurately [34, 52].
- **Strategic Decision-Making:** CI, enriched by Industry 4.0 technologies, supports strategic decision-making by providing a comprehensive view of the competitive landscape. This includes insights into operational efficiencies, supply chain management, and innovation opportunities.
- **Enhanced Collaboration:** Industry 4.0 promotes enhanced collaboration across departments and with external partners. By integrating diverse data sources and perspectives, CI benefits from this, leading to more robust competitive analyses and strategies.

Thus, CI in the context of Industry 4.0 is about leveraging advanced digital technologies to gather, analyze, and act upon information in a way that is faster, more accurate, and more actionable than ever before. This synergy enables companies to navigate the complexities of the modern industrial landscape more effectively, driving innovation, optimizing operations, and securing competitive advantages. In addition, CI can play a crucial role in implementing the six strategies (for more information, see subsection 8.5.5) for creating a resilient supply chain. Here is how CI can be integrated into each strategy:

1. **Inventory and Capacity Buffers:** CI can provide insights into industry benchmarks and best practices regarding inventory levels and capacity management. Understanding how competitors manage

their buffers can inform a company's approach to balancing cost against the need for resilience, helping them justify investments in additional capacity or inventory to stakeholders.

2. **Manufacturing Network Diversification:** CI can help identify trends in manufacturing location shifts, including regions competitors are exiting or entering. It can also uncover potential risks and opportunities in different geopolitical regions, guiding strategic decisions about diversifying manufacturing networks in response to trade wars or other disruptions.
3. **Multisourcing:** CI can be used to analyze the supplier base of competitors, identifying critical suppliers and how others implement diversification strategies in the industry. This insight can help a company identify potential new suppliers or assess the resilience of its supply chain compared to competitors.
4. **Nearshoring:** By monitoring the nearshoring trends among competitors, CI can reveal shifts towards regional supply chains and the reasons behind these changes, such as cost considerations, cycle time reductions, or response to tariff impacts. This information can guide strategic decisions about whether and how to implement nearshoring.
5. **Platform, Product, or Plant Harmonization:** Competitive intelligence can provide insights into how competitors achieve standardization and harmonization within their operations. This includes understanding the benefits and challenges they face, which can inform a company's strategy for technology standardization and component harmonization to enhance resilience.
6. **Ecosystem Partnerships:** CI can identify successful partnerships and collaborations within the industry, offering models for structuring relationships with suppliers, contract manufacturers, and logistics partners. Understanding competitors' strategic alliances and partnerships can guide the development of a more resilient and diversified supply chain ecosystem.

Incorporating competitive intelligence into these strategies enhances a company's ability to build a resilient supply chain. It ensures that strategic decisions are informed by a deep understanding of the competitive landscape, potential threats, and opportunities. This comprehensive approach enables a company to navigate disruptions more effectively and maintain a competitive edge in its industry.

8.2 Competitive Intelligence During the Life Cycle Engineering

Integrating competitive intelligence (CI) into the Life cycle Engineering (LCE) framework is essential for maintaining a competitive edge in the industry. CI involves systematically collecting, analyzing, and utilizing information about competitors, market trends, and external factors to make informed strategic decisions. Within the LCE framework, CI can provide crucial insights at every stage, from material development and product design to manufacturing processes and supply chain management. By leveraging CI, companies can anticipate market shifts, innovate effectively, and ensure their products meet evolving consumer demands and regulatory standards.

In the concept and design phase, CI gathers and analyzes information about market needs, technological advancements, and competitive threats. By integrating CI, companies can ensure that their designs are innovative, meet customer expectations, and incorporate sustainable practices from the outset, setting a strong foundation for the entire life cycle [68].

In the initial stages of the LCE framework, such as material and product development, CI is vital in identifying emerging technologies and consumer preferences. Companies can tailor their research and development efforts to create superior products by understanding competitors' advancements in material science and product features. For example, in material development, CI can highlight competitors' use of advanced materials or sustainable practices, enabling companies to adopt similar or better approaches. In product development, insights into competitors' design philosophies and market strategies can guide the creation of innovative products that meet or exceed market expectations.

CI helps optimize production methods and enhance operational efficiency during the process development and manufacturing system stages. Companies can identify best practices, streamline operations, and reduce costs by analyzing competitors' manufacturing processes. For instance, understanding how competitors utilize digital technologies like 3D printing, robotics, and AI-driven process optimization can inform a company's strategy to adopt similar technologies, improving production speed and quality. Additionally, CI can reveal potential bottlenecks or inefficiencies in competitors' processes, allowing companies to address these issues in their own operations proactively [131].

In the supply chain development stage, CI is crucial for building a resilient and efficient supply chain. Companies can enhance their supply chain resilience by monitoring competitors' supplier networks, logistics strategies, and inventory management practices. For example, CI can identify reliable suppliers, optimal transportation methods, and competitors' innovative warehousing practices. This knowledge enables companies to develop robust supply chains capable of withstanding disruptions and efficiently delivering products to market. Moreover, understanding competitors' strategies for managing end-of-life products can inform sustainable practices, enhancing the company's corporate image and compliance with environmental regulations.

CI's role extends to the strategic management of the entire product life cycle, ensuring that companies remain competitive from ideation to disposal. Companies can make informed decisions at every stage of the product life cycle by continuously gathering and analyzing data on market trends, technological advancements, and competitor actions. This includes integrating eco-design principles during product development, optimizing manufacturing processes for sustainability, and implementing effective end-of-life management strategies. Such comprehensive CI practices enable companies to innovate continuously, improve operational efficiency, and maintain a competitive advantage in a rapidly evolving market.

Incorporating CI within the LCE framework also helps companies navigate the challenges of rapid technological changes and cybersecurity risks. By staying informed about competitors' adoption of new technologies and potential vulnerabilities, companies can proactively address security concerns and safeguard their innovations. Additionally, understanding regulatory trends and standards, such as ISO 14000 for environmental management, ensures that companies remain compliant while pursuing competitive strategies. Thus, CI supports strategic decision-making and enhances overall operational resilience and sustainability.

The process is a strategic approach that involves gathering, analyzing, and utilizing information about competitors and market trends throughout a product or service's entire life cycle. This approach ensures that strategic decisions are informed by a comprehensive understanding of the competitive landscape, from the initial concept and design phases to production, operation, and eventual disposal or recycling [68].

Thus, competitive intelligence is a crucial component of the Life cycle Engineering framework, providing valuable insights at every product development and manufacturing stage. By leveraging CI, companies can anticipate market trends, optimize operations, enhance sustainability, and maintain a competitive edge. Integrating CI with LCE practices ensures that companies are well-equipped to innovate, meet consumer demands, and navigate the complexities of the modern industrial landscape. This strategic approach enables companies to achieve long-term success and sustainability in an increasingly competitive and dynamic market environment.

8.3 Competitive Intelligence within the 5S Systems Framework

CI enables organizations to gather, analyze, and leverage information to maintain a competitive edge. By integrating CI into the 5S principles—Social, Sustainable, Sensing, Smart, and Safe—organizations can enhance their strategic decision-making, foster innovation, and ensure long-term sustainability and success. Here is how competitive intelligence can be incorporated into each principle of the 5S systems:

- **Social:**
  - ***Market and Stakeholder Analysis:*** Conduct thorough analyses of market trends, customer needs, and stakeholder expectations. Understanding social dynamics helps design socially relevant and impactful products and systems.
  - ***Community Engagement Insights:*** Utilize CI to gather insights on community needs and expectations. Engaging with communities can reveal opportunities for new products or improvements that can enhance social well-being and corporate reputation.
- **Sustainable:**
  - ***Sustainability Benchmarking:*** Use CI to benchmark against industry leaders in sustainability practices. This helps identify gaps and opportunities for adopting best practices that reduce environmental impact and improve resource efficiency.
  - ***Regulatory Monitoring:*** Stay informed about evolving environmental regulations and standards through CI. Adapting proactively to regulatory changes ensures compliance and positions the organization as a leader in sustainability.
- **Sensing:**
  - ***Technology Trends:*** Monitor advancements in sensing technologies through CI to integrate cutting-edge solutions into products and processes. Staying ahead in technology adoption can enhance product functionality and operational efficiency.
  - ***Competitive Product Analysis:*** Analyze competitor products with integrated sensing capabilities. Understanding competitor strengths and weaknesses can inform the development of superior sensing features in your products.

- **Smart:**
  - ***Innovation Tracking:*** Use CI to track technological innovations and emerging trends in AI, IoT, and other smart technologies. This enables the organization to incorporate the latest advancements and maintain a competitive edge.
  - ***Data Analytics:*** Leverage CI to analyze data on market demands, customer preferences, and operational performance. Smart use of data can drive continuous improvement and innovation in products and processes.
- **Safe:**
  - ***Safety Standards and Best Practices:*** Monitor industry standards and best practices in safety through CI. Ensuring that your products and systems meet or exceed these standards can prevent risks and enhance trust with customers and regulators.
  - ***Risk Intelligence:*** Use CI to identify potential risks and threats in the market, such as cybersecurity risks or supply chain vulnerabilities. Proactively addressing these risks ensures the safety and resilience of your systems.

By embedding competitive intelligence into the 5S systems framework, organizations can enhance their ability to innovate, respond to market changes, and maintain a competitive advantage while adhering to social responsibility, sustainability, and safety principles. This comprehensive approach drives business success and contributes to broader societal and environmental goals.

8.4 Ethical and Legal Considerations

CI, when conducted within ethical boundaries, involves tracking and collecting publicly available data. The ethical concerns arise from the methods used to gather this intelligence. While hiring private investigators or consultants to find hard-to-source information may still be legal, it must comply with a company's corporate code of conduct, the Strategic Consortium of Intelligence Professionals (SCIP) Code, and all applicable local laws. Activities that cross ethical lines can still be legal, but illegal CI often involves industrial espionage.

Unethical CI practices include methods that violate legal and moral standards. These practices often involve infringing on privacy, copyrights, and intellectual property rights. Some examples of unethical CI practices include [144]:

- **Industrial Espionage:** This is both illegal and unethical and involves the theft of business trade secrets to gain a competitive advantage. It can include spying, bribery, cyber-attacks, or hiring competitor employees to disclose confidential information. This practice can severely damage the economy, costing billions of dollars and millions of jobs annually.
- **Misrepresentation:** This involves deceitful tactics such as posing as a client, jobseeker, or other trusted party to obtain insider information. Another example is posting fake job ads to extract information from applicants about their current employers. While not illegal in many jurisdictions, these tactics are unethical.
- **Corruption:** Offering bribes or other incentives to employees of competing organizations or state institutions to gain confidential information is highly unethical and often illegal.

Unethical CI can lead to serious consequences, including legal penalties and business failure. Ethical CI practices are essential for maintaining credibility, stakeholder trust, and a positive company reputation. Ethical CI involves sourcing information from publicly available resources and conducting business transparently. Hence, CI must be communicated to all employees and consider the following:

- **Policy Formation:** Develop clear policies outlining ethical CI practices and permissible activities.
- **Training:** Conduct regular training to ensure all employees understand and adhere to ethical guidelines.
- **Oversight:** Implement monitoring systems to ensure compliance with ethical standards.
- **Review and Update:** Regularly review and update policies to reflect changes in laws, regulations, and the business environment.

Understanding and adhering to legal requirements is vital for conducting CI. For example, in the United States, federal statutes like the Uniform Trade Secrets Act (UTSA) and the Economic Espionage Act of 1996 (EEA) address the theft or misappropriation of trade secrets, with severe penalties for violations. Other relevant statutes include antitrust regulations and fraud laws. Internationally, legal CI statutes may be more restrictive, necessitating additional care [80].

Beyond legal compliance, ethical considerations are equally important. Companies must navigate the gray zone of gathering competitor information that is not openly shared but not a corporate secret. Protecting the brand and adhering to corporate and CI-specific ethical guidelines is essential. Hence, when engaging third-party CI vendors, it is essential to ensure they adhere to the same legal standards. The company remains responsible for any legal violations committed by vendors on its behalf. Vetting the history and reputation of CI vendors can help ensure compliance with high ethical standards.

8.5 Building the LCE Competitive Intelligence Framework

The diagram in FIGURE 8.2 shows the integration of CI within the LCE framework at different stages and entities of the product life cycle. This LCE framework illustrates how CI informs each stage—ranging from Ideation/Imagination to End-of-Life—across five key entities: Material Development, Product Development, Process Development, Manufacturing System Development, and Supply Chain Development. Each element specifies the application of CI to guide decisions, improve competitive alignment, and foster innovation within each respective phase and entity. The framework underscores the incorporation of the 5S principles—Social, Sustainable, Sensing, Smart, and Safe—ensuring that all activities are conducted with strategic intelligence and sustainability considerations. By systematizing the use of CI, the figure highlights how organizations can effectively anticipate market and competitor movements, optimize their processes, and maintain a competitive edge throughout the entire product life cycle.

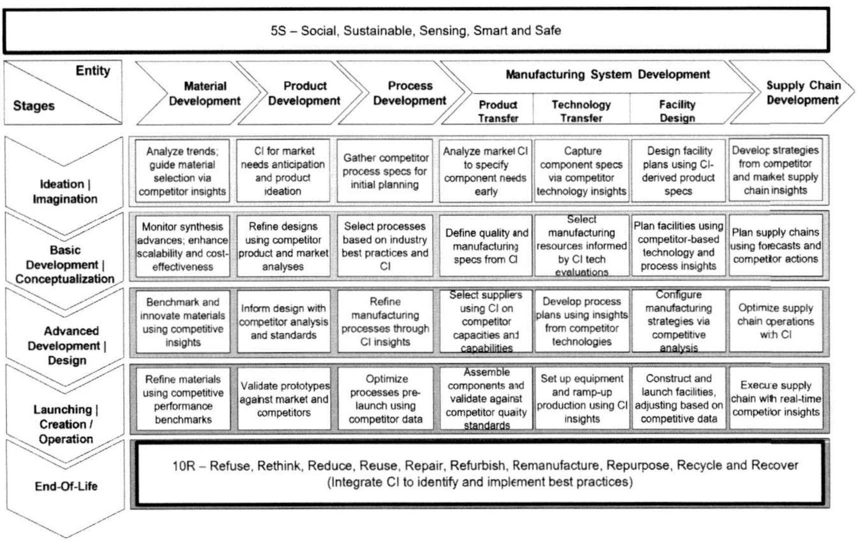

FIGURE 8.2
Competitive intelligence integration within the life cycle engineering framework.

The development of a robust LCE Competitive Intelligence Framework is critical for fostering innovation, improving operational efficiency, and enhancing organizations' competitive edge across various life cycle stages of materials, products, processes, manufacturing systems, and supply chains. This section explores the application of CI within these key development entities, focusing on its role in strategic decision making, innovation, and adaptability to evolving industry landscapes.

By systematically gathering, analyzing, and applying information about competitors, market trends, technological advancements, and regulatory environments, CI empowers organizations to anticipate challenges, identify opportunities, and make data-driven decisions. The framework outlined here provides a structured approach to integrating CI across different phases of the life cycle—from ideation and conceptualization to advanced development and operational launch—ensuring alignment with emerging market demands, sustainability goals, and best industry practices.

Each of the subsequent subsections examines CI within a specific entity of the LCE Framework:

- **Material Entity Competitive Intelligence:** Focuses on guiding the life cycle of material development, from ideation and molecular design to synthesis, testing, and market launch. CI in this entity enables the benchmarking of material performance, cost-effectiveness, and sustainability while inspiring innovation and identifying new applications.
- **Product Entity Competitive Intelligence:** Emphasizes the transformation of product concepts into tangible, market-ready solutions. CI supports decision making at all stages of product development: design, development, prototyping, and launch, ensuring competitive functionality, design quality, and customer alignment.
- **Process Entity Competitive Intelligence:** Centers on optimizing manufacturing processes through insights into industry best practices, cost efficiencies, and technological innovations. CI informs decisions on process design, material selection, and operational enhancements to improve scalability, quality, and flexibility.
- **Manufacturing System Entity Competitive Intelligence:** Examines the creation and refinement of manufacturing systems, including product transfer, technology adoption, and facility design. CI identifies technological trends and operational strategies that streamline manufacturing processes and ensure systems are efficient, competitive, and scalable.
- **Supply Chain Competitive Intelligence:** Highlights the importance of CI in building resilient and adaptive supply chains. By analyzing competitor strategies, logistics operations, and market demands, CI enables organizations to optimize sourcing, distribution, inventory management, and overall supply chain performance.
- **End of Life:** 5S Framework Integration: Discusses the application of CI in ensuring responsible end-of-life management through the 5S principles—Social, Sustainable, Sensing, Smart, and Safe. CI supports innovation in eco-design, recycling processes, and smart technologies, aligning end-of-life strategies with sustainability and regulatory requirements.

Finally, this section concludes with an exemplification of CI in Mexico through nearshoring, showcasing how Competitive Intelligence can address the challenges and opportunities associated with nearshoring initiatives. By using CI in materials, products, processes, and supply chains, organizations can position themselves as leaders in the nearshoring landscape, driving sustainable economic growth and improving their global competitiveness.

8.5.1 Material Entity Competitive Intelligence

The Material Development Entity guides the creation and refinement of new materials from the conceptual stage to their practical application and mass production. This entity encompasses a material's life cycle journey, from understanding user needs and functional requirements through the molecular engineering and design phase to the material's synthesis, testing, and launching into the market. It combines scientific research, engineering, design principles, and operational planning to ensure that the developed materials meet specific performance criteria, are cost-effective, scalable, and have a manageable environmental impact. Each process stage is critical in determining the material's final properties, suitability, and overall success in the intended applications.

Complementing, ***Material Entity Competitive Intelligence*** refers to the strategic process of gathering and analyzing information regarding competitors' activities and market trends, specifically within material development. It encompasses scrutinizing the competitive landscape to inform decisions at every stage of a material's life cycle, from initial concept to production. This intelligence gathering focuses on understanding how competitors and the broader industry approach new materials' design, synthesis, and application and how they manage cost, scalability, and environmental considerations. The aim is to leverage this knowledge to enhance one's own material innovation process, ensuring that new developments are competitive in terms of quality and functionality and aligned with market demands and emerging industry standards.

- **Ideation—Imagination:** CI helps understand user needs, functionality, and performance requirements by analyzing market trends, competitor product offerings, and technological advancements. This stage benefits from the systematic collection and analysis of data to identify material choices and performance criteria that meet product purposes like structural strength or conductivity. For example, the study on Lean Design for Six Sigma emphasizes the importance of orienting the New Product Development (NPD) process toward customer needs to ensure product quality and reliability while minimizing manufacturing costs [8].
  - CI can identify emerging customer needs and market trends, ensuring that new material development aligns with user demands and anticipated future requirements.

  - Benchmarking against competitors can help set performance criteria that aim not just to meet but to exceed industry standards.
  - Insight into new applications and uses of materials in different industries can inspire innovation and identify untapped opportunities.
  - Monitor trends in material technology to stay ahead of other competitors.

- **Basic Development—Conceptualization:** During the conceptualization stage, CI plays a crucial role in the design and engineering of the molecular makeup of materials to achieve desired properties and functions. By leveraging insights from CI, companies can create materials with specific characteristics to match the product's requirements, considering scalability, cost-effectiveness, and environmental impact. The Product Innovation Design Method Based on BP Neural Network illustrates how advanced algorithms and innovative design methods driven by data can improve product design capability using information technology [63].
  - CI can provide insights into how competitors and other industry players approach molecular structure and synthesis, potentially highlighting new methods or materials that have yet to be considered.
  - Understanding cost structures and sourcing strategies for raw materials in the industry can inform more cost-effective and scalable material development approaches.
  - Analysis of regulatory trends and environmental impact assessments conducted by competitors can guide the integration of sustainability into the development process.

- **Advanced Development—Design:** CI contributes to establishing the materials' form, structure, and dimensions. It ensures that the materials match production capabilities by analyzing competitor manufacturing strategies and technological innovations. The Formation and Development of Intelligent Processing Technology overview highlights the role of intelligent manufacturing technologies in enhancing product performance and optimizing control, foundational for intelligent processing in this stage [14].
  - CI can reveal best practices in material engineering design, including how competitors manage the transition from molecular design to actual production-ready materials.
  - Competitor patent filings and intellectual property strategies can inform the creation of drawings and specifications that are both innovative and legally defensible.
  - Market analysis of similar materials can ensure that the design is not only manufacturable but also competitively positioned.

- **Launching—Creation/Operation:** CI supports the synthesis, analysis, characterization, and prototyping of new materials by providing insights into the latest technological advancements and market needs. It enables the evaluation of materials' physical, chemical, and mechanical properties to ensure they meet design specifications. The Automating the Process of Energy-Technology Invention paper discusses AI-enabled design capabilities being developed for energy applications, showcasing how CI can accelerate material development and prototyping processes [135].
  - Monitoring how similar materials are introduced to the market can provide valuable insights into effective launch strategies.
  - Competitive benchmarking can inform the analysis and characterization phase, ensuring that the new materials meet or exceed the competitive offerings regarding quality and performance.
  - CI on prototyping and testing methods can reveal more efficient ways to test materials in real-world conditions and predict potential production challenges.

Competitive intelligence (CI) provides knowledge regarding the competitive landscape, customer expectations, regulatory environment, and technological advancements in each stage. By leveraging CI, a company can enhance its material development process, anticipate market movements, innovate more effectively, and create well-positioned materials to succeed in the marketplace.

8.5.2 Product Entity Competitive Intelligence

The Product Development Entity serves to bring new products from a mere concept to a tangible item on the market. It outlines transforming innovative ideas into successful products through a structured series of stages. These stages include the initial ideation, where product concepts are generated and evaluated; the development phase, which involves detailed planning, conceptual design, and technical specifications; followed by an advanced design phase, where the product is meticulously engineered; and finally, the launching phase, which includes prototyping and testing to resolve any issues and ensure the product meets its intended functionality and design goals. This entity is integral to a company's innovation and growth, requiring cross-functional collaboration and a keen understanding of market demands, customer needs, and technological possibilities.

Additionally, ***Product Entity Competitive Intelligence*** gathers, analyzes, and applies information about competitors' products, market trends, consumer preferences, and technological advancements, specifically within the product development cycle. It supports the strategic decision-making process at each stage of product creation—from ideation and conceptualization to detailed design and prototyping. CI for product entities aims to uncover actionable insights that can influence product features, design, functionality, and market positioning, ensuring that the end product meets the market needs and possesses a competitive edge in its segment. This type of intelligence is essential for companies looking to innovate effectively and capture market share in an environment where consumer preferences and technology are constantly evolving.

- **Ideation—Imagination:** At this initial stage, CI aids in the systematic search for, selection, and development of promising product ideas. It involves analyzing both technically and economically viable applications. CI gathers and analyzes data on market trends, competitor products, and customer feedback to define the project's scope and create a detailed project plan. For instance, the integration of Artificial Intelligence (AI) into the design process of Minimum Viable Products (MVPs) can enhance efficiency, quality, and innovation, highlighting the potential advantages of AI in MVP design [146].
  - CI can identify market gaps and customer needs that haven't been met, guiding the ideation process toward viable product ideas.
  - It can analyze competitive product strategies to determine the scope and potential for new products, ensuring economic viability.
  - CI can help define the project scope and plan by providing insights into the competitive landscape, including risks and opportunities.
- **Basic Development—Conceptualization:** CI collects information about constraints and customer requirements during conceptualization. This information is crucial for developing a conceptual design and target specifications that embody the solution. CI tools, such as Lean design for Six Sigma, can accelerate the achievement of reliable product quality, cost-effective implementation, and desired time-to-market by incorporating customer-oriented approaches into the New Product Development (NPD) process [8].
  - CI gathers information on competitor product features, customer feedback, and market trends to inform the target specifications.
  - It can help identify and understand constraints by looking at how competitors have addressed similar challenges.
  - CI can refine customer requirements for the product by analyzing successful product attributes in the market.
- **Advanced Development—Design:** In the detailed and engineering design stage, CI contributes to developing the arrangement, form, dimensions, and surface properties of all individual parts. It ensures that the design meets manufacturing capabilities and customer expectations. For example, the study on Design for Additive Manufacturing and Advanced Development Methods applied to an innovative multifunctional fan demonstrates how modern models and technologies can be leveraged for efficient product design and prototyping [44].

  - CI can provide benchmarks for arrangement, form, and dimensions by studying competitor products.
  - It can inform about the latest manufacturing technologies and materials competitors use, which can be considered in the detailed design.
  - CI can help avoid patent infringement by ensuring the product design does not violate existing intellectual property rights.

- **Launching—Creation/Operation:** CI plays a crucial role in the prototyping stage by verifying any remaining problems in the product. It involves testing the functionality and potential design modifications. The Optimizing Automotive Product Development paper discusses the integration of Electronic Control Units and virtual prototypes using virtual reality to enhance the efficiency of product development processes, showcasing how CI can streamline project stages and reduce the use of physical prototypes [15].
  - CI can assess competitor prototyping stages, providing insights into common issues and effective solutions.
  - It can identify testing standards and processes that ensure functionality and compliance with industry standards.
  - CI can reveal potential design modifications made by competitors that enhance product performance or user experience.

8.5.3 Process Entity Competitive Intelligence

The Process Development Entity helps to optimize the creation and implementation of manufacturing processes for new products or improvements to existing products. This approach spans from the initial ideation phase, where product designs are broken down to identify component specifications through selecting appropriate manufacturing processes, to the detailed planning and evaluation of these processes to ensure they meet production, cost, quality, and flexibility goals. It involves a meticulous examination of individual component requirements, decision-making on manufacturing in-house or outsourcing, and the use of advanced simulation tools to validate manufacturing plans. This entity plays a pivotal role in ensuring that manufacturing processes are efficient, cost-effective, and capable of producing high-quality products that meet market demands and customer expectations.

Hence, ***Process Entity Competitive Intelligence*** refers to the strategic gathering and analysis of information about competitors' manufacturing and production processes, technological advancements, and industry benchmarks within process development. This form of CI aims to understand how competitors optimize their production process to improve efficiency, reduce costs, and enhance product quality. It involves analyzing the manufacturing strategies, process innovations, supply chain logistics, and quality control measures employed by leading players in the industry. The goal is to leverage this intelligence to inform and improve one's own process development entity, ensuring that manufacturing processes are competitive and aligned with the industry's best practices and latest trends. This ensures the production of high-quality products most efficiently and cost-effectively possible, enhancing the company's competitive edge in the market.

- **Ideation—Imagination:** CI helps identify and specify individual components in this stage by analyzing external data on similar manufacturing processes, market demands efficiency and material innovations. CI can uncover insights into emerging trends in geometry, material choices, and production rates required by the market, guiding the project scope and plan definition. This involves a systematic search for information that could influence the selection and development of component specifications, ensuring they are both technically and economically viable [116].
  - CI can provide insights into how competitors and market leaders define specifications for similar components, which can inform the specification process.
  - It can reveal customer preferences and expectations, helping to ensure that the component specifications align with market demands.
  - CI can guide the project scope and plan by benchmarking against similar successful projects to identify best practices and avoid potential pitfalls.
- **Basic Development—Conceptualization:** CI is crucial for selecting the most appropriate manufacturing processes during the conceptualization stage. It involves evaluating competitors' manufacturing strategies and technological advancements. Thus, it is relevant to manage information for decision-making, from the company's strategy to manufacturing strategy, emphasizing quick and flexible responses to market situations [116].
  - CI can identify the manufacturing processes used by competitors for similar parts, which can inform whether to manufacture in-house or transfer components.
  - It can assess the viability of standard parts in the industry, helping to decide if standardization or customization is more competitive.
- **Advanced Development—Design:** CI contributes to the development of a detailed manufacturing process plan by offering competitive insights into time and cost optimization strategies, quality control measures, and flexibility in production. It involves analyzing best practices, technological innovations, and efficiency measures adopted by leading competitors and industries. This stage requires a deep dive into the operational aspects of manufacturing, where CI can identify potential bottlenecks and opportunities for process improvement.

- CI can provide data on the efficiency and cost-effectiveness of different manufacturing processes used within the industry.
- It can benchmark time and cost estimates against competitors, ensuring the plan is competitive.
- CI can offer insights into the flexibility of competitors' manufacturing processes, which can inform how to structure the manufacturing plan to achieve similar flexibility.

• **Launching—Creation/Operation:** In the final stage, CI supports evaluating the manufacturing process plan through benchmarking against industry standards and competitor practices. By leveraging simulation tools, CI can help verify the feasibility of the process in achieving desired production volumes and identify areas for potential design modifications. This evaluation is critical to ensure the manufacturing process is optimized for efficiency, cost-effectiveness, and quality before full-scale operation.

  - CI can analyze competitors' use of simulation tools and their outcomes, providing a reference point for evaluating one's manufacturing process plan.
  - It can highlight the effectiveness of different simulation methodologies used in the industry, potentially suggesting improvements to the evaluation process.

In each stage, CI provides a strategic advantage by enabling data-driven decisions that enhance the manufacturing process's efficiency, cost-effectiveness, and market responsiveness. Companies can adapt their process development strategies to maintain a competitive edge by continuously monitoring the competitive landscape and technological advancements.

8.5.4 Manufacturing System Entity Competitive Intelligence

The Manufacturing System Development Entity aims to guide companies through creating, refining, and implementing a manufacturing system. It encapsulates the journey from the initial conceptualization of product and process specifications through the selection and integration of the appropriate technology and equipment to the design and operationalization of the manufacturing facility itself. This entity is structured around three main pillars: Product Transfer, which ensures the precise specification and quality of components; Technology Transfer, focusing on the adoption and adaptation of manufacturing technologies; and Facility Design, which involves the strategic layout and operational planning of the manufacturing environment. The entity aims to optimize manufacturing operations, enhance efficiency, ensure product quality, and drive competitive advantage in the marketplace through this comprehensive approach.

Furthermore, ***Manufacturing System Competitive Intelligence*** refers to the strategic process of gathering, analyzing, and applying detailed information about competitors' manufacturing systems, technologies, processes, and operational strategies. It aims to understand the strengths and weaknesses of competitors' manufacturing capabilities, technology adoption, efficiency of production processes, quality control measures, and facility design and layout. This intelligence helps organizations identify best practices, innovative technologies, and operational efficiencies that can be leveraged to improve their manufacturing systems. The goal is to enhance production efficiency, reduce costs, ensure product quality, and ultimately achieve a competitive edge in the market by making informed decisions based on a comprehensive understanding of the manufacturing landscape and competitor activities.

- **Product Transfer:**
  - ***Ideation—Imagination:*** CI can benchmark against competitors' Bill of Materials (BOM) strategies to identify best practices in material selection, component standardization, and delivery times, ensuring competitive quality and efficiency.
  - ***Basic Development—Conceptualization:*** CI can provide insights into industry standards for manufacturing and quality control, helping to define competitive product attributes and control variables.
  - ***Advanced Development—Design:*** CI can aid in evaluating potential suppliers by analyzing competitors' partnerships, focusing on capacities, capabilities, and historical performance to select the best partners.
  - ***Launching—Creation/Operation:*** CI can track and analyze competitor assembly processes and quality control measures, optimizing manufacturing and assembly operations.
- **Technology Transfer:**
  - **Ideation—Imagination:** CI can gather information on the latest technological trends and competitor component specifications, ensuring the selection of cutting-edge materials and production rates.
  - ***Basic Development—Conceptualization:*** CI can evaluate available technologies used by competitors, assisting in selecting the most efficient manufacturing resources.
  - ***Advanced Development—Design:*** CI can provide insights into the process planning strategies of leading manufacturers, including tool and fixture selection, to inform the development of effective process charts.
  - ***Launching—Creation/Operation:*** CI can assess how competitors manage equipment setup and production ramp-up, identifying best practices for efficient production start-up and scaling.

- **Facility Transfer:**
  - ***Ideation—Imagination:*** CI can analyze competitors' facility designs based on their product and manufacturing process specifications, guiding the initial design of a new facility.
  - ***Basic Development—Conceptualization:*** CI can benchmark available technologies and equipment selections against those industry leaders use to ensure the facility is equipped with state-of-the-art technology.
  - ***Advanced Development—Design:*** CI can inform the manufacturing system design by examining competitors' manufacturing strategies, shop-floor configurations, and capacity planning approaches.
  - ***Launching—Creation/Operation:*** CI can review how competitors have successfully constructed, set up, and ramped up production in new facilities, offering insights into best practices for facility performance optimization.

By leveraging Competitive Intelligence throughout the Manufacturing System Development Entity, organizations can make informed decisions that enhance their manufacturing efficiency and product quality and ensure that their manufacturing systems are competitive and aligned with industry benchmarks and best practices.

8.5.5 Supply Chain Competitive Intelligence

The Supply Chain Development Entity constructs, optimizes, and continuously improves its supply chains from the initial strategy formulation to the operational execution stage. It involves developing a robust framework that ensures the efficient flow of materials, information, and finances from suppliers to end consumers. This process encompasses strategic planning to address technological integration, demand forecasting, inventory management, and risk mitigation, followed by the conceptualization and design of supply chain operations, including procurement, production, distribution, and logistics. The entity culminates in the practical execution of these plans, focusing on order fulfillment, logistics management, and supplier collaboration to achieve operational efficiency and customer satisfaction. This entity is essential for businesses to enhance their competitive advantage by ensuring their supply chain is resilient, responsive, and aligned with market demands.

In addition, **Supply Chain Entity Competitive Intelligence** is the strategic process of collecting, analyzing, and applying information related to competitors' supply chain strategies, operations, and performance. It aims to understand how rival organizations manage their supply chains, from sourcing and procurement to production, distribution, and logistics. This intelligence covers technological integration, inventory and logistics management efficiency, supplier relationships, and responsiveness to market changes. Supply Chain Entity CI aims to inform and guide an organization's supply chain development efforts, ensuring they are competitive, resilient, and aligned with industry best practices. By leveraging this intelligence, companies can identify opportunities for improvement, innovate their supply chain processes, and achieve superior customer satisfaction and market positioning.

- **Ideation—Imagination:** CI aids in planning the efficient flow of materials and information from sourcing to consumer. It focuses on technology integration and adaptability to demand and market shifts. The study by Monnagaaratwe and Mathu explores SCM as a competitive advantage for SMEs, emphasizing the role of supply chain collaboration in enhancing the constant flow of products for business continuity and cost-saving [96].
  - CI can provide insights into how competitors structure their supply chain strategies to manage the flow of materials and information efficiently, highlighting opportunities for technology integration and adaptability.
  - It can identify trends and shifts in the market, enabling the formulation of a strategy resilient to demand fluctuations and other risks.
- **Basic Development—Conceptualization:** During the conceptualization stage, CI is crucial for integrating forecasting, inventory, production, and distribution to meet customer demand. Sharabati et al. discuss how green supply chain practices affect competitive strategy development, highlighting the importance of green operations in enhancing the competitiveness of the Jordanian pharmaceutical manufacturing sector [128].
  - CI can offer benchmarks for forecasting accuracy, inventory management, and production planning, allowing for a comparison with competitors and identifying areas for improvement.
  - It can reveal strategies used by competitors for managing stock levels, negotiating with suppliers, and balancing costs with service goals, guiding more effective planning decisions.
- **Advanced Development—Design:** CI addresses daily supply chain management by focusing on procurement, production, distribution, and logistics. Reyes, Mula, and Díaz-Madroñero develop a conceptual model for lean supply chain planning in Industry 4.0, merging digital technologies with lean manufacturing tools to reduce waste and minimize costs, thereby streamlining processes and managing inventory [118].
  - CI can uncover operational efficiencies and best practices in procurement, production, distribution, and logistics from leading competitors, providing a roadmap for streamlining processes.

  - It can highlight innovative approaches to inventory management and adaptation to market changes, which can be incorporated into the supply chain operations design.

- **Launching—Creation/Operation:** CI supports the execution of the established supply chain, including order fulfillment and logistics management. Almatarneh, Jarah, and Jarrah establish the role of management accounting in developing supply chain performance in logistics manufacturing companies, underlining the importance of real-time tracking and collaboration with all supply chain partners for efficient operations and customer satisfaction [7].
  - CI can assess competitor capabilities in order fulfillment, logistics management, and supplier performance monitoring, offering insights into effective execution strategies.
  - It can identify technologies and platforms used by competitors for real-time tracking and collaboration, suggesting areas where automation and technology integration can enhance execution efficiency and customer satisfaction.

By employing Competitive Intelligence at each stage of the Supply Chain Development Entity, industries can make informed decisions that improve efficiency, adaptability, and customer service. CI enables a proactive approach to supply chain management, ensuring that strategies and operations are aligned with current industry standards and positioned to anticipate and respond to future market dynamics effectively.

8.5.6 End of Life: 5S—Social, Sustainable, Sensing, Smart, Safe within the LCE Framework

The End of Life stage is critical for managing a product's disposal, recycling, and repurposing once it has reached the end of its useful life. Incorporating the 5S principles—Social, Sustainable, Sensing, Smart, and Safe—ensures that this stage is handled responsibly, ethically, and efficiently.

Integrating competitive intelligence into end-of-life processes involves systematically gathering and analyzing data on competitor strategies, market trends, and regulatory changes related to disposal, recycling, and repurposing. This data informs decisions on eco-design, community engagement, safety, and smart technologies, ensuring that each 5S principle is thoroughly addressed. For the ***Material Entity***, CI helps identify sustainable materials and recycling methods. In the ***Product Entity***, it guides the design of products for easy disassembly and recycling. CI offers insights into efficient manufacturing processes that minimize waste for the ***Process Entity***. The ***Manufacturing System Entit***y informs facility design and equipment selection for safe and efficient end-of-life management. Finally, for the ***Supply Chain Entity***, CI supports the development of logistics and supplier strategies that facilitate responsible disposal and recycling. By embedding CI into the 5S framework, organizations ensure that end-of-life management is compliant, responsible but also innovative, and competitive, contributing to overall sustainability and market leadership.

- **Social:** This aspect ensures that end-of-life processes address broader societal needs and engage with stakeholders effectively.
  - ***Applications in CI:***
    * Understanding competitor efforts in educating consumers about responsible disposal and recycling.
    * Identifying successful community engagement strategies to adopt similar or improved approaches.
    * Monitoring regulatory compliance and ethical practices across the industry.
- **Sustainable:** This involves using eco-design principles and ensuring that end-of-life processes enhance environmental performance throughout the product life cycle.
  - ***Applications in CI:***
    * Benchmarking against competitors' sustainability metrics and goals.
    * Identifying innovative recycling and reuse strategies in the industry.
    * Staying ahead of regulatory changes and ensuring compliance.
- **Sensing:** This incorporates integrating sensors and data utilization to monitor and optimize the end-of-life processes.
  - ***Applications in CI:***
    * Analyzing competitor implementations of sensing technologies to improve end-of-life processes.
    * Using data analytics to optimize recycling and disposal operations.
    * Monitoring environmental impact through data to develop competitive and sustainable practices.
- **Smart:** This aspect integrates advanced technologies such as connectivity features and artificial intelligence to optimize end-of-life processes.
  - ***Applications in CI:***
    * Identifying how competitors use smart technologies for end-of-life management.
    * Leveraging AI and IoT for predictive maintenance and efficient decommissioning.
    * Developing smart features that enhance the life cycle and recyclability of products.

- **Safe:** This ensures that end-of-life processes prioritize safety, including the handling and disposal of hazardous materials.
  - ***Applications in CI:***
    * Monitoring safety protocols and standards adopted by competitors.
    * Developing comprehensive safety plans for product end-of-life management.
    * Ensuring compliance with the latest safety regulations and best practices.

By embedding competitive intelligence into the 5S end-of-life processes, organizations can develop compliant, responsible but also innovative, and competitive strategies. This comprehensive approach ensures that end-of-life management is a critical component of the product life cycle, contributing to sustainability, safety, and overall competitive advantage.

8.5.7 Exemplification of CI through Nearshoring in Mexico

CI, in the context of nearshoring, is crucial for making informed decisions that align with strategic business goals. CI involves systematically collecting, analyzing, and applying information about competitors, market trends, technologies, and regulatory frameworks. This intelligence supports companies in maintaining or gaining a competitive edge by anticipating market shifts, identifying emerging opportunities, and mitigating potential threats [103].

The global shift towards nearshoring, driven by disruptions like the COVID-19 pandemic and geopolitical tensions (e.g., US-China relations), has positioned Mexico as a key player due to its proximity to the United States, the world's largest market. Mexico offers a compelling case for investment from global companies seeking to reduce their dependence on distant supply chains [103].

To capitalize on this opportunity, Mexico must overcome various technological and policy challenges that impede strengthening its supply chains. As Mexico positions itself as a strategic hub for nearshoring due to its proximity to the United States and other North American markets, it faces many challenges that must be addressed to capitalize on this opportunity fully. These challenges span strategic planning, infrastructure, energy, human capital, and technological advancement. This analysis outlines these key challenges with a sustainable focus, aiming to provide insights for academia, industry, and manufacturing stakeholders [103].

1. **Strategic Planning and Coordination:** One of the most significant challenges is the lack of a cohesive, focused strategic plan for nearshoring efforts. Currently, no unified roadmap addresses the specific needs of different regions, productive sectors,

and manufacturing chains within Mexico. The absence of a comprehensive public-private partnership plan exacerbates this issue, leading to fragmented efforts and inefficiencies. Moreover, there is a critical need for real-time data and competitive intelligence to inform decision-making processes effectively.

2. **Energy Supply:** The capacity to supply energy, particularly clean and renewable energy, is a major hurdle. Mexico faces an insufficient energy infrastructure to meet the potential demand from new and expanding investments. Additionally, there is no clear public policy that incentivizes the use of renewable energies, which is crucial for attracting sustainable investments. Ensuring a stable, sufficient, and green energy supply is essential for the country's competitiveness in the nearshoring landscape.
3. **Water Scarcity:** Water scarcity is another pressing issue. Mexico lacks a national plan to address the current and future distribution and availability of water, especially considering the increased demand from relocated businesses. Modern, efficient systems for water collection and reuse are not widely implemented, posing a significant risk to sustainable development and the attractiveness of Mexico as a nearshoring destination.
4. **Security and Infrastructure:** Generalized insecurity and a lack of rule of law create an environment of uncertainty for investors. High levels of violence and insecurity, particularly on major transportation routes, pose substantial risks. Moreover, Mexico's transportation infrastructure is inadequate, with insufficient and poor-quality roads, railways, ports, and airports. This infrastructure deficit hampers efficient mobility and logistics, which are critical for integrating into global supply chains.
5. **Human Capital:** There is a notable shortage of skilled technicians with updated knowledge and competencies. The profiles of graduates often do not meet the evolving demands of modern production processes. This skills gap includes both technical expertise and essential soft skills such as leadership, communication, and teamwork. Furthermore, there is a significant deficit in English language proficiency among technicians, limiting their ability to engage with international partners and technologies effectively.
6. **Government Support:** The support provided by the government to businesses is often insufficient and unclear. Existing programs and incentives need more clarity and accessibility. Additionally, there is a disparity in the support offered to foreign versus domestic investors, which can deter local businesses from expanding or modernizing. Ensuring equitable and robust support

mechanisms is vital for fostering a conducive environment for both domestic and international investments.

7. **Technological Infrastructure:** The technological infrastructure of many Mexican manufacturing plants is outdated, reducing their competitiveness. There is a pressing need for modernization to align with global standards and technological advancements. This includes adopting Industry 4.0 technologies, which integrate digital technologies into manufacturing processes to create smart factories that are more efficient and sustainable.
8. **SME Integration:** Small and medium-sized enterprises (SMEs) face significant barriers to integrating into global supply chains. Many SMEs lack the necessary certifications and compliance with international standards. There is also a general lack of strategic planning and investment in technology and long-term projects among SMEs, often due to immediate financial pressures and limited resources.

Addressing these challenges requires a multifaceted approach that incorporates strategic planning, infrastructure investment, energy, human capital development, and technological modernization. By leveraging competitive intelligence and fostering collaboration among academia, industry, and government, Mexico can create a sustainable and competitive environment that attracts and retains nearshoring investments. This comprehensive strategy will not only boost the country's economic growth but also enhance its global standing as a preferred nearshoring destination.

In addition, CI involves systematically collecting, analyzing, and applying information about competitors, market trends, technologies, and regulatory frameworks. By leveraging CI, stakeholders can make informed decisions, anticipate market shifts, identify emerging opportunities, and mitigate potential threats. Here is how CI can be integrated into various entities involved in the nearshoring process:

- **Material Entity:**
  - ***Understanding Competitor Materials and Technologies:*** CI helps gather and analyze information on the material choices and technological advancements made by competitors. By benchmarking against best practices, Mexican companies can ensure their material development processes are competitive. This includes understanding the properties, costs, and sustainability of different materials used in similar products.
  - ***Benchmarking for Innovation and Sustainability:*** CI enables companies to benchmark their material development against industry leaders. This helps identify innovative and sustainable materials that meet or exceed market expectations. For instance, if a competitor uses

a novel eco-friendly material, CI can inform Mexican companies about its benefits and potential applications.

- **Product Entity:**
  - ***Market Trends and Customer Insights:*** CI allows companies to analyze competitor products, market trends, and customer feedback. This intelligence helps define product features, design, and functionality that align with current market demands. Companies can tailor their products to meet or exceed these expectations by understanding what competitors are offering and how customers respond.
  - ***Guiding Product Design and Innovation:*** Companies can gain insights into competitor product development processes, including design philosophies and technological integrations through CI. This information is crucial for creating innovative products that are competitive and cater to evolving consumer preferences and technological advancements.
- **Process Entity:**
  - ***Optimizing Manufacturing Processes:*** CI helps understand the manufacturing strategies and process innovations employed by competitors. This includes insights into efficiency measures, cost reductions, and quality control practices. By analyzing these aspects, Mexican manufacturers can optimize their own production methods to improve operational efficiency and reduce costs.
  - ***Enhancing Operational Efficiency:*** CI provides detailed information on how competitors handle their production processes, from raw material sourcing to final product assembly. This intelligence helps identify potential bottlenecks and opportunities for process improvement, ensuring that manufacturing operations are streamlined and efficient.
- **Manufacturing System:**
  - ***Technology Adoption and Facility Design:*** CI gathers data on competitors' technology adoption and facility design. This includes understanding the equipment, technologies, and layout strategies leading competitors use to optimize their manufacturing systems. By leveraging this intelligence, Mexican companies can design and implement manufacturing systems that are both efficient and scalable.
  - ***Ensuring Competitive Manufacturing Systems:*** CI enables companies to benchmark their manufacturing systems against industry standards. This helps identify gaps and areas for improvement, ensuring that their systems are aligned with best practices and capable of competitively producing high-quality products.

- **Supply Chain:**
  - ***Strategic Supply Chain Management:*** CI provides insights into competitors' supply chain strategies and performance. This includes analyzing how competitors manage sourcing, procurement, production, distribution, and logistics. By understanding these strategies, companies can develop supply chains that are resilient, responsive, and aligned with market demands.
  - ***Enhancing Supply Chain Resilience and Responsiveness:*** CI helps identify best practices in supply chain management, including inventory control, supplier relationships, and logistics efficiency. This intelligence supports the development of supply chains that can adapt to market changes and disruptions, ensuring continuous and efficient operations.
- **End of Life:**
  - ***Social:*** CI helps understand competitor efforts in community engagement and regulatory compliance. By adopting best practices, companies can enhance their social responsibility and stakeholder relations, which is crucial when setting up nearshoring operations that require community and regulatory support.
  - ***Sustainable:*** CI identifies industry leaders in sustainability and benchmarks their practices. This helps companies adopt eco-friendly materials, recycling methods, and waste management strategies, aligning nearshoring operations with sustainable practices to meet environmental standards and customer expectations.
  - ***Sensing:*** CI provides insights into the use of sensing technologies for monitoring and optimizing end-of-life processes. This ensures that companies can efficiently track and manage the disposal and recycling of products, which is crucial for nearshoring, where local regulations and sustainability practices must be adhered to.
  - ***Smart:*** CI analyzes competitors' use of advanced technologies such as AI and IoT for end-of-life management. This information guides the integration of smart features that enhance life cycle and recyclability, making nearshoring operations more efficient and environmentally friendly.
  - ***Safe:*** CI monitors safety standards and protocols adopted by competitors. By implementing these best practices, companies can ensure safe handling, disposal, and recycling of products, protecting both employees and the environment in nearshoring operations.

Integrating Competitive Intelligence into various entities involved in nearshoring enables Mexican companies to make informed, strategic decisions that enhance their competitiveness. Companies can innovate effectively, optimize their operations, and build resilient supply chains by systematically gathering, analyzing, and applying information about competitors and market trends. This comprehensive approach to CI supports Mexico's nearshoring goals and fosters sustainable growth and long-term success in the global market.

8.5.8 The Ten Core Technologies for the CI Integration within the LCE Framework

To specifically integrate the ten core technologies within the LCE framework and outline their role in enhancing CI, FIGURE 8.3 delineates how the ten core technologies can be strategically employed to optimize the processes across the LCE framework:

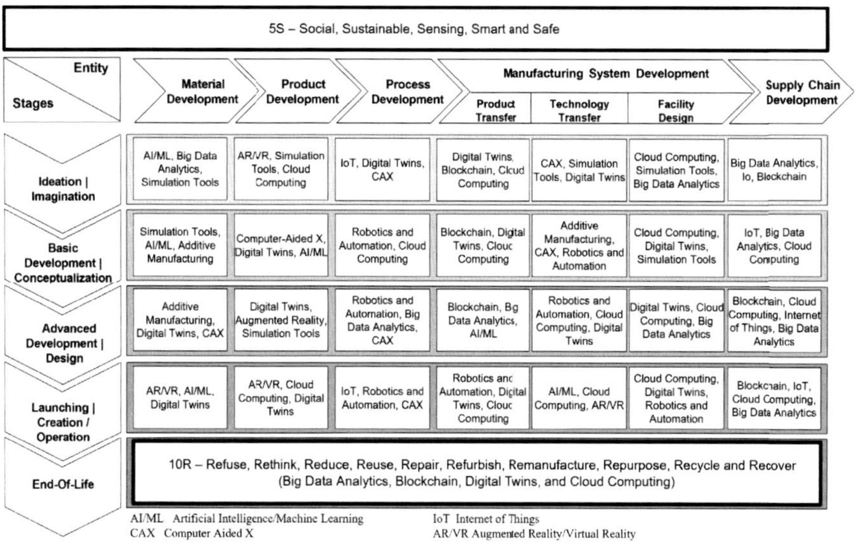

FIGURE 8.3
Core technologies for the CI integration applied to LCE framework.

1. **Ideation—Imagination:** This stage focuses on gathering, analyzing, and utilizing competitor and market insights.
   - ***Material Development:*** AI/ML and Big Data Analytics predict and analyze material trends. Simulation Tools test theoretical material properties.
   - ***Product Development:*** AR/VR visualize initial product concepts, enhancing design through immersive experiences. Simulation Tools and Cloud Computing facilitate collaborative design and simulation.
   - ***Process Development:*** IoT tracks real-time data from initial process trials. Digital Twins simulate these processes, while Computer-Aided X refines the process design.

- ***Manufacturing System Development:***
    i. *Product Transfer:* Digital Twins visualize component integration, Blockchain ensures data integrity, and Cloud Computing coordinates across platforms.
    ii. *Technology Transfer:* Computer-Aided X defines precise component specifications, aided by Simulation Tools and Digital Twins for accuracy.
    iii. *Facility Design:* Cloud Computing and Simulation Tools plan and optimize the layout; Big Data Analytics supports decision-making based on CI.
- ***Supply Chain:*** Big Data Analytics and IoT develop initial supply chain frameworks. Blockchain ensures security and traceability of early supply chain data.

2. **Basic Development—Conceptualization:** Enhances designs and processes based on deeper insights into technology, market demands, and operational feasibility.
    - ***Material Development:*** Simulation Tools, AI/ML, and Additive Manufacturing enable rapid prototyping and testing of new materials.
    - ***Product Development:*** Computer-Aided X enhances design detail, Digital Twins and AI/ML for advanced product simulations.
    - ***Process Development:*** Robotics and Automation streamline process setups, enhanced by Cloud Computing for data management.
    - ***Manufacturing System Development:***
        i. *Product Transfer:* Blockchain for secure data management, Digital Twins for precise component integration, and Cloud Computing for real-time tracking.
        ii. *Technology Transfer:* Additive Manufacturing, Computer-Aided X, Robotics and Automation accelerate prototype production.
        iii. *Facility Design:* Cloud Computing, Digital Twins, and Simulation Tools for detailed facility modeling.
    - ***Supply Chain:*** IoT and Big Data Analytics refine logistics planning, and Cloud Computing enhances data-driven decision-making.
3. **Advanced Development—Design:** Focuses on finalizing product and process designs by integrating complex CI data to optimize performance and efficiency.
    - ***Material Development:*** Additive Manufacturing, Digital Twins, and Computer-Aided X allow complex material designs and simulations.

- ***Product Development:*** Digital Twins, Augmented Reality, and Simulation Tools refine product designs and user interactions.
- ***Process Development:*** Robotics and Automation, Big Data Analytics, and Computer-Aided X finalize manufacturing processes.
- ***Manufacturing System Development:***
  - i. *Product Transfer:* Blockchain, Big Data Analytics, and AI/ML ensure high-quality, compliant component integration.
  - ii. *Technology Transfer:* Robotics and Automation, Cloud Computing, and Digital Twins implement advanced manufacturing technologies.
  - iii. *Facility Design:* Digital Twins, Cloud Computing, and Big Data Analytics finalize facility designs and systems.
- ***Supply Chain:*** Blockchain, Cloud Computing, IoT, and Big Data Analytics ensure an efficient, traceable, and responsive supply chain.

4. **Launching—Creation/Operation:** Utilizes CI to ensure the seamless introduction of products and systems into the market, focusing on quality, compliance, and operational efficiency.
   - ***Material Development:*** AR/VR, AI/ML, and Digital Twins ensure materials meet operational requirements.
   - ***Product Development:*** AR/VR, Cloud Computing, and Digital Twins prepare products for market launch.
   - ***Process Development:*** IoT, Robotics and Automation, and Computer-Aided X monitor and optimize the launch process.
   - ***Manufacturing System Development:***
     - i. *Product Transfer:* Robotics and Automation, Digital Twins, and Cloud Computing ensure efficient production scaling.
     - ii. *Technology Transfer:* AI/ML, Cloud Computing, and AR/VR finalize system integrations.
     - iii. *Facility Design:* Cloud Computing, Digital Twins, and Robotics and Automation adjust and optimize operational workflows.
   - ***Supply Chain:*** Blockchain, IoT, Cloud Computing, and Big Data Analytics manage full-scale operations for optimal delivery and customer satisfaction.

The strategic integration of these ten core technologies within the LCE framework optimizes each stage of the product development process and significantly enhances Competitive Intelligence capabilities. By meticulously applying technologies such as AI/ML, Digital Twins, AR/VR, and Blockchain across various stages, from Ideation to Launch and Operation, the framework ensures a comprehensive, data-driven approach to product life cycle management.

This integration allows for a deeper and more accurate analysis of market and competitor data, fostering innovation and improving decision-making processes. Consequently, organizations equipped with this technologically enhanced LCE framework are better positioned to anticipate market trends, respond to environmental and competitive challenges, and succeed in delivering products that meet the evolving needs of consumers and regulatory standards.

The utilization of these core technologies not only streamlines operations but also builds a robust foundation for sustainable competitive advantage, making the LCE framework a pivotal model for future-focused industries aiming to thrive in dynamic markets.

8.6 Case Study Applying CI in Industry

The Organization for Economic Co-operation and Development (OECD) defines green growth as "promoting economic growth and development, while ensuring that natural assets continue to provide the resources and environmental services on which our well-being depends." [102] The United Nations Environment Programme (UNEP) maintains that green growth is related to an economy that "results in improved human well-being and social equity, while significantly reducing environmental risks and ecological constraints." [115]

Green growth and decarbonization can be a catalyst for the diversification of economies, moving them away from carbon-intensive activities and promoting sectors such as sustainable agriculture or eco-tourism. Ricardo Hausmann, professor at the Harvard Kennedy School and director of the Harvard Growth Lab, argues that decarbonization will transform global production and trade patterns so radically that new growth opportunities will emerge for developing countries. According to Hausmann, developing countries must carve out a role for themselves in a world economy that is becoming greener and that will create new industries and markets. The governments of these countries must study the value chains emerging behind industries that generate the green production necessary to reduce emissions and develop explicit strategies to attract investments from emerging green industries [58].

Hausmann suggests that decarbonization will reduce demand for dirty goods and services and increase demand for cleaner, greener ones. The growth opportunities for developing countries will not lie so much in reducing their emissions, but in how to underpin their development by tapping into fast-growing sectors that will enable the world to reduce emissions and achieve net-zero emissions. But these opportunities will be best seized by those countries that actively intervene and are experts in the capabilities that will sustain their future comparative advantage [59].

Any definition of green growth highlights the main elements that must be in place to achieve a substantial change in the way we produce/consume economic goods and natural resources. These include:

- Increasing efficiency in the use of resources by implementing technologies and practices that reduce the consumption of materials and energy.
- Technological innovation, such as renewable energy or cleaner production processes, that drive the decarbonization of the economy and minimize greenhouse gas emissions and environmental pollution.
- Designing public policies (economic, fiscal and social) that incorporate the sustainability component.
- Ensuring a fair transition that does not leave vulnerable sectors behind.

In this context, aligned with the Framework for Life cycle Engineering (LCE), specifically with the development of supply chains, and as a result of the Challenge-Based Research Funding Program 2022 of the Tecnológico de Monterrey, researchers from the Regional Development Laboratory (LadDeR) and the Decision Science Research Center (CDD) of the School of Government and Public Transformation, and the Institute of Advanced Materials for Sustainable Manufacturing (IAMSM), developed the Observatory of Sustainable Manufacturing Industries (OSMI) [106], with the objective of being a node of applied research and providing tools for the design of technologies and public policies that promote industrial decarbonization and green growth in Mexico and the world. FIGURE 8.4 depicts the four core elements within the OSMI ecosystem.

The OSMI website provides relevant information and data to understand the current challenges and opportunities of manufacturing industries and the different regions of Mexico. Below is a general description of the information contained in the current sections of the Observatory.

8.6.1 Manufacturing Industries

This section analyzes regional and industrial structures of Mexico with the aim of providing a clear overview of the location, concentration and industrial sustainability in Mexico, as well as the conformation of manufacturing supply chains and regional industrial structures. The section is divided into 4 subsections, each with dynamic data visualization tools to analyze the industrial and regional characteristics of the country.

8.6.1.1 Regional Structure

Each region has particular economic structures. Understanding their characteristics and the economic importance of each industry in a region is

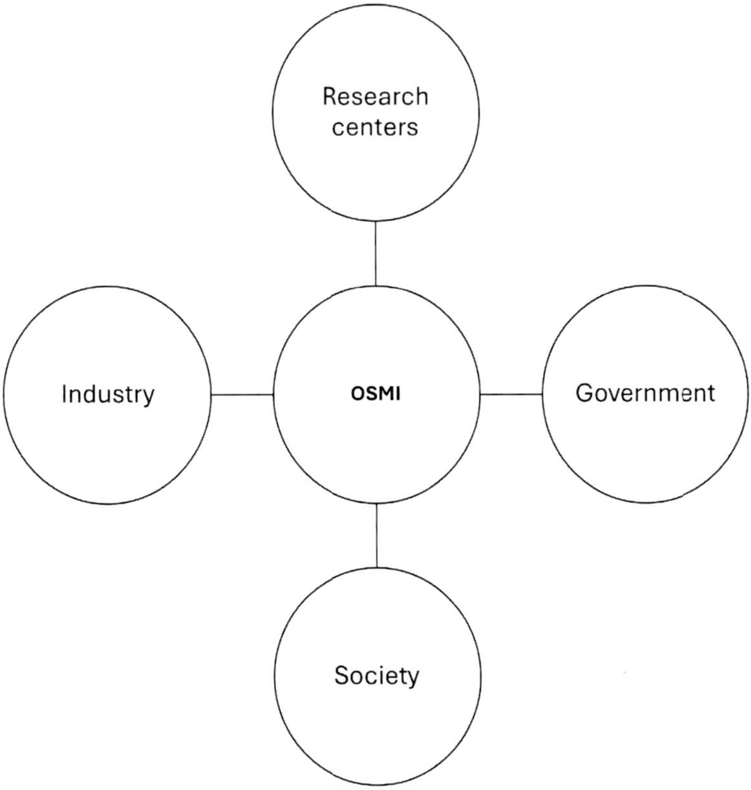

FIGURE 8.4
OSMI ecosystem.

the starting point for the design of productive development policies that promote structural transformation.

This subsection presents the conformation of the manufacturing productive structures in the different regions of Mexico. Using the tool, the relative importance of each subsector and industry group, based on the 2018 North American Industrial Classification System (NAICS, 2018), can be explored in each of the 32 federal entities, 70 metropolitan areas and more than 2,400 municipalities. For instance, FIGURE 8.5 shows the industrial structure of the Metropolitan Area of Monterrey in Mexico, composed by different economic industry groups.

8.6.1.2 Industrial Location

Industries tend to be located and concentrated in those regions that provide them with better productive ecosystems that allow them to increase their competitiveness. Knowing the location of industries and/or regional

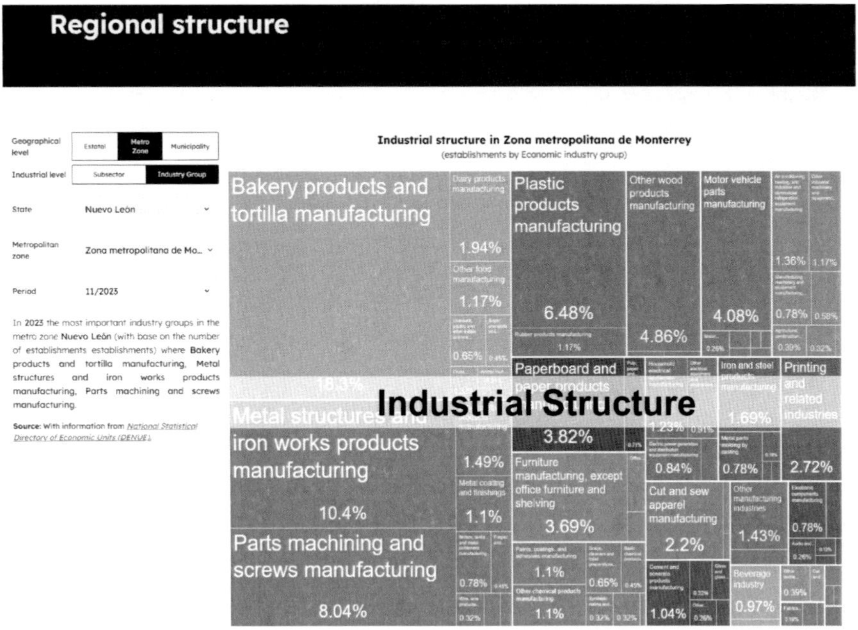

FIGURE 8.5
Industrial structure of the metropolitan area of monterrey.

specialization allows us to identify opportunities to attract investments and analyze routes for productive diversification.

This subsection presents the concentration of manufacturing industries in the different regions of Mexico. The tool allows us to explore, through the location coefficient, the state, metropolitan or municipal specialization of Mexico, in 21 subsectors and 86 manufacturing industry groups [106].

As an exemplification, FIGURE 8.6 presents the aerospace equipment manufacturing industry in Mexico.

8.6.1.3 Supply Chains

Industries, as complex systems, are made up of different stages, processes, suppliers, etc., that allow them to generate the goods that we trade and consume daily. Understanding the particular system in which each industry is inserted is important to identify the strategic links to increase its competitiveness and/or opportunities for new business.

This subsection presents the structures of the supply chains of the manufacturing subsectors and economic branches of Mexico. Through network analysis we can explore the relative importance of the different suppliers of products and services that make up the manufacturing supply chains at a national level, as well as the level of development of these in each

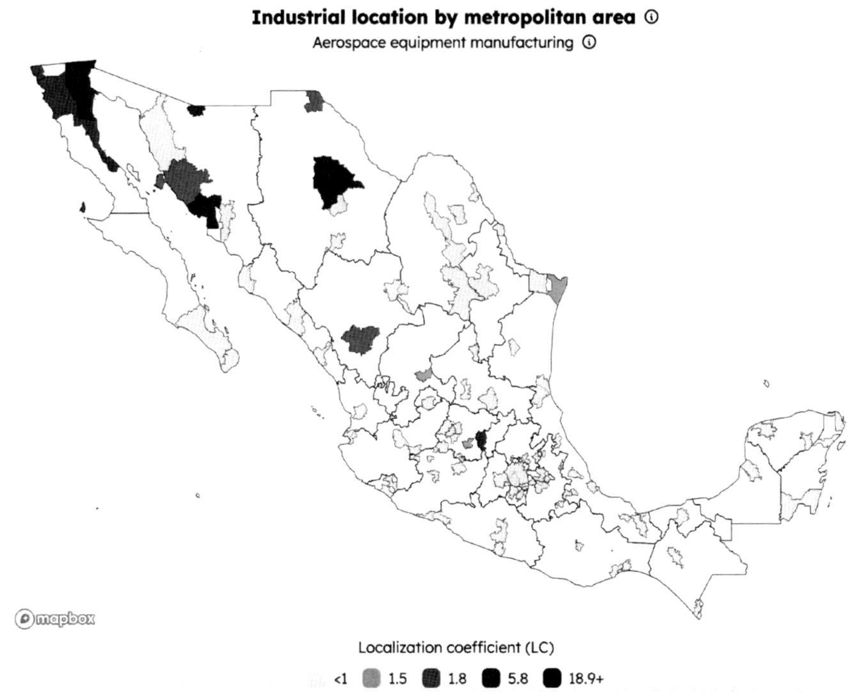

FIGURE 8.6
Location of the aerospace equipment manufacturing industry.

of the states of the country. For instance, FIGURE 8.7 shows the industry supply chain of automobiles and trucks manufacturing industry.

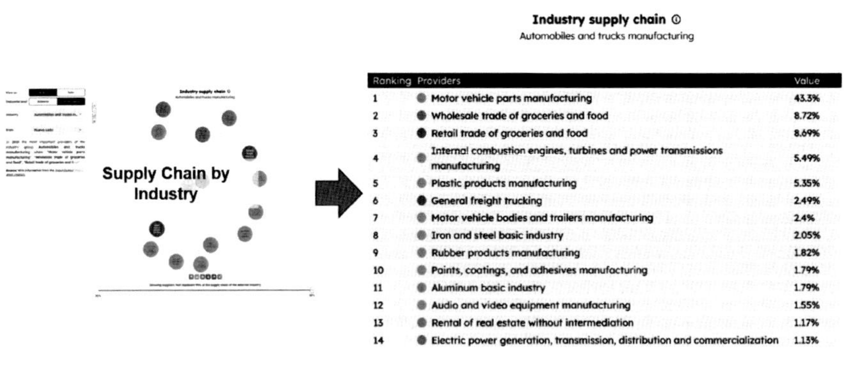

FIGURE 8.7
Supply chain of the automobile and truck manufacturing industry.

8.6.1.4 Industrial Sustainability

Industries use natural, energy, and material resources in different production processes. Analyzing each industry's productive efficiency in these areas allows us to identify the technologies necessary to promote sustainability and green growth.

In this subsection, it is possible to analyze the productive efficiency of 21 subsectors and 86 manufacturing industry groups [101], in terms of the use of electricity, fuels, water and packaging, at the national and state level. FIGURE 8.8 depicts the productivity of electricity from 2008 to 2018.

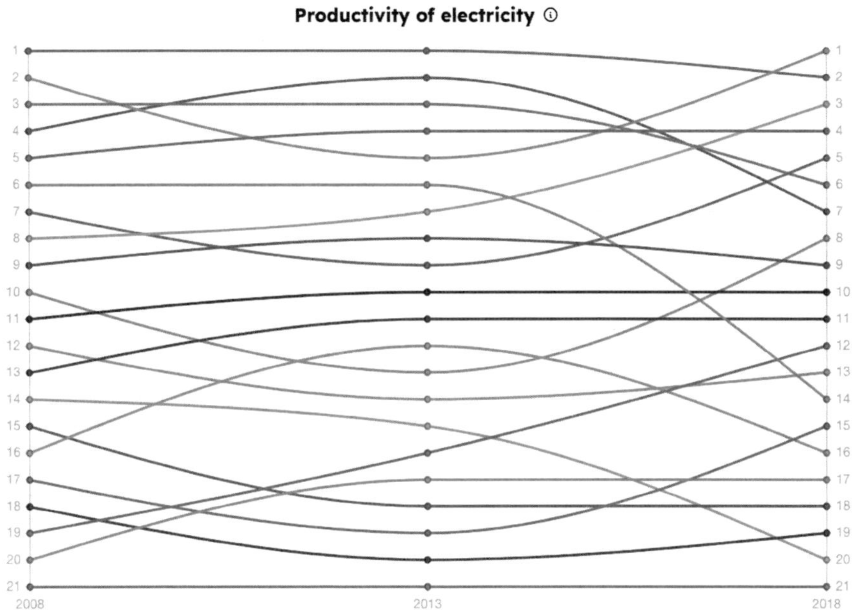

FIGURE 8.8
Productivity of the use of electricity by subsector at the national level.

8.6.2 Decarbonization Routes

In this section, we explore Mexico's options for solidly meeting the goal of achieving net zero emissions, which is in line with the objectives of the Paris Agreement. To this end, we have developed SiSePuede (Simulation of Sectoral Pathways and Uncertainty Exploration for Decarbonization), a reference framework for designing development strategies and analyzing their emissions, costs, benefits and alignment with development objectives and the goals of the Paris Agreement in the midst of uncertainty. We have used SiSePuede to evaluate decarbonization strategies in Mexico under conditions of uncertainty.

SiSePuede models emissions in six integrated economic sectors based on the Intergovernmental Panel on Climate Change (IPCC) methodology: agriculture, industry, buildings, transport, waste, and electricity and fuel production [19, 105]. SiSePuede's emissions reduction modeling process can be understood through the ASIF framework, originally designed for the transport sector [125]. ASIF stands for Activity, Mode Share, Emissions Intensity. First, activity levels (e.g. demand for food, transport, buildings, industrial production, and energy) are based on underlying factors such as gross domestic product (GDP) and population. The variables that determine emissions are calibrated from publicly available datasets on economic activity, energy consumption, and emissions from sources such as the IDB, the World Bank, the International Monetary Fund (IMF), the International Energy Agency (IEA), and the Food and Agriculture Organization of the United Nations (FAO). Given the large number of uncertainties influencing emissions, costs, and benefits, we couple SiSePuede with Robust Decision Making (RDM), a method for managing deep uncertainties [53, 71, 75, 76]. RDM is an iterative, stakeholder-supported process for conducting comprehensive "what-if" analyses. Using RDM's vulnerability analysis techniques, we evaluate strategies under a range of future conditions and assumptions to understand the potential range of outcomes and conditions that lead strategies to meet (or fail to meet) decarbonization and economic goals.

The Decarbonization Pathways section features 9 subsections, each explaining Mexico's potential path to 2050. The images refer to the Business as Usual (BAU) scenario in the "Base" tab of each chart. The "Mitigation" tab in each chart shows the potential path to 2050 when decarbonization actions are implemented.

The "Base" tab of each chart shows a Business as Usual (BAU) path for Mexico, which serves as a backdrop against which other decarbonization actions (called transformations) are compared in the "Mitigation" tab. These transformations refer to actions such as:

- **Incremental Improvements:** these achieve marginal benefits over current practices, without making significant structural changes. For example, increasing the energy efficiency of vehicles, increasing the carrying capacity of livestock, and expanding basic waste and wastewater management.
- **Supply-side Solutions:** These may involve significant structural changes in production, such as pursuing nearly emissions-free electricity, shifting to industrial processes such as hydrogen-based steel production, and transitioning transport to electricity. These changes are typically made by producers and are not under significant control of consumers.
- **Changing Consumption:** These include transformations that are observable by consumers and often involve their participation. Examples include using public transport instead of the personal car for commuting or

changing food production and consumption patterns to reduce the relative importance of high-carbon foods and increase that of low-carbon foods.

- **All Transformations:** This refers to all transformations to identify how maximum effort can achieve net zero emissions by 2050.

8.6.2.1 Business as Usual Leads to High Emissions

Business as usual reflects a future largely unconcerned with decarbonization and aligned with historical development trajectories. Mexico's Nationally Determined Contributions (NDC) establish a commitment to reduce greenhouse gas (GHG) emissions by 35% with respect to the business-as-usual (BAU) scenario and foresees that 30% will be achieved with domestic resources and the remaining 5% with international cooperation. Unfortunately, the overall rating of Mexico's climate policies and commitments by the Climate Action Tracker is critically insufficient. The country estimates that with current policies it will be able to meet this 35% reduction target; however, emissions will continue to increase until 2030 [136].

In 2020, Mexico's total emissions reached 670 MtCO2eq, representing 1.35% of global emissions. The main drivers of emissions in its NDC BAU scenario include energy, which contributes 66% of emissions, followed by agriculture with 16%, waste with 8%, industrial processes with 7%, and land-use change and forestry with almost 3%. Emissions have grown by 18% compared to the 2000 baseline year.

8.6.2.2 Achieving Carbon Neutrality is Feasible

What each of the 9 sections of the Decarbonization Pathways section shows in detail is an analysis that complements Mexico's long-term strategies (LTS) by providing insights on how to combine economy-wide transformations to achieve carbon neutrality.

The strategy aligns various sectors, such as land use planning, urban development, sustainable buildings, energy, transport, waste management and water policies. A review of this LTS identifies weaknesses, including the lack of a net-zero emissions target and the need for more targeted mitigation and adaptation actions. Our analysis complements Mexico's LTS by providing insights on how to combine economy-wide transformations to achieve carbon neutrality.

In this analysis, we set a threshold of 250 MtCO2e by 2050 (below the more than 1,200 MtCO2e of the BAU pathway) to identify the strategies that reduce emissions the most. Approximately 60% of these variants reduce emissions below the threshold, while 40% of the pathways exceed it.

To the extent that Business as Usual reflects Mexico's NDC BAU scenario, all individual categories of transformations would achieve the 35% emissions reduction targets expressed in the NDC. However, no single category alone achieves net-zero emissions by 2050. If Mexico were to decide to achieve net-zero emissions by 2050, a more comprehensive approach closer to All Transformations/All Actions would likely be needed.

We find that while there are thousands of different paths to net-zero, there are three changes that are critical:

- Producing electricity (and, to a lesser extent, hydrogen) from renewable energy sources. Economy-wide fuel shifts reduce emissions only to the extent that electricity and hydrogen—the main replacement fuels—are produced from renewables rather than fossil fuels. So cleaning up the grid and producing green hydrogen are critical to decarbonization.
- Use electricity (and to a lesser extent hydrogen) instead of fossil fuels to power transport. Transport is one of the most challenging sectors to reduce GHG emissions in, given the many diffuse sources of emissions, the growing demand for private vehicle travel, and the built infrastructure that perpetuates road-based travel patterns. Our analysis points to fuel shifting in transport as key, as it reduces emissions even if modal and activity shifts turn out to be limited.
- Turn land into a carbon sink. This is necessary to offset any residual emissions from the rest of the economy. It involves protecting forests and enabling afforestation through a combination of increasing agricultural productivity and shifting agricultural production and consumption from higher-carbon foods to lower-carbon foods, which are typically healthier.

These three actions must be accompanied by other, potentially more modest, economy-wide changes, such as energy efficiency in buildings, switching to heat pumps, more efficient use of materials, and phasing out potent fluorinated gases.

8.6.3 Applied Research

This section presents the scientific research work and applied research projects with governments and industries that researchers from the School of Government and Public Transformation and the Institute of Advanced Materials for Sustainable Manufacturing (IAMSM) of the Tecnológico de Monterrey carry out on green growth and decarbonization. By disseminating research, we seek to generate collaborations and promote strategic projects between government, industry and academia, as well as inform and raise awareness in society in general about the importance of transforming the way we currently produce and consume.

Green growth and decarbonization will create new development and markets opportunities, but this requires identifying them and quickly taking action. With this mission, OSMI generates tools and research to develop the technologies, knowledge and public policies necessary to promote a change in the productive paradigm that leads to sustainable and inclusive development in Mexico and the world.

8.6.4 Integrating LCE and CI in the OSMI

The OSMI harmonizes LCE and CI to support Mexico's transition towards sustainable and competitive manufacturing sectors. Using these frameworks, OSMI provides actionable insights into the current state and future potential of the manufacturing industry in scenarios of green growth and decarbonization.

1. **LCE framework in OSMI:** Life cycle Engineering involves understanding and optimizing the life cycle of products and services from conception to disposal, guiding on effective end-of-life management strategies to ensure products are disposed of or recycled in an environmentally friendly manner. In OSMI, this framework is applied through several key activities:
   (a) ***Material Analysis:*** OSMI investigates the types of materials used in different manufacturing industries and assesses their environmental impacts, promoting the use of sustainable materials.
   (b) ***Product Life cycle Management:*** Tools and data are provided to analyze how products can be designed, manufactured, used, and disposed of in a way that minimizes environmental impact while maximizing resource efficiency.
   (c) ***Process Development:*** OSMI identifies innovative manufacturing processes that reduce waste, energy consumption, and emissions. It also promotes the integration of renewable energy sources into manufacturing processes.
   (d) ***Manufacturing System Optimization:*** Designed in a manner that is efficient, manufacturing systems that incorporate advanced technologies such as IoT and AI to optimize the use of resources and energy.
   (e) ***Supply Chain Analysis:*** OSMI analyzes the supply chains of key manufacturing sectors to identify opportunities to reduce carbon footprints, improve material recyclability, and ensure more sustainable logistics practices.
2. **Competitive Intelligence in OSMI:** CI involves gathering, analyzing, and using information about competitors, markets, and environmental trends to make informed strategic decisions. In the context of OSMI, CI is pivotal in several ways:
   (a) ***Market and Competitor Analysis:***
      - *Industrial Trends:* By monitoring and analyzing global and regional trends in manufacturing, OSMI provides insights into how competitive dynamics are shaping industry practices towards sustainability.
      - *Competitor Strategies:* Information on how other countries and companies are advancing in sustainable manufacturing

helps Mexican industries adopt best practices and innovative technologies.

(b) ***Technological Innovation and Adoption:***

- *Technology Scouting:* OSMI uses CI to identify emerging technologies that can revolutionize manufacturing processes, such as additive manufacturing, advanced robotics, and smart automation.
- *Innovation Benchmarking:* It benchmarks Mexican industries against global standards to identify gaps in technology use and management practices, guiding investments in technology adoption and innovation.

(c) ***Policy and Strategic Planning:***

- *Regulatory Insights:* CI provides an understanding of the regulatory environment, helping industries anticipate and adapt to changes in environmental laws and standards.
- *Strategic Decision-Making:* OSMI uses CI to help manufacturers and policymakers make informed decisions about where to focus efforts and resources to drive growth and sustainability simultaneously.

3. **Practical Applications and Outcomes:**

(a) ***Manufacturing Industry Analysis Tools:*** OSMI provides dynamic data visualization tools that allow stakeholders to examine the regional and industrial structure of manufacturing in Mexico. This analysis helps to understand the economic impact, sustainability, and potential areas of green growth within the industry.

(b) ***Decarbonization Pathways Modeling:*** Using the SiSePuede tool, OSMI simulates different decarbonization scenarios to understand their potential impacts on emissions, costs, and economic growth. This helps in strategizing national approaches to achieving Paris Agreement goals.

(c) ***Applied Research and Collaboration:*** OSMI promotes collaboration among academia, government, and industry to apply research findings in real world settings, promoting sustainable practices and technologies in Mexican manufacturing sectors.

The integration of the LCE framework and CI within OSMI not only enhances the competitiveness of Mexico's manufacturing sectors but also aligns them with global sustainability and green growth objectives. This holistic approach ensures that strategic planning, technological innovation, and policy-making are well-informed, targeted, and effective in promoting sustainable industrial development.

8.7 Future Trends

The landscape of Competitive Intelligence (CI) is rapidly evolving, driven by technological advancements and increasing global competition. Future trends in CI emphasize data-driven insights, real-time monitoring, integration with Industry 4.0, enhanced collaboration, and ethical considerations. These trends significantly impact how CI is conducted and utilized, providing businesses with the tools and strategies to maintain a competitive edge.

- **Increased Use of AI and Machine Learning:**
  - ***Predictive Analytics:*** AI and machine learning will be crucial in predicting market trends, competitor actions, and customer preferences, enabling businesses to make more informed decisions.
  - ***Automated Data Collection and Analysis:*** Automation will streamline the collection and analysis of large datasets, making it faster and more efficient to gain actionable insights.
- **Big Data and Advanced Analytics:**
  - ***Enhanced Data Integration:*** The ability to integrate and analyze diverse data sources, including social media, IoT devices, and public records, will provide a more comprehensive view of the competitive landscape.
  - ***Real-Time Insights:*** Real-time data analytics will become more prevalent, allowing companies to respond more quickly to market changes and competitor actions.
- **Ethical and Legal Considerations:**
  - ***Privacy and Data Security:*** As data collection methods become more advanced, there will be an increased emphasis on ensuring CI practices comply with privacy laws and data protection regulations.
  - ***Ethical AI Usage:*** Companies will need to navigate the ethical implications of using AI and machine learning in CI to avoid bias and ensure fair competition.
- **Social Media and Digital Footprint Analysis:**
  - ***Social Listening:*** Analyzing social media trends and consumer sentiment will become integral to CI, providing insights into public perception and competitor marketing strategies.
  - ***Digital Footprint Tracking:*** Monitoring the online activities of competitors, including website traffic, online advertisements, and digital campaigns, will offer valuable competitive insights.

- **Collaboration and Knowledge Sharing:**
  - ***Cross-Functional Collaboration:*** CI will increasingly involve collaboration across different departments, such as marketing, sales, R&D, and strategy, to create a holistic understanding of the competitive environment.
  - ***Knowledge Management Systems:*** Implementing advanced knowledge management systems will facilitate the sharing and utilization of CI insights across the organization.
- **Specialized CI Tools and Platforms:**
  - ***CI Software:*** The development and adoption of specialized CI software and platforms that offer advanced analytics, visualization, and reporting capabilities will become more widespread.
  - ***Customized Solutions:*** Tailored CI solutions that address different industries' specific needs and challenges will gain traction.
- **Focus on Competitive Benchmarking:**
  - ***Performance Metrics:*** More organizations will use competitive benchmarking to measure their performance against industry standards and competitors.
  - ***Strategic Benchmarking:*** Companies will focus on strategic benchmarking to identify best practices and areas for improvement in their business processes.
- **Integration with Strategic Planning:**
  - ***Strategic Insights:*** CI will be more deeply integrated into strategic planning processes, providing critical insights that inform long-term business strategies and decisions.
  - ***Scenario Planning:*** Leveraging CI for scenario planning will help organizations anticipate and prepare for various future scenarios and market conditions.
- **Human-Centered CI:**
  - ***Human Intelligence:*** Despite technological advances, human intelligence will remain essential for interpreting CI data and making strategic decisions.
  - ***Training and Development:*** Investing in training and development programs for CI professionals to enhance their analytical and strategic skills will be crucial.
- **Globalization and Market Expansion:**
  - ***Global CI Networks:*** As businesses expand globally, there will be a need for more extensive CI networks that provide insights into international markets and competitors.

– ***Cultural Intelligence:*** Understanding cultural differences and market dynamics in different regions will be an important aspect of global CI efforts.

8.8 Conclusion

Integrating CI into the LCE framework—such as material, product, process, manufacturing systems, and supply chains—provides a comprehensive approach to maintaining a competitive edge. Hence CI within the LCE framework and the 5S principles—Social, Sustainable, Sensing, Smart, and Safe—enables companies to enhance strategic decision-making, foster innovation, and optimize operations. CI supports industries by systematically collecting, analyzing, and applying information about competitors, market trends, technologies, and regulatory frameworks, anticipating market shifts, identifying emerging opportunities, and mitigating potential threats.

Incorporating CI into material and product development processes drives innovation and efficiency, providing insights into competitor advancements, customer preferences, and technological trends. This allows companies to develop superior products and materials that meet or exceed market expectations. Moreover, CI optimizes manufacturing processes by analyzing competitors' strategies, technological innovations, and efficiency measures, leading to improved operational efficiency, cost reduction, and enhanced product quality.

CI also plays a crucial role in building resilient and responsive supply chains. By understanding competitors' supply chain strategies and performance, companies can develop robust supply chain frameworks that adapt to market changes and disruptions, ensuring continuous and efficient operations. Integrating CI within the end-of-life processes ensures sustainable and ethical practices, allowing companies to adopt eco-friendly and compliant disposal, recycling, and repurposing strategies.

The synergy between CI and Industry 4.0 technologies, such as AI, IoT, and big data analytics, provides a significant competitive advantage. These technologies enable real-time data collection and analysis, predictive analytics, and smart manufacturing processes, enhancing the effectiveness of CI. Strategic planning for nearshoring, particularly in regions like Mexico, is bolstered by CI, addressing challenges related to energy supply, water scarcity, security, infrastructure, human capital, government support, technological infrastructure, and SME integration. This supports the development of a conducive environment for nearshoring investments.

Future trends in CI emphasize increased use of AI and machine learning, big data and advanced analytics, ethical and legal considerations, social media and digital footprint analysis, enhanced collaboration, specialized CI tools and platforms, competitive benchmarking, integration with strategic planning, human-centered CI, and globalization and market expansion. These trends will shape the evolution of CI, making it more data-driven, real-time, collaborative, and globally focused.

The case study of nearshoring to Mexico highlights CI's importance in addressing challenges related to energy supply, water scarcity, security, infrastructure, human capital, government support, technological infrastructure, and SME integration. CI supports the development of a conducive environment for nearshoring investments by providing insights that help overcome these challenges. The future of CI is marked by increased use of AI and machine learning, big data and advanced analytics, ethical and legal considerations, social media and digital footprint analysis, enhanced collaboration, specialized CI tools and platforms, competitive benchmarking, integration with strategic planning, human-centered CI, and globalization and market expansion. These trends will shape the evolution of CI, making it more data-driven, real-time, collaborative, and globally focused.

Thus, the comprehensive integration of CI across various business entities and functions within the LCE framework and the 5S principles is essential for maintaining a competitive edge in today's dynamic industrial landscape. Leveraging CI for strategic decision-making, innovation, efficiency, sustainability, and ethical practices enables industries to navigate market complexities, drive growth, and achieve long-term success. The future of CI, shaped by technological advancements and a global perspective, will be critical to business strategy and operations.

References

[1] It's time to reinvent CPGs' product development and operations. How digital helps unlock profitable growth. https://www.accenture.com/content/dam/accenture/final/a-com-migration/r3-3/pdf/pdf-139/accenture-cgs-time-reinvent-product.pdf, 2020. Accessed: 2024-06-06.

[2] AccTek. CNC router control systems comprehensive guide. https://acctekcnc.com/cnc-router-control-systems-comprehensive-guide. Accessed: 2024-11-07, 2024.

[3] ACEEE, Alliance to Save Energy. 2022. The Business Council for Sustainable Energy. Energy efficiency impact report. https://energyefficiencyimpact.org/. Accessed: 2024-06-06.

[4] Adobe Communications Team. 2022. Product life cycle stages and how to use them. https://business.adobe.com/blog/basics/product-life-cycle-stages-and-how-to-use-them. Accessed: 2024-06-06.

[5] Rita Quetziquel Fuentes Aguilar, Alejandro Garcia Gonzalez, Daniel Aragon Han, Yoku Sashida Mendez and Juan Carlos Vazquez Fuentes. Sistema, método y aparato para la evaluación y seguimiento del rendimiento muscular con retroalimentación autoajustable, April 25, 2019. WO2019209098A1 and PCT/MX2019/000014.

[6] Rita Quetziquel Fuentes Aguilar, Alejandro Garcia Gonzalez, Daniel Aragon Han, Yoku Sashida Mendez and Juan Carlos Vazquez Fuentes. 2023. System, method and apparatus for assessing and monitoring muscle performance with self-adjusting feedback. US Patent 11,633,140.

[7] Zeyad Almatarneh, B. Jarah and M. Jarrah. 2022. The role of management accounting in the development of supply chain performance in logistics manufacturing companies. Uncertain Supply Chain Management. 10(1): 13–18.

[8] Jesus Cruz Alvarez. 2015. Lean design for Six Sigma: An integrated approach to achieving product reliability and low-cost manufacturing. International Journal of Quality and Reliability Management. 32(8): 895–905.

[9] David A. Arcos-Novillo and David Güemes-Castorena. 2017. Development of an additive manufacturing technology scenario for opportunity identification—the case of Mexico. Futures. 90: 1–15.

[10] M.F. Ashby. 2011. Materials selection in mechanical design, Butterworth-Heinemann.

[11] Mike Ashby. 2016. Material property data for engineering materials. Engineering Department and Granta Design. 27.

[12] Mohsen Attaran, Sharmin Attaran and Bilge Gokhan Celik. 2023. The impact of digital twins on the evolution of intelligent manufacturing and Industry 4.0. Advances in Computational Intelligence. 3(3): 11.

[13] Irina Badurashvili. 2023. Achievement of sustainable manufacturing from Industry 4.0 technologies–future perspective.

[14] D.N. Bai, P.F. Gao, X.G. Yan and Yao Wang. 2021. Intelligent forming technology: State-of-the-art review and perspectives. Journal of Advanced Manufacturing Science and Technology. 1(3): 2021008.

[15] Fernando H. Bedulli, Rodrigo Leite, Fabio Freitas, Thiago Murari and Ingrid Winkler. 2024. Optimizing automotive product development: Integration of electronic control units and virtual prototypes using virtual reality. Technical Report, SAE Technical Paper.

[16] Morad Behandish, Saigopal Nelaturi, Chaman Singh Verma and Mats Allard. 2019. Automated process planning for turning: A feature-free approach. Production and Manufacturing Research. 7(1): 415–432.

[17] Douja Ben Aissia. 2022. Potential of digitalization to improve the risk management in supply chain processes. PhD thesis, Technische Hochschule Ingolstadt.

[18] Brinda, B., Mortlock D. and Dixon, J. 2020. Product management in the age of disruption. Accessed: 2024-06-06.

[19] Eduardo Buendia, Kiyoto Tanabe, Andrej Kranjc, Baasansuren Jamsranjav, Maya Fukuda, Akira Osako, et al. 2019. Refinement to the 2006 IPCC Guidelines for National Greenhouse Gas Inventories. 11.

[20] Felipe Cerdas, Max Juraschek, Sebastian Thiede and Christoph Herrmann. 2017. Life cycle assessment of 3D printed products in a distributed manufacturing system. Journal of Industrial Ecology. 21(51): 580–593.

[21] Danni Chang, CKM Lee and Chun-Hsien Chen. 2014. Review of life cycle assessment towards sustainable product development. Journal of Cleaner Production. 83: 48–60.

[22] Dante Chavarría-Barrientos, Rafael Batres, Paul K. Wright and Arturo Molina. 2018. A methodology to create a sensing, smart and sustainable manufacturing enterprise. International Journal of Production Research. 56(1–2): 584–603.

[23] Ming-Chuan Chiu and Chih-Hsing Chu. 2012. Review of sustainable product design from life cycle perspectives. International Journal of Precision Engineering and Manufacturing. 13: 1259–1272.

[24] Sunil Chopra. 2018. Supply Chain Management: Strategy, Planning, and Operation. Pearson. New York.

[25] CIMdata. 2014. Accenture unveils digital product development framework: Every business is a digital business (commentary). https://www.cimdata.com/de/resources/complimentary-reports-research/commentaries/item/2691-accenture-unveils-digital-product-development-framework-every-business-is-a-digital-business-commentary/2691-accenture-unveils-digital-product-development-framework-every-business-is-a-digital-business-commentary. Accessed: 2024-06-06.

[26] Andreas Ciroth, Matthias Finkbeiner, Marzia Traverso, Jutta Hildenbrand, Walter Kloepffer, Bernard Mazijn, et al. 2011. Towards a life cycle sustainability assessment: making informed choices on products.

[27] William C. Copacino. 2019. Supply chain management: The basics and beyond. Routledge.

[28] Daniel Cortés, José Ramérez, Jorge Medina and Arturo Molina. 2022. Integrated product, process and manufacturing system development for multifunctional micromachine tool. The International Journal of Advanced Manufacturing Technology. 120(11): 7477–7521.

[29] Daniel Cortés, José Ramérez, Pedro Ponce and Arturo Molina. 2021. S3 manufacturing process taxonomy. Journal of Manufacturing Processes. 67: 579–610.

[30] Ricardo Henrique Da Silva, Paulo C Kaminski and Fabiano Armellini. 2020. Improving new product development innovation effectiveness by using problem solving tools during the conceptual development phase: Integrating design thinking and TRIZ. Creativity and Innovation Management. 29(4): 685–700.

[31] Jim Davis, Thomas Edgar, James Porter, John Bernaden and Michael Sarli. 2012. Smart manufacturing, manufacturing intelligence and demand-dynamic performance. Computers and Chemical Engineering. 47: 145–156.

[32] Jos, de la O Campos. 2019. Critical data: Teaching design through critical design, physical computing, and digital data. *In*: Bohemia, E., Kovacevic, A., Buck, L., Brisco, R., Evans, D., Grierson, H., et al. (eds). DS 95: Proceedings of the 21st International Conference on Engineering and Product Design Education (E&PDE 2019). University of Strathclyde, Glasgow. 12th–13th September 2019.

[33] Diaz, A., Schöggl, J.-P., Reyes, T. and Baumgartner, R.J. 2021. Sustainable product development in a circular economy: Implications for products, actors, decision-making support and lifecycle information management. Sustainable Production and Consumption. 26: 1031–1045. https://doi.org/10.1016/j.spc.2020.12.044

[34] Yuxuan Du and Kaining Ge. 2024. Embracing the digital intelligence: A strategic approach to optimizing mass customization. *In*: SHS Web of Conferences. 181: 04011. EDP Sciences.

[35] Mahboob Elahi, Samuel Olaiya Afolaranmi, Jose Luis Martinez Lastra and Jose Antonio Perez Garcia. 2023. A comprehensive literature review of the applications of AI techniques through the lifecycle of industrial equipment. Discover Artificial Intelligence. 3(1): 43.

[36] Schneider Electric. 2022. What are the 3 emissions scopes and how do you manage them? https://perspectives.se.com/blog-stream/what-are-the-3-emissions-scopes-and-how-do-you-manage-them. (Accessed on 07/04/2024).

[37] Schneider Electric. 2023. Supply chain decarbonization: An essential step towards net-zero. https://perspectives.se.com/sustainability/supply-chain-decarbonization-an-essential-step-towards-net-zero. (Accessed on 07/04/2024).

[38] Armando Elizondo-Noriega, Mario G Beruvides, Naveen Tiruvengadam and David Guemes-Castorena. 2019. An improvement in robustness of a system dynamics model for simulating cost of quality in manufacturing operations. *In*: Proceedings of the American Society for Engineering Management, Philadelphia, Pennsylvania.

[39] Elizondo-Noriega, A., Tiruvengadam, N. and Guemes-Castorena, D. 2021. Impact of RFID implementation on operational productivity of a manufacturing firm: a simulation-based study. International Journal on Interactive Design and Manufacturing (IJIDeM). 15: 555–575.

[40] Ellen MacArthur Foundation. 2021. The circular economy: A wealth of flows. https://www.ellenmacarthurfoundation.org/. Accessed: 2023-11-07.

[41] Yasamin Eslami, Mario Lezoche, Hervé Panetto and Michele Dassisti. 2021. On analysing sustainability assessment in manufacturing organisations: A survey. International Journal of Production Research. 59(13): 4108–4139.

[42] Anna Maria Ferrari, Lucrezia Volpi, Davide Settembre-Blundo and Fernando E García-Muiña. 2021. Dynamic life cycle assessment (LCA) integrating life cycle inventory (LCI) and enterprise resource planning (ERP) in an Industry 4.0 environment. Journal of Cleaner Production. 286: 125314.

[43] FICTIV. Sustainability in manufacturing report. 2023. https://www.fictiv.com/pdf/sustainability-in-manufacturing-report-2023. Accessed: 2024-06-06.

[44] Leonardo Frizziero, Giampiero Donnici, Alfredo Liverani and Karim Dhaimini. 2020. Design for additive manufacturing and advanced development methods applied to an innovative multifunctional fan. *In*: Additive manufacturing: Breakthroughs in research and practice. 52–85. IGI Global.

[45] Gartner. 2024. Future of supply chain. https://www.gartner.com/en/supply-chain/topics/future-of-supply-chain. Accessed: 2024-02-04.

[46] Gartner. 2024. Gartner announces ranking of the Gartner healthcare supply chain top 25 for 2023. https://www.gartner.com/en/newsroom/press-releases/2023-12-13-gartner-announces-ranking-of-the-gartner-healthcare-supply-chain-top-25-for-2023. Accessed: 2024-02-04.

[47] Gartner. 2024. The Gartner supply chain top 25 for 2023. https://www.gartner.com/en/supply-chain/research/supply-chain-top-25. Accessed: 2024-02-04.

[48] Gartner. 2024. How to build a supply chain ecosystem? https://www.gartner.com/en/supply-chain/trends/supply-chain-ecosystem. Accessed: 2024-02-04.

[49] Gartner. 2024. Supply chain technologies and digital transformation. https://www.gartner.com/en/supply-chain/topics/supply-chain-digital-transformation. Accessed: 2024-02-04.

[50] Gartner. 2024. We help clients make faster, smarter decisions. https://www.gartner.com/en/about. Accessed: 2024-02-04.

[51] SDG Goal. 9: Build resilient infrastructure, promote sustainable industrialization and foster innovation. United Nations. Link available at: https://www.un.org/sustainabledevelopment/infrastructureindustrialization/(20.12.2019).

[52] Sandra Grabowska, Bożena Gajdzik and Sebastian Saniuk. 2020. The role and impact of Industry 4.0 on business models. Sustainable Logistics and Production in Industry 4.0: New Opportunities and Challenges. 31–49.

[53] David G. Groves and Robert J. Lempert. 2007. A new analytic method for finding policy-relevant scenarios. Global Environmental Change. 17(1): 73–85.

[54] David Güemes-Castorena and Karina Blando. 2015. Methodology to develop a technological front route of exploration. *In*: 2015 Portland International Conference on Management of Engineering and Technology (PICMET). 2218–2222. IEEE, 2015.

[55] Sudheer Gupta and Omkar D. Palsule-Desai. 2011. Sustainable supply chain management: Review and research opportunities. IIMB Management Review. 23(4): 234–245.

[56] Michael Z. Hauschild, Sami Kara and Inge Røpke. 2020. Absolute sustainability: Challenges to life cycle engineering. CIRP Annals. 69(2): 533–553.

[57] Michael Z. Hauschild, Ralph K. Rosenbaum and Stig Irving Olsen. 2018. Life Cycle Assessment. Springer.

[58] Ricardo Hausmann. ¿el desarrollo verde es un oxímoron? https://www.project-syndicate.org/commentary/decarbonization-green-technologies-developing-countries-growth-by-ricardo-hausmann-2021-06/spanish, 06 2021. (Accessed on 07/04/2024).

[59] Ricardo Hausmann. 2022. Oportunidades de crecimiento verde. https://www.imf.org/es/Publications/fandd/issues/2022/12/green-growth-opportunities-ricardo-hausmann. (Accessed on 07/04/2024).

[60] Katsundo Hitomi. 1990. Manufacturing systems engineering: the concept, its context and the state of the art. International Journal of Computer Integrated Manufacturing. 3(5): 275–288.

[61] John Horn. 2023. Inside the Competitor's Mindset: How to Predict Their Next Move and Position Yourself for Success. MIT Press.

[62] Jianjun Hu, Stanislav Stefanov, Yuqi Song, Sadman Sadeed Omee, Steph-Yves Louis, Edirisuriya MD Siriwardane, et al. 2022. Org: A materials informatics web app platform for materials discovery and survey of state-of-the-art. NPJ Computational Materials. 8(1): 65.

[63] Shihui Huangl. 2022. Product innovation design method based on BP neural network. Advances in Multimedia. 6830892.

[64] Guo Hui, Rose Dahlina Rusli and Asliza bt Aris. 2023. Remodeling the traditional fashion industry in the era of Industry 4.0. International Journal of Global Optimization and Its Application. 2(3): 165–178.

[65] Bonsa Regassa Hunde and Abraham Debebe Woldeyohannes. 2022. Future prospects of computer-aided design (CAD)—A review from the perspective of artificial intelligence (AI), extended reality, and 3D printing. Results in Engineering. 14: 100478.

[66] Hyekyung Imottesjo and Jaan-Henrik Kain. 2022. The Urban CoCreation Lab—An integrated platform for remote and simultaneous collaborative urban planning and design through web-based desktop 3D modeling, head-mounted virtual reality and mobile augmented reality: prototyping a minimum viable product and developing specifications for a minimum marketable product. Applied Sciences. 12(2): 797.

[67] International Energy Agency. 2023. Industry. https://www.iea.org/energy-system/industry. Accessed: 2023-11-07.

[68] Amit Kumar Jain, Sandeep Kumar and Shubham Tayal. 2022. Life cycle engineering in the era of Industry 4.0. Frontiers in Manufacturing Technology, 2: 1008987.

[69] Mohd Javaid, Abid Haleem, Ravi Pratap Singh, Shahbaz Khan and Rajiv Suman. 2022. Sustainability 4.0 and its applications in the field of manufacturing. Internet of Things and Cyber-Physical Systems. 2: 82–90.

[70] Xinxiang Ji and Shiva Abdoli. 2023. Challenges and opportunities in product life cycle management in the context of Industry 4.0. Procedia CIRP. 119: 29–34.

[71] Kalra, N., Groves, D.G., Bonzanigo, L., Molina Perez, E., Ramos, C. and Rodriguez Cabanillas, I. 2015. Republic of Peru Robust Decision-Making in the water sector: A strategy for implementing Lima's long-term water resources master plan. World Bank. 2015. https://hdl.handle.net/10986/22493

[72] Elena Kazakova and Joosung Lee. 2022. Sustainable manufacturing for a circular economy. Sustainability. 14(24): 17010.

[73] Farzad Kazemi and Fatemeh Soltani. 2023. The energy industry's response to competitive intelligence in green marketing (case study of Qatar). Journal of Open Innovation: Technology, Market, and Complexity, 100192.

[74] Alexis Laurent, Christine Molin, Mikołaj Owsianiak, Peter Fantke, Wim Dewulf, Christoph Herrmann, et al. 2019. The role of life cycle engineering (LCE) in meeting the sustainable development goals–report from a consultation of LCE experts. Journal of Cleaner Production. 230: 378–382.

[75] Robert J. Lempert. 2003. Shaping the next one hundred years: New methods for quantitative, long-term policy analysis.

[76] Robert J. Lempert, Benjamin P. Bryant, Myles T. Collins, Andrew Hackbarth, Tom LaTourrette, Robert T. Reville, et al. 2013. Making good decisions without predictions: Robust decision making for planning under deep uncertainty.

[77] Exploit the product life cycle. https://hbr.org/1965/11/exploit-the-product-life-cycle, 1965. Accessed: 2024-06-06.

[78] Chen Li and Kun Zheng. 2023. Methods, progresses, and opportunities of materials informatics. InfoMat. 5(8): e12425.

[79] Kendrik Yan Hong Lim, Pai Zheng and Chun-Hsien Chen. 2020. A state-of-the-art survey of digital twin: techniques, engineering product

lifecycle management and business innovation perspectives. Journal of Intelligent Manufacturing. 31(6): 1313–1337.

[80] Limacher, M. and L. Fauconnet. 2024. The legal and ethical guardrails for sound competitive intelligence. https://www.pragmaticinstitute.com/resources/articles/product/the-legal-and-ethical-guardrails-for-sound-competitive-intelligence. (Accessed on 07/04/2024).

[81] Marcello Lorusso, Marco Rossoni, Giorgio Colombo, et al. 2020. Conceptual modeling in product design within virtual reality environments. Computer-Aided Design and Applications. 18(2): 383–398.

[82] Luís Madureira, Aleš Popovič and Mauro Castelli. 2021. Competitive intelligence: A unified view and modular definition. Technological Forecasting and Social Change. 173: 121086.

[83] Edwin Mauricio Martinez, Pedro Ponce, Israel Macias and Arturo Molina. 2021. Automation pyramid as constructor for a complete digital twin, case study: A didactic manufacturing system. Sensors. 21(14): 4656.

[84] Matmatch. 2024. The complete guide to sustainable materials selection. https://matmatch.com/guide/sustainable-materials-selection. Accessed: 2024-06-06.

[85] McKinsey & Company. Product managers for the digital world, 2017. Accessed: 2024-06-06.

[86] McKinsey & Company. What separates top product managers from the rest of the pack. https://www.mckinsey.com/industries/technology-media-and-telecommunications/our-insights/what-separates-top-product-managers-from-the-rest-of-the-pack#/, 2023. Accessed: 2024-06-06.

[87] Z. Hauschild Michael, K. Rosenbaum Ralph and Irving Olsen Stig. 2018. Life cycle assessment: Theory and practice.

[88] J. Miranda, R. Pérez-Rodríguez, V. Borja and P.K. Wright. 2017. Integrated product, process and manufacturing system development reference model to develop cyber-physical production systems-the sensing, smart and sustainable microfactory case study. IFAC-PapersOnLine. 50(1): 13065–13071.

[89] Issam AR Moghrabi, Sameer Ahmad Bhat, Piotr Szczuko, Rawan A AlKhaled and Muneer Ahmad Dar. 2023. Digital transformation and its influence on sustainable manufacturing and business practices. Sustainability. 15(4): 3010.

[90] Arturo Molina and R. Bell. 1999. A manufacturing model representation of a flexible manufacturing facility. *In*: Proceedings of the Institution

of Mechanical Engineers. Part B: Journal of Engineering Manufacture. 213(3): 225–246.

[91] Arturo Molina, Daniel Cortés, Isaac Chairez, Mariel Alfaro-Ponce, Mario Moisés Alvarez and Grisell Trujillo de Santiago. 2024. Comprehensive sustainable development of a multifunctional machine: 3D food printer and didactic platform. International Journal of Sustainable Engineering. 17(1): 1–16.

[92] Arturo Molina, Pedro Ponce, Jhonattan Miranda and Daniel Cortés. 2021. Enabling Systems for Intelligent Manufacturing in Industry 4.0, Volume 277. Springer.

[93] Arturo Molina, Pedro Ponce, Jhonattan Miranda, Daniel Cortés, Arturo Molina, Pedro Ponce, et al. 2021. Integrated product, process and manufacturing system development reference model framework. Enabling Systems for Intelligent Manufacturing in Industry 4.0: Sensing, Smart and Sustainable Systems for the Design of S3 Products, Processes, Manufacturing Systems, and Enterprises. 29–69.

[94] Arturo Molina, Marcela Velandia and Nathalie Galeano. 2007. Virtual enterprise brokerage: A structure-driven strategy to achieve build to order supply chains. International Journal of Production Research. 45(17): 3853–3880.

[95] Maria E. Mondejar, Ram Avtar, Heyker Lellani Baños Diaz, Rama Kant Dubey, Jesús Esteban, Abigail Gómez-Morales, et al. 2021. Digitalization to achieve sustainable development goals: Steps towards a smart green planet. Science of The Total Environment. 794: 148539.

[96] Kgosietsile F. Monnagaaratwe and Kenneth Mathu. 2022. Supply chain management as a competitive advantage for grocery small and medium-sized enterprises in Mahikeng, South Africa. Journal of Transport and Supply Chain Management. 16: 791.

[97] Juliana Montoya, Jorge Medina, Arturo Molina, Janet Gutiérrez, Blanca Rodríguez and Rodrigo Marín. 2021. Impact of viscoelastic and structural properties from starch-mango and starch-arabinoxylans hydrocolloids in 3D food printing. Additive Manufacturing. 39: 101891.

[98] Fredrick Mumali and Joanna Kałkowska. 2024. Intelligent support in manufacturing process selection based on artificial neural networks, fuzzy logic, and genetic algorithms: Current state and future perspectives. Computers and Industrial Engineering. 110272.

[99] M. Adeel Munir, M. Salman Habib, A. Hussain, Muhammad A. Shahbaz, Adnan Qamar, Tariq Masood, et al. 2022. Blockchain adoption for sustainable supply chain management: Economic, environmental, and social perspectives. Frontiers in Energy Research. 10: 899632.

[100] Alberto Navajas, Leire Uriarte and Luis M. Gandía. 2017. Application of eco-design and life cycle assessment standards for environmental impact reduction of an industrial product. Sustainability. 9(10): 1724.

[101] NEGI. 2023. Sistema de clasificación industrial de América del norte, México Scian 2023. https://www.inegi.org.mx/scian/. Accessed on 07/04/2024.

[102] OECD. 2011. OECD green growth studies towards green growth. Organization for Economic Co-operation and Development (OECD), Paris Cedex, France.

[103] Institute of Advanced Materials for Sustainable Manufacturing. Retos tecnológicos y políticas de apoyo necesarias para el fortalecimiento de los encadenamientos productivos requeridos por las inversiones de la relocalización de empresas a México. https://tec.mx/sites/default/files/inline-files/reporte%20IAMSM_nearshoring%20v6.1.pdf, 05 2024. Accessed on 07/04/2024.

[104] Okechukwu Okorie, Konstantinos Salonitis, Fiona Charnley, Mariale Moreno, Christopher Turner and Ashutosh Tiwari. 2018. Digitisation and the circular economy: A review of current research and future trends. Energies. 11(11): 3009.

[105] IPCC (Intergovernmental Panel on Climate Change). 2006. Guidelines for national greenhouse gas inventories.

[106] OSMI. 2024. Observatory of sustainable manufacturing industries. https://www.osmi.mx/en, 2024. Accessed on 07/04/2024.

[107] Stig Ottosson. 2002. Virtual reality in the product development process. Journal of Engineering Design. 13(2): 159–172.

[108] Vural Özdemir and Nezih Hekim. 2018. Birth of industry 5.0: Making sense of big data with artificial intelligence,"the internet of things" and next-generation technology policy. Omics: a journal of integrative biology. 22(1): 65–76.

[109] Andreas Tri Panudju, Sapta Rahardja, Mala Nurilmala, et al. 2023. Supply chain performance evaluation using the SCOR® model and fuzzy-TOPSIS. Operational Research in Engineering Sciences: Theory and Applications. 6(2).

[110] Bhaveshkumar N Pasi, Subhash K. Mahajan and Santosh B. Rane. 2020. Smart supply chain management: A perspective of Industry 4.0. Supply Chain Management. 29(5): 3016–3030.

[111] Pedro Ponce, Brian Anthony, Russell Bradley, Javier Maldonado-Romo, Juana Isabel Méndez, Luis Montesinos and Arturo Molina. 2024. Developing a virtual reality and AI-based framework for advanced digital manufacturing and nearshoring opportunities in Mexico. Scientific Reports. 14(1): 11214.

[112] Pedro Ponce, Therese Peffer, Juana Isabel Méndez Garduño, Ursula Eicker, Arturo Molina, Troy McDaniel, et al. 2023. Data and AI Driving Smart Cities. Studies in Big Data. Springer International Publishing.

[113] Pedro Ponce, Citlaly Pérez, Aminah Robinson Fayek and Arturo Molina. 2022. Solar energy implementation in manufacturing industry using multi-criteria decision-making fuzzy TOPSIS and S4 framework. Energies. 15(23): 8838.

[114] Pedro Ponce, Kenneth Polasko, Arturo Molina, Miguel Ramírez and Patricia Mora. 2016. Intellectual property basic manual for researchers in universities. Bentham Science Publishers.

[115] United Nations Environment Programme. 2011. Towards a green economy: Pathways to sustainable development and poverty eradication.`https://sustainabledevelopment.un.org/content/documents/126GER _ synthesis _ en.pdf`. Accessed on 07/04/ 2024.

[116] Poblano-Ojinaga, E.R., López, R.R., Gómez, J.A.H. and Torres-Arguelles, V. 2019. Effect of competitive intelligence on innovation capability: An exploratory study in Mexican companies. Journal of Intelligence Studies in Business. 9(3): 62–67.

[117] Denise Reike, Walter J.V. Vermeulen and Sjors Witjes. 2018. The circular economy: new or refurbished as CE 3.0?—Exploring controversies in the conceptualization of the circular economy through a focus on history and resource value retention options. Resources, Conservation and Recycling. 135: 246–264.

[118] John Reyes, Josefa Mula and Manuel Díaz-Madroñero. 2023. Development of a conceptual model for lean supply chain planning in Industry 4.0: multidimensional analysis for operations management. Production Planning and Control. 34(12): 1209–1224.

[119] Theresa Riedelsheimer, Sabrina Neugebauer and Kai Lindow. 2021. Progress for life cycle sustainability assessment by means of digital lifecycle twins—A taxonomy. EcoDesign and Sustainability II: Social Perspectives and Sustainability Assessment. 329–345.

[120] E.E. Rodriguez-Garcia, A. Elizondo-Noriega, N. Tiruvengadam, D. Güemes-Castorena and M. Rodriguez-Salvador. 2023. Evaluation of decentralized manufacturing via system dynamics simulation: A review of the state-of-the-art. *In*: 2023 Portland International Conference on Management of Engineering and Technology (PICMET). 1–9. IEEE.

[121] Yesim Can Saglam, Sibel Yildiz Çankaya, Ismail Golgeci, Bulent Sezen and Selim Zaim. 2022. The role of communication quality, relational commitment, and reciprocity in building supply chain resilience: A social exchange theory perspective. Transportation Research Part E: Logistics and Transportation Review. 167: 102936.

[122] A., Sartal, R. Bellas, Ana M. Mejías and Alberto García-Collado. 2020. The sustainable manufacturing concept, evolution and opportunities within industry 4.0: A literature review. Advances in Mechanical Engineering. 12(5): 1687814020925232.

[123] Donald Alford Sayre. 2014. INSDE ISO 14000: The competitive advantage of environmental management. CRC Press.

[124] Peter Scallan. 2003. Process planning: the design/manufacture interface. Elsevier.

[125] Lee Schipper and Celine Marie-Lilliu. 1999. Transportation and CO_2 emissions: Flexing the link—a path for the World Bank. World Bank, Environment Department.

[126] Julie M. Schoenung, Elsa A. Olivetti and Guest Editors. 2023. Sustainable development of materials: Broadening stakeholder engagement. MRS Bulletin. 48(4): 362–367.

[127] Zhiming Shang and Liming Zhang. 2022. The sustainable digitalization in the manufacturing industry: a bibliometric analysis and research trend. Mobile Information Systems. 2022: 1–11.

[128] A. Sharabati, N. Almokdad, A. Marei and H. Abusaimeh. 2023. Competitive strategy development through green supply chain practices. Uncertain Supply Chain Management. 11(4): 1507–1518.

[129] Thibault Simon, Pierre Rust, Romain Rouvoy and Joël Penhoat. 2023. Uncovering the environmental impact of software life cycle. *In*: 2023 International Conference on ICT for Sustainability (ICT4S). 176–187. IEEE.

[130] Sinis, Georgios. 2024. Recycling carbon footprint savings calculator. `https://recyclewits.com/tools-calculators/carbon-footprint-savings/`. Accessed: 2024-06-06.

[131] S.P. Sundar Singh Sivam, S. Rajendra Kumar, Sathiyamoorthy Karuppiah and A. Rajasekaran. 2017. Competitive study of engineering change process management in manufacturing industry using product life cycle management—A case study. *In*: 2017 International Conference on Inventive Computing and Informatics (ICICI). 76–81. IEEE.

[132] Mohsen Soori and Mohammed Asmael. 2021. Classification of research and applications of the computer aided process planning in manufacturing systems. Independent Journal of Management and Production. 12(5): 1250–1281.

[133] Christian Spreafico and Davide Russo. 2021. A sustainable cheese packaging survey involving scientific papers and patents. Journal of Cleaner Production. 293: 126196.

[134] Kenneth G. Swift and Julian D. Booker. 2003. Process selection: From design to manufacture. Elsevier.

[135] David E. Tew, Mihaela Quirk and Lipi Acharya. 2021. Automating the process of energy-technology invention. *In*: AAAI 2022 Workshop on AI for Design and Manufacturing (ADAM).

[136] Climate Action Tracker. 2024. Cat climate target update tracker-mexico. `https://climateactiontracker.org/climate-target-update-tracker/mexico/`. Accessed on 07/04/2024.

[137] Karl T. Ulrich and Steven D. Eppinger. 2016. Product Design and Development. McGraw-Hill.

[138] SETAC UNEP. 2015. Guidance on organizational life cycle assessment, 2015.

[139] United Nations Environment Programme. 2021. Cleaner production: A guide to information sources. `http://www.uneptie.org/pc/cp/network/cp_worldwide.htm`. Accessed: 2023-11-07.

[140] United Nations Environment Programme. 2021. Sustainable product development. `https://www.unep.org/`. Accessed: 2023-11-07.

[141] United States Environmental Protection Agency. 2021. Sustainable materials management: The road ahead. `https://www.epa.gov/sites/default/files/2015-09/documents/vision2.pdf`. Accessed: 2024-06-06.

[142] Flore Vallet, Benoît Eynard, Dominique Millet, Stéphanie Glatard Mahut, Benjamin Tyl and Gwenola Bertoluci. 2013. Using eco-design tools: An overview of experts' practices. Design Studies. 34(3): 345–377.

[143] Ben T. Wittbrodt, Alexandra G. Glover, J. Laureto, G.C. Anzalone, D. Oppliger, J.L. Irwin, et al. 2013. Life-cycle economic analysis of distributed manufacturing with open-source 3-D printers. Mechatronics. 23(6): 713–726.

[144] WMC. 2024. Ethical competitive intelligence: A complete guide. `https://www.watchmycompetitor.com`. Accessed on 07/04/2024.

[145] World Economic Forum. 2023. Reducing the carbon footprint of the manufacturing industry through data sharing. `https://www.weforum.org/impact/carbon-footprint-manufacturing-industry`. Accessed: 2024-06-06.

[146] Lianfan Wu. 2023. Agile design and AI integration: Revolutionizing MVP development for superior product design. International Journal of Education and Humanities. 9(1): 226–230.

[147] Wai Sze Yip, HongTing Zhou and Suet To. 2023. A critical analysis on the triple bottom line of sustainable manufacturing: Key findings and implications. Environmental Science and Pollution Research. 30(14): 41388–41404.

[148] Leila Zehtaban, Omar Elazhary and Dieter Roller. 2016. A framework for similarity recognition of CAD models. Journal of Computational Design and Engineering. 3(3): 274–285.

[149] Guoqing Zhang, Yiqin Yang and Guoqing Yang. 2023. Smart supply chain management in Industry 4.0: the review, research agenda and strategies in North America. Annals of Operations Research. 322(2): 1075–1117.

[150] Jianchao Zhang and Zhihong Yu. 2016. Overview of 3D printing technologies for reverse engineering product design. Automatic Control and Computer Sciences. 50(2): 91–97.

[151] Jiwei Zhou and Jorge D. Camba. 2021. Computer-aided process planning in immersive environments: A critical review. Computers in Industry. 133: 103547

Glossary

10R Framework: A set of strategies integrated into the 5S principles to maximize resource efficiency and sustainability through actions like Refuse, Rethink, Reduce, Reuse, Repair, Refurbish, Remanufacture, Repurpose, Recycle, and Recover to minimize environmental impact throughout a product's lifecycle.

5S Framework: A holistic framework that integrates a reference model, methodology, and tools based on the principles of 5S (Social, Sustainable, Sensing, Smart, and Safe) to design, develop, and create systems such as Materials, Products, Processes, Manufacturing Systems, and Supply Chains.

5S Principles: A set of principles focusing on Social, Sustainable, Sensing, Smart, and Safe dimensions used to guide the design, development, and creation of systems such as Materials, Products, Processes, Manufacturing Systems, and Supply Chains.

5S Systems: Industry 5.0 systems are designed using the principles of Social, Sustainable, Sensing, Smart, and Safe to create human-centric, environmentally responsible, intelligent, and secure industrial solutions.

Additive Manufacturing (3D Printing): A process that builds materials layer by layer, allowing for precise control of material structure and composition, reducing waste, and creating complex geometries.

Adjacent Invaders: Companies from adjacent industries that leverage their capabilities to enter new markets.

Advanced Characterization Techniques: Methods such as electron microscopy, X-ray diffraction, and spectroscopy provide detailed insights into the microstructure and composition of materials.

Advanced Process Control (APC): Techniques and control strategies used to optimize manufacturing processes, maintaining product quality, and reducing variations.

Agile Methodologies: Approaches in project management, typically in software development, involve continual planning, learning, improvement, team collaboration, and a flexible response to change.

Artificial Intelligence (AI): The simulation of human intelligence in machines programmed to think and learn, used to optimize production and predict failures in manufacturing.

Augmented Reality (AR): An interactive experience where digital information is overlaid on the real-world environment, aiding in design visualization and modifications.

Basic Development—Conceptualization: A stage in the LCE framework focused on transforming initial ideas into tangible concepts with specified characteristics.

Big Data Analytics: The complex process of examining large and varied data sets to uncover hidden patterns, correlations, market trends, and customer preferences.

Bio-based Materials: Materials derived from renewable resources that offer reduced environmental footprints and biodegradability.

Blockchain: A decentralized ledger technology that ensures secure, transparent, and traceable transactions, enhancing supply chain management.

Chief Supply Chain Officer (CSCO): An executive responsible for overseeing the supply chain operations of an organization.

Circular Economy: A sustainability concept aiming to minimize waste through the reuse, recycling, and repurposing of materials, supporting long-term resource efficiency and environmental conservation. An economic system to minimize waste and maximize resources through reuse, recycling, and remanufacturing.

Collaborative Robotics (Cobots): Robots designed to work alongside human beings in a shared workspace, assisting in tasks and improving efficiency to enhance manufacturing processes.

Competitive Intelligence (CI): The systematic and ethical collection, analysis, and application of information about the business environment to support strategic decision-making and help companies maintain or gain a competitive edge.

Computational Materials Science: The use of computer simulations to study and predict the properties of materials before they are synthesized.

Computer-Aided Design (CAD): Software designers use to create precise drawings and technical illustrations in 2D or 3D.

Computer-Aided Manufacturing (CAM): The use of software to control machine tools and related machinery in the manufacturing of components.

Computer-Aided Process Planning (CAPP): Software that translates design models into actionable manufacturing process plans, bridging the gap between CAD and CAM. Therefore, a system that assists in manufacturing planning by automating the selection and sequencing of production processes.

Core Digital Technologies: The foundational set of digital tools and systems, aligned with the LCE framework, enable innovation, efficiency, and sustainability across the stages of each entity.

Data-Driven Insights: Insights are derived from analyzing large datasets generated by technologies like IoT, providing valuable information about competitor activities, market trends, and technological advancements.

Digital Technologies: Refer to electronic tools, systems, devices, and resources that generate, store, or process data.

Digital Thread: Integrates data across the entire product lifecycle, from design to manufacturing to after-sales service, ensuring continuity and consistency.

Digital Twins: Virtual representations of physical devices allow manufacturers to simulate production processes and identify inefficiencies without physical testing.

Disruptive Innovators: Entities that create significant market changes through groundbreaking innovations.

Eco-design: Designing products with consideration for their environmental impacts throughout their lifecycle.

Embodied Energy: The total energy required for the extraction, processing, production, and delivery of a material or product, excluding operational and disposal energy costs.

End-of-Life Management: Refers to strategies and processes at the end of a product's useful life, focusing on recycling, upcycling, and disposal to minimize environmental impact, ensure sustainable resource use, and maintain social responsibility.

Entrepreneurs and Start-ups: New market entrants bring fresh ideas, often characterized by agility and a higher risk-taking approach.

Environmental, Social, and Governance (ESG): Criteria used to evaluate a company's operations and performance on environmental sustainability, social responsibility, and corporate governance.

Extended Producer Responsibility (EPR): Legislative frameworks hold producers accountable for the entire lifecycle of their products, particularly their take-back, recycling, and disposal.

Facility Design: The detailed planning and layout of manufacturing plants and equipment to optimize the production process, enhance efficiency, and ensure worker safety and environmental compliance.

Fast-Moving Innovators: New entrants or existing players typically employ disruptive innovations to gain a market edge.

Finite Element Analysis (FEA): A simulation technique used to predict how products will react to real-world forces and conditions.

Foreign Direct Investment (FDI): Investment made by a company into business interests located in another country, often involving establishing business operations or acquiring business assets.

Generative CAPP: Automatically generates process plans using algorithms and a knowledge base, providing flexibility for various parts and processes.

Generative Design Models: Advanced models that propose new material compositions and structures based on specified criteria, often using algorithms like Variational Autoencoders (VAEs) and Generative Adversarial Networks (GANs).

Greenhouse Gas Emissions: Emissions of gases such as CO_2 that trap heat in the atmosphere, contributing to global warming and climate change.

Group Technology (GT): A method for categorizing parts with similar features and processes, streamlining production by grouping and standardizing manufacturing plans.

High-Throughput Screening: Techniques that combine automated synthesis with rapid characterization, enabling the evaluation of many material compositions simultaneously.

Holistic Product Development: A future approach to product development that incorporates eco-design from the ideation phase, ensuring every product is designed with its entire lifecycle in mind.

Hybrid CAPP: Combines Variant and Generative CAPP to retrieve standard plans and apply generative techniques for customization and flexibility.

Industrial Internet of Things (IIoT): The use of interconnected sensors and devices in industrial applications to collect and analyze data, optimizing manufacturing processes.

Industry 4.0: The integration of digital technologies, including IoT, AI, and big data analytics, into manufacturing processes to create smart, efficient, and autonomous production systems.

Integrated Product Lifecycle Management (PLM): Software that manages all stages of a product's life, from conception to disposal, enhancing collaboration and data management.

Internet of Things (IoT): A network of physical objects embedded with sensors, software, and other technologies to connect and exchange data with other devices and systems over the internet.

Life Cycle Assessment (LCA): A technique to assess the environmental impact of a material throughout its entire lifecycle, from raw material extraction to disposal.

Life Cycle Engineering (LCE): An integrative approach that focuses on optimizing the entire life cycle of products, materials, processes, systems, and supply chains for sustainability and cost-effectiveness.

Machine Learning (ML): A type of AI that allows computers to learn from and make data-based decisions, used to optimize manufacturing processes.

Manufacturing System Development: The design and operation of manufacturing systems prioritizing efficiency, modularity, and adaptability, with sustainable practices integral to daily operations.

Materials Informatics: Integrating big data, AI, and machine learning into materials science to accelerate the discovery and development of sustainable materials.

Micromachine: A modular CNC machine tool used as a case study to illustrate the application of LCE and 5S principles across its development lifecycle.

Mio-Tracker: A medical device designed to evaluate muscular activity through a single-channel electromyograph system, used in rehabilitation and sports settings to provide real-time data on muscle function.

Nearshoring: Relocating business processes to a nearby country, typically to reduce travel times, improve cultural alignment, or take advantage of similar time zones.

Offshoring: The relocation of business processes from a company's home country to a foreign country, often to leverage cost advantages.

Opitz Coding System: A classification system used in Group Technology to categorize parts by geometry, material, and manufacturing attributes, facilitating efficient production planning.

Predictive Maintenance: Techniques designed to help determine the condition of in-service equipment in order to predict when maintenance should be performed, thereby preventing unexpected equipment failures.

Predictive Modeling: Using computational modeling to predict material properties and behaviors before physical testing, enhancing the efficiency of material development.

PRIMA Matrix: A process selection tool that maps product design requirements to appropriate manufacturing processes, aiding in the visualization and decision-making in manufacturing.

Process Development: The stage in LCE focused on creating sustainable manufacturing methods, including waste minimization and energy conservation.

Process Optimization: The continuous improvement of manufacturing processes to increase efficiency, quality, and productivity, often guided by data analytics and real-time feedback.

Product Lifecycle Management: The process of managing the entire lifecycle of a product from inception, through engineering design and manufacture, to service and disposal.

Product Transfer: The process of transitioning a product from the development phase into mass production, ensuring that all specifications and quality standards are met throughout the manufacturing and distribution stages.

Prototyping: Creating an early sample, model, or product release for testing in real-world conditions before full-scale production.

Public vs. Private Competitors: Differentiation between public companies, which disclose more due to regulatory requirements, and private companies, which keep their strategies more confidential.

Raw Material Extraction: The process of extracting natural resources used in manufacturing, with a focus on reducing environmental and social impacts.

Reconfigurable Manufacturing Systems: Systems designed to be flexible and modular, allowing quick adaptation to new manufacturing methods or changes in product design.

Recycling: The process of converting waste materials into new materials and objects reduces the need for virgin resources and minimizes environmental impact.

Relocation: The process of moving business operations either back to the home country (reshoring) or to a third country, which could be nearshoring if the third country is closer to the home country.

Reshoring: Bringing previously offshored business operations back to the home country to enhance control, improve quality, or respond to consumer preferences for local products.

Resilience: The ability of a supply chain to quickly recover from disruptions and maintain continuous operations.

Resource Efficiency: The practice of optimizing the use of materials, water, energy, and other resources to reduce waste and lower environmental impacts in manufacturing.

Safe Dimension: Ensures worker safety and compliance with environmental regulations, reducing accident risks and protecting workers and the environment.

Scenario Planning: Developing best-case and worst-case scenarios based on possible competitor actions and simulating competitive responses to test different strategies.

Scope 1 Emissions: Direct emissions from owned or controlled sources by a company.

Scope 2 Emissions: Indirect emissions from generating purchased electricity, heat, or steam a company consumes.

Scope 3 Emissions: All other indirect emissions not covered under Scope 2, including those from the supply chain, business travel, and waste disposal.

Sensing Dimension: Involves using sensors and data systems for real-time monitoring, ensuring continuous process optimization and predictive maintenance.

Simulation Software: Tools like FEA and CFD are used to model and predict the performance of manufacturing processes, reducing the need for physical prototypes.

Smart Dimension: Integrates advanced technologies like AI, IoT, and automation to enhance manufacturing efficiency, flexibility, and responsiveness.

Smart Factories: Highly digitized manufacturing facilities that rely on Industry 4.0 technologies to optimize production processes, improve operational efficiency, and enhance product quality.

Smart Manufacturing: The use of AI, machine learning, and IIoT to create adaptive, efficient manufacturing systems that optimize operations and predict maintenance needs.

Social Dimension: Evaluates manufacturing processes based on their impact on workers, communities, and ethical standards, promoting fair labor practices and community engagement.

Supply Chain Development: Strategic planning and management of materials, components, and finished products to ensure timely production, cost efficiency, and market demand alignment.

Supply Chain Execution: Equivalent to Drivers, it involves implementing strategies to ensure the efficient flow of goods and information. This includes managing inventory, diversifying the manufacturing network, multisourcing, nearshoring, harmonizing technology, and building ecosystem partnerships.

Supply Chain Management: The management of the flow of goods and services involves the movement and storage of raw materials, work-in-process inventory, and finished goods from the point of origin to the point of consumption.

Supply Chain Operations: Encompasses Production/Service Strategy, Core Process Selection/Definition, and KPIs (Key Performance Indicators). It involves executing daily supply chain activities, managing core processes, and measuring performance to ensure efficiency and customer satisfaction.

Supply Chain Planning: Equivalent to Value Chain Strategy involves organizing the value chain to optimize resource utilization, enhance flexibility, and improve communication. This includes vertical integration, horizontal integration, and collaboration.

Supply Chain Planning: Equivalent to Value Chain Strategy involves organizing the value chain to optimize resource utilization, enhance flexibility, and improve communication. This includes vertical integration, horizontal integration, and collaboration.

Supply Chain Strategy: Equivalent to Competitive Strategy involves developing a comprehensive plan that aligns the supply chain with business goals to gain a competitive advantage. This includes innovation, operational excellence, and customer focus.

Supply Chain Strategy: Equivalent to Competitive Strategy involves developing a comprehensive plan that aligns the supply chain with business goals to gain a competitive advantage. This includes innovation, operational excellence, and customer focus.

Sustainability Benchmarking: Using competitive intelligence to benchmark against industry leaders in sustainability practices to identify gaps and opportunities for adopting best practices.

Sustainability Dimension: Focuses on minimizing environmental impacts through resource efficiency, waste management, and carbon footprint reduction.

Sustainable Materials Management: A systematic approach that addresses environmental impacts at every stage of the material lifecycle, focusing on the most productive and sustainable use of resources.

Sustainable Operations: Integrating sustainable practices into supply chain operations to drive revenue growth, cost savings, and environmental goals.

Technology Transfer: The process of disseminating technology, skills, knowledge, methodologies, or manufacturing techniques among universities, government organizations, private sector companies, or across geographies.

Traditional Competitors: Well-established entities competing on similar products and services within the same market.

Triggers for Reaction: Market changes that could prompt a competitive reaction, monitored to anticipate and respond to competitors' actions.

Virtual Reality (VR): An immersive digital environment used for training, process visualization, and product testing, enhancing the understanding and optimization of manufacturing processes.

Index

D

E

H

I

L

About the Authors

Dr. Arturo Molina holds a PhD in Manufacturing Engineering from Loughborough University, a University Doctorate degree in Mechanical Engineering from Technical University of Budapest, and MSc and BSc degrees in Computer Science from Tecnológico de Monterrey. Arturo is a member of various prestigious scientific academies and working groups, such as the National Researchers System of Mexico (SNI-Nivel III). He has authored 20 books, over 150 scientific papers, and holds 5 patents and 3 patent solicitations. Additionally, he has founded 3 technology-based companies, including IECOS, Albiomar, and SMES, which provide innovative solutions for manufacturing enterprise systems.

Dr. Juana Isabel Méndez is an architect and an engineer in electronic design with a Master's in Energy Management and a laureate PhD in Engineering Sciences from Tecnológico de Monterrey, with a doctoral stay at UC Berkeley, who received the best thesis award at TecScience Summit 2023. She is a COMEXUS Fulbright Scholar. Isabel is the *2024 Premio Mujer Tec* awardee under the category of *Environment.* Her research focuses on building energy simulations, thermal comfort, tailored dynamic interfaces, personalized gamification strategies, AI-based decision systems, and parametric architecture applied to the energy savings field. Isabel has around 50 scientific publications and has presented at national and international conferences and a patent application.

Dr. Dora Iliana Medina earned her Bachelor´s degree in Chemical Engineering from the National Polytechnic Institute in Mexico City, graduating summa cum laude. She then moved to Riverside, CA, to pursue an MSc in Chemical and Environmental Engineering which she completed at the University of California. Afterwards, Dr. Medina relocated to Cambridge, UK, to pursue her PhD degree and received her PhD in Chemical Engineering and Biotechnology from the University of Cambridge. For her outstanding work, Dr. Medina was highly commended in the Sellafield Ltd Award for Engineering Excellence (IChemE awards) and was invited as a Keynote speaker at the Annual Meeting of the Polymer Processing Society. Subsequently, she was selected as a UNESCO-L'ORÉAL "For Women in Science" International Fellow at MIT.

Dr. Fernando Gómez-Zaldívar holds a PhD in Public Policy from Tecnológico de Monterrey. He is a professor at the School of Government and Public Transformation at Tecnológico de Monterrey, a researcher at the Regional Development Lab, and a member of Mexico's National System of Researchers (SNI). His research focuses on the use of economic complexity principles and tools to design development policies tailored to each region. As a consultant, he has worked with governments, businesses, and international institutions, including the IDB and the World Bank, on designing productive development strategies in Mexico and Latin America.

Dr. Alex Elías-Zúñiga has published more than 165 articles in International Journals (JCR) and 120 papers in International Conferences and Proceedings—inventor of six patents granted in Mexico and one in the USA. Dr. Elías has led several scientific-technology developments with global companies such as Vitro, Cemex, Prolec GE, Honeywell Aerospace, Safran Aerospace, Whirlpool, and Siemens, to name a few, with deliverables that exceed TRL6 level. He is a member of the National Research System of Mexico (SNI-Level III) and

the Mexican Academy of Sciences. Co-founder and CTO of ZOA 4U—Technology-based Company (www.zoa4U.com). Currently, he is leading the Accelerated Materials Discovery Research Unit at the Institute of Advanced Materials for Sustainable Manufacturing and is also a professor at the Department of Mechanical Engineering and Advanced Materials Department.

Dr. Pedro Ponce from Xalapa, Veracruz, Mexico, is an expert in automation, control systems, manufacturing systems, and electric drives, with a Master's and PhD in Electrical Engineering. He has over 14 years of industrial

experience as a field and design engineer and has worked with robotics, power electronics, advanced manufacturing systems, and intelligent systems. Since 2002, Pedro has been a professor at Tecnológico de Monterrey, where he served as Director of the Graduate Program in Science in Engineering from 2006 to 2014. Now, he is a leader researcher in enabling technology at the Institute of Advanced Materials for Sustainable Manufacturing in Mexico City. He also has over 15 years of experience advising international industries. His academic contributions include 19 books, 19 patents, and more than 200 research papers. He enjoys reading, basketball, swimming, and traveling in his free time.

Dr. Rita Q. Fuentes-Aguilar is a researcher at Tecnológico de Monterrey and leader of the Enabling Technologies Unit at the Institute of Advanced

Materials for Sustainable Manufacturing. Her work focuses on biomedical engineering, medical device design, and artificial intelligence. She graduated with honors in Biomedical Engineering from IPN and holds a master's and PhD in Automatic Control from CINVESTAV. She is a member of Mexico's National System of Researchers and a distinguished professor at Guadalajara campus. She leads the Advanced Cyber-Physical Systems Lab, with projects supported by national and international sponsors. In 2023, she was named among Forbes' "100 Most Powerful Women in Mexico" and received two Diseña México awards for innovation in medical technology. Dr. Fuentes has multiple patents, and a technology transfer, and has given over 20 talks promoting STEAM women's education.

Dr. Alejandro Montesinos-Castellanos holds a PhD and a Master's degree in Chemical Engineering from the Universidad Autónoma Metropolitana

Iztapalapa and a Bachelor's degree in Chemical Engineering from the Instituto Tecnológico de Orizaba. Dr. Montesinos has distinguished himself with over 46 published articles and six national and international patents. His experience includes national and international collaborations with institutions such as the National Polytechnic Institute and the University of California at Berkeley, the Delhi Institute of Technology, the Southwest Research Institute, and the Massachusetts Institute of Technology. He is an expert in inorganic materials, specializing in their synthesis and characterization for various applications, from catalysts to CO_2 absorbers, covering topics from fundamental research to industrial applications.

Made in the USA
Las Vegas, NV
15 September 2026

43b3a480-6855-4277-b76c-9745444bdbeaR01